Waste to Sustainable Energy

MFCs – Prospects through Prognosis

Editors

Lakhveer Singh
Biological and Ecological Engineering
Oregon State University, Corvallis, Oregon, USA

Durga Madhab Mahapatra
Biological & Ecological Engineering
Oregon State University, Corvallis, Oregon, USA

CRC Press
Taylor & Francis Group
Boca Raton London New York

CRC Press is an imprint of the
Taylor & Francis Group, an **informa** business

A SCIENCE PUBLISHERS BOOK

Cover illustrations reproduced by kind courtesy of Supriyanka Rana, Lakhveer Singh, Zularisam bin ab Wahid.

CRC Press
Taylor & Francis Group
6000 Broken Sound Parkway NW, Suite 300
Boca Raton, FL 33487-2742

First issued in paperback 2021

© 2019 by Taylor & Francis Group, LLC
CRC Press is an imprint of Taylor & Francis Group, an Informa business

No claim to original U.S. Government works

Version Date: 20181206

ISBN-13: 978-0-367-78018-0 (pbk)
ISBN-13: 978-1-138-32821-1 (hbk)

Library of Congress Cataloging-in-Publication Data

Names: Singh, Lakhveer, editor. | Mahapatra, Durga Madhab, editor.
Title: Waste to sustainable energy : MFCs, prospects through prognosis / editors, Lakhveer Singh (Biological and Ecological Engineering, Oregon State University, Corvallis, Oregon, USA), Durga Madhab Mahapatra (Biological & Ecological Engineering, Oregon State University, Corvallis, Oregon, USA).
Description: Boca Raton, FL : CRC Press, Taylor & Francis Group, 2019. | "A science publishers book." | Includes bibliographical references and index.
Identifiers: LCCN 2018054085 | ISBN 9781138328211 (hardback)
Subjects: LCSH: Microbial fuel cells. | Fuel cells. | Clean energy. | Climate change mitigation.
Classification: LCC TK2931 .W37 2019 | DDC 621.31/2429--dc23
LC record available at https://lccn.loc.gov/2018054085

Visit the Taylor & Francis Web site at
http://www.taylorandfrancis.com

and the CRC Press Web site at
http://www.crcpress.com

Preface

Extensive use of fossil fuels for energy have negatively contributed to the environment owing to the emission of carbon dioxide and other harmful gases as serious atmospheric pollutants and particulates, and has resulted in soaring global warming. At the same time, various waste products from domestic, agricultural, animal facilities, refineries and industries also cause a tremendous environmental burden that needs treatment and recycle. Energy systems from MFC's can be used for combating both environmental problems and offsetting the pollution loads. This creates more opportunities for renewable and green fuel production that can substitute fossil fuels and generate commercially important coproducts from waste substrates. MFC's has a plethora of benefits, over other kinds of energy production routes with a) ceased emissions (such as SOx, NOx, CO_2 and CO), b) higher efficiency, c) no mobile parts and d) least sound pollution. Although there have been significant attempts to produce bioelectricity from bacterial electrolytic systems right from early 1900's, there were limitations in the yield and feasibility of the process. However, early 1990's witnessed innovations in MFC designs and incorporation of various electron donor, electrode apparatus and biocatalysts (biological agents: algae etc.) and thus have become far more appealing. This book will focus on the state of the art of MFC's with various combinatories of substrates yielding bioelectricity with valued co-products. Essentially the book will provide fundamental ideas and basics of MFC technologies, entailing various design and modelling aspects with examples. Various sections of the book will deal with unique aspects of basic sciences, reactor configuration, application, market feasibility with lucid illustrations and explanations.

The techno-economics and life cycle section will critically assess the feasibility of waste-powered MFC's for sustainable bioenergy production and essentially highlight the tradeoff between resource needs and energy production. This section will help academicians, entrepreneurs, industrialists to understand the scope and challenges and select unique, and specific integrated approaches in unit processes.

Lakhveer Singh
Durga Madhab Mahapatra

Contents

MFCs - From Lab to Field

Abhilasha Singh Mathuriya

Department of Biotechnology, School of Engineering and Technology, Sharda University, Knowledge Park III, Greater Noida - 201306, India
Email: imabhilasha@gmail.com

Introduction

In order to develop a green world in the future, research initiatives are focused on alternate, renewable, and carbon neutral energy sources which pose minimal or no negative environmental impact. Microbial fuel cells (MFCs) are fairly contributing in this attempt. MFCs are bioelectrochemical systems, used for the direct conversion of chemical energy of chemical compounds into electricity using microorganisms (Allen and Bennetto 1993, Kim et al. 1999, Shukla et al. 2004, Mathuriya and Yakhmi 2014).

A conventional two-chamber MFC contains an anodic and a cathodic chamber separated by an ion exchange membrane (Bond and Lovley 2003, Mathuriya and Sharma 2010a, Mathuriya 2016a). At the anode, electrons are generated by metabolic activities of microorganisms when they metabolize organic compounds. Those electrons travel through external circuit and protons through membrane to cathode, where both react with oxygen to form water (Bennetto 1990).

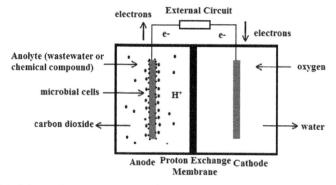

Figure 1.1. Schematic representation of typical two-chambered microbial fuel cell.

The basic reactions involved can be represented as (Jiang et al. 2013)

$$C_6H_{12}O_6 + 6H_2O \rightarrow 6CO_2 + 24H^+ + 24e^- \ (E^0 = \textbf{0.014 V}) \qquad (1)$$

$$(\text{Intermediate})_{ox} + e^- \rightarrow (\text{Intermediate})_{re} \qquad (2)$$

$$6O_2 + 24H^+ + 24e^- \rightarrow 12H_2O \ (E^0 = \textbf{1.23 V}) \qquad (3)$$

There are no set parameters to measure the performance of an MFC, since the performance depends on many factors like the nature of electrochemical reactions, type of electrode used, organic substrates, composition of the microbial population, and architecture, etc. Therefore, different measures might be taken in order to improve the MFC performance. Over the past few years, researchers across the globe are engaged in an attempt to enhance the efficiency and reliability of MFCs in different ways. This includes MFC architecture, development of high efficiency electrode materials, operational costs, and robust microbial community (Kaewkannetra et al. 2011, Qian et al. 2011, Huang et al. 2012, Buitrón and Cervantes-Astorga 2013, Mathuriya 2016 a,b).

MFCs offer several attractive features, viz. (a) direct electricity generation from chemical compounds which leads to high efficiency; (b) the chemical to electricity conversion by MFCs is not limited by the Carnot cycle because it does not involve the conversion of energy into heat—instead directly into electricity and, a much higher conversion efficiency can be achieved (70%) (Mathuriya and Yakhmi 2014); (c) owing to microbially operated technology, operate effectively at ambient temperature; (d) quiet and safe performance (Rabaey and Verstreate 2005), and (e) does not requires off-gas treatment (Mathuriya and Sharma 2009). It is hypothesized that the amount of electricity generated by MFCs in the wastewater treatment process can reduce the power required in a conventional treatment process that involves aerating of the activated sludge (Watanabe 2008).

Potential Areas for Microbial Fuel Cells Application

MFCs are witnessing rapid growth and their application aura is continuously increasing. Some areas are discussed here.

Wastewater treatment: Wastewater generates from any combination of domestic, commercial, industrial, or agricultural activities. Wastewater retains complex chemical and biological matter, which can be invaluable sources of energy, but due to random degradation create serious health, sanitation, and environmental problems. Moreover, wastewater also hosts innumerable extremophilic microbial flora, which can degrade those complex matter. Developing a route to produce electricity from degradation of complex matter in wastewater by microorganisms is the mover for the development of MFCs (Mathuriya and Sharma 2009).

Wastewater treatment is the prime area of MFC application and MFCs have proved themselves as efficient wastewater treatment systems (Kim et al. 2004, Huang and Logan 2008, Mathuriya and Sharma 2009, Ieropoulos et al. 2013). Only this system can convert organic matter of waste directly into electricity. MFCs have been reported to treat a wide range of wastewaters. The list is too large to accommodate in the word limit of this chapter. Some remarkable mentions are human feces wastewater (Fangzhou et al. 2011), food processing wastewater (Mathuriya and Sharma 2010a, Velasquez-Orta et al. 2011), oilfield wastewater (Gong and Qin 2012), landfill leachate (Sonawane et al. 2017), domestic wastewater (Ren et al. 2014), paper industry wastewater (Mathuriya and Sharma 2009), swine wastewater (Ding et al. 2017), tannery wastewater (Mathuriya 2014), brewery wastewater (Mathuriya and Sharma 2010b, Huang et al. 2011), refinery waste (Agarry 2017), synthesis gas (Mehta et al. 2010), paracetamol (Zhang et al. 2015), several dyes *viz.* azo dyes (Sun et al. 2011, Fang et al. 2013, Fang et al. 2017, Yuan et al. 2017), wastewaters containing various heavy metals *viz.* Chromium (Ryu et al. 2011, Song et al. 2016), Vanadium (Qiu et al. 2017), Copper (Wu et al. 2016), Arsenic (Wang et al. 2014), Silver (Choi and Cui 2012), Cobalt (Huang et al. 2013), and even Uranium (Gregory et al. 2004).

Powering low energy devices: Electricity generation from any chemical compound is the most fascinating feature of MFCs. MFCs have been reported for the generation of electricity since 1910 (Potter 1910, Bond and Lovley 2003, You et al. 2010, Mathuriya 2016a, Krieg et al. 2017). MFCs are an attractive option as sustainable lower scale power supplies. Many studies have successfully demonstrated the applications of MFCs in powering low-power-consuming devices (Winfield et al. 2014) like sensors (Tommasi et al. 2014, Khaled et al. 2016), as biochemical oxygen demand (BOD) sensors (Kim et al. 2003a, Kharkwal et al. 2017), telecommunication systems (Tender et al. 2008, Thomas et al. 2013), mobiles and smart phones (Ieropoulos et al. 2013).

MFC-type BOD sensor developed by Kim et al. (2003a) was reported to operate for over five years without remarkable maintenance, which is longer than other types of BOD sensors in the market. MFCs can also be used to monitor the presence of chemical toxicants (Jiang et al. 2017) *viz.* formaldehyde (Davila et al. 2011), anaerobic digestion fluid (Liu et al. 2011), heavy metal and phenol in the test water (Kim et al. 2003b, Tao et al. 2017), toxic matter in potable water (Stein et al. 2012), and to detect illegal dumping or pollution (Chang et al. 2004). Various microbial activity-monitoring sensors (Lovley and Nevin 2016) were also developed in recent years. Moreover, demonstration to apply MFCs in powering gadgets like mobile phones (Ieropoulos et al. 2013, Walter et al. 2017) is making MFCs popular among public.

Robotics: Autonomous robotics is a recent attention of scientists. Conventionally robots are run by standard battery (Kluger 1997), which require periodical manual input to change battery. MFCs can be used to power such robots, especially to those assigned 'start and forget' missions (Melhuish et al. 2006). Various generations of robots are reported to use MFCs as their power source, such as Green bug robots (Wilkinson and Campbell 1996), Gastrobots (Kelly et al. 1999, Wilkinson 1999, 2000a), Gastronome

(Wilkinson 2000b), EcoBot (Melhuish et al. 2006), Slugbot (Melhuish and Kubo 2007), and EvoBot (Theodosiou et al. 2017).

Journey of MFCs-Lab

In 1911, Michael C. Potter noticed an electric force between electrodes, which were immersed in microbial culture suspension in a battery-type setup (Potter 1910, 1911) and reported that electricity can be generated from the microbial metabolism of organic compounds. Two decades later Cohen reported a stacked bacterial fuel cell capable of producing a voltage of 35 V at a current of 0.2 mA (Cohen 1931). However, MFCs were alien to the scientific community until 1960s, when the idea of electricity production was used by NASA space program—as a system to recycle human wastes to electricity during space flights (Canfield et al. 1963, Bean et al. 1964, Canfield and Goldner 1964, Del Duca and Fuscoe 1964). In early 2000s their application for wastewater treatment were explored (Kim et al. 2001). Since then, MFCs have been well-studied in laboratories across the globe (Bond and Lovley 2003, Huang and Logan 2008, Pant et al. 2010, Ieropoulos et al. 2013, Mathuriya and Yakhmi 2014, Mathuriya 2016a, Ding et al. 2017).

Over the past few years, researchers across the globe are in attempt to advance the MFCs in different ways in order to increase its reliability and efficiency. Major areas of MFC advancement include, a) MFC designs (Kaewkannetra et al. 2011, Qian et al. 2011, Huang et al. 2012, Mathuriya 2016a, Monasterio et al. 2017), b) cost reduction (Buitrón and Cervantes-Astorga 2013, Mathuriya 2014), c) development of high efficiency electrode materials *viz.* graphene/TiO$_2$ hybrids (Zhao et al. 2014), monolithic three-dimensional graphene (Chen et al. 2016), binder-free carbon black/stainless steel mesh composite (Zheng et al. 2015), gold-sputtered carbon paper (Sun et al. 2010), d) efficient cathode catalyst *viz.* manganese (IV) dioxide (MnO$_2$) (Cao et al. 2003), zirconium oxycarbonitride (Ichihashia et al. 2016), iron oxide and partly graphitized carbon from waste cornstalks and pomelo skins (Ma et al. 2014); nickel oxide and carbon nanotube composite (NiO/CNT) (Huang et al. 2015), cobalt oxide (Kumar et al. 2016); biochar derived from bananas (Yuan et al. 2014), carbon nanotube/polypyrrole nanocomposite (Ghasemi et al. 2016), composite catalysts (Jadhav 2017), e) high strength and permeable separator *viz.* Nafion (Mathuriya and Sharma 2010b, Fangzhou et al. 2011), ultra-filtration membranes (Kim et al. 2007), clayware (Behera et al. 2010, Jadhav et al. 2014), natural rubber (Winfield et al. 2013), glass wool (Mohan et al. 2008), J-cloth (Fan et al. 2007), nylon and glass fiber (Zhang et al. 2010), mixed cellulose easter (Wang and Lim 2017), and natural eggshell (Ma et al. 2016), and f) utilization of efficient microbial community (Niessen et al. 2006, Parameswaran et al. 2010, Ewing et al. 2017, Paitier et al. 2017).

Journey of MFCs-Pilot Study

Enormous attempts have been made worldwide by the scientific community to develop the candidature of MFCs in real world. Worldwide universities, governments, and companies are also investing on their R&D resources for the future commercialization of MFCs. Kamaraj et al. (2015) digested the Nopal biogas effluent (NPE) by a clay cup MFC. This group succeeded in powering the digital clock with MFC of 2-cell connected in series. Bombelli et al. (2016) reported a bryophyte MFC (bryoMFC), which was able to power a commercial radio receiver or an environmental sensor (LCD desktop weather station). Mathuriya (2016b) developed a large scale continuous auto flow and power-free large scale microbial auto flow fuel cell (MAFFC) reactor. This system eliminated the need of external agitation and power requirements, therefore is self sustained and net power generating system, and is efficient for long-term continuous operation with minimal maintenance. Ieropoulos et al. (2016) conducted the MFC based pee power urinal field trials for internal lighting at Frenchay Campus (UWE, Bristol) from February–May 2015, and reported the possibility of modular MFCs for lighting, with University staff and students as the users.

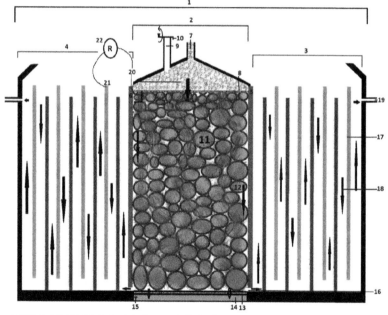

1. Main reactor assembly, 2. Anode tank, 3. Cathode chamber 1, 4. Cathode chamber 2, 5. Separator 1, 6. Separator 2, 7. Wastewater inlet point, 8. Inert sieve, 9. Gas pipe, 10. Butterfly valve, 11. Anodes packing in anode tank, 12. Direction of anolyte flow, 13. Lower inert sieve, 14. Sludge settling space, 15. Sludge removal port, 16. Anode outlet point, 17. Cathode plates, 18. Direction of catholyte flow, 19. Outlet for treated water, 20. Electrical fitting at anode, 21. Electrical fitting at cathode, 22. External resistance.

Figure 1.2. Schematic presentation of novel micobial auto flow fuel cell (MAFFC) (Mathuriya 2016c).

Mathuriya (2016c) disclosed a 25L tubular trickling bio-electrochemical filter reactor, with peripheral air cathode arrangement (Fig. 1.3). In this design, wastewater to be treated moved from the inlet to the outlet through a pathway with less hydraulic resistance, which occupied the part of the interior space of the anodic chamber. The feed supplied from the bottom of the reactor passed upward through the anode and exited at the top. The compact reactor is suitable for rural electrification. Ghadge and Ghangrekar (2015) developed a low cost 26L air cathode MFC with a clayware separator and observed stable performance up to 14 months.

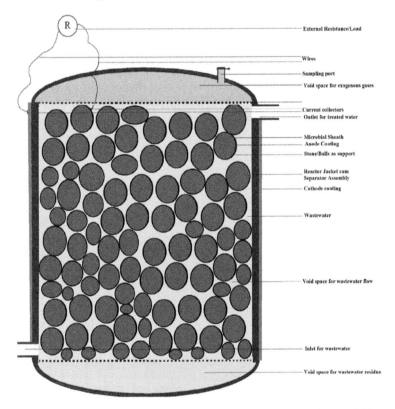

Figure 1.3. Schematic representation of trickling bioelectrochemical filter reactor and arrangement of various electrodes (Mathuriya 2016b).

Recently Jadhav (2017) developed 100L stacked bioelectric toilet MFC, capable of treating 87-92% of organic matter present in waste received from septic tank and capable of charging mobile phone batteries and operating different electronic appliances using advanced power management system. Bioelectric toilet MFC fabricated with 36 pairs of separator electrode assemblies (SEAs) and PVC battery/modified clayware separator was operated in an upflow manner with continuous substrate flow in S-shape direction to allow uniform distribution of wastewater.

Journey of MFCs-Field

No technology can bolster the scientific inquisitiveness, until it is practically feasible. MFCs have the potentials to be practically applicable in various areas of modern interest *viz.* for sensors (Jiang et al. 2017), for small electronic gadgets, mobile phones (Ieropoulos et al. 2013), and robots (Ieropoulos et al. 2010), as well as pee power urinals (Ieropoulos et al. 2016). Therefore, researchers and industries are coming forward to apply this technology in the real field. Over the last decade, MFCs have experienced significant scientific and technological development, to the point of becoming close to commercialization (Trapero et al. 2017). Key players of the global MFC market include Fluence Corporation, Cambrian Innovation Inc., MICROrganic Technologies Inc., Microbial Robotics, Prongineer, and Triqua International, etc.

The US army is developing a 2250L sewage run MFC to power itself and even have an excess to be stored for future use. MFC would help reduce large supplies of fuel and water to isolated army bases in Afghanistan and Iraq. The army is working alongside a Boston company, Cambrian Innovation, which has developed the 'BioVolt', electrically active microbes that are used as catalysts to treat wastewater and generate electricity (https://www.engerati. com/article/viewing-waste-another-energy-resource, accessed on 10.3.18). Some companies like MICROrganic Technologies (http://microrganictech. com/) and Keego Technologies (https://www.magicalmicrobes.com/) are developing kits and science materials for classrooms, high-school students, and researchers. The kits are also available on online shopping sites and the company's store. Pilus Energy, a wholly-owned subsidiary of Tauriga Sciences, Inc. a USA based life sciences company, develops MFC based bacterial robots, or BactoBots. The BactoBot is used for water remediation during direct current (DC) electricity, and economically important gases and chemicals production. BactoBots are anaerobically and aerobically active, even with low BOD and chemical oxygen demand (COD) (http://www. taurigasciences.com/). Cambrian Innovation Inc. Boston, Massachusetts, USA (http://cambrianinnovation.com/) deals with MFCs for industrial applications under the brand name 'EcoVolt', a wastewater treatment technology; and 'EcoVolt Mini', an all-in-one water reuse container. The company also provides 'EcoVolt MBR' (membrane bioreactor) to serve the food and beverage industries.

Many other companies like Hy-SyEnce Inc. USA (http://www.hy-syence.com/index.php); Zigco LLC, USA (http://www.zigcollc.com); Emefcy, Israel; Fluence Corporation Limited, Israel (Emefcy); Pron-gineer, Canada (http://prongineer.com/) and CASCADE (Computer-Assisted Strain Construction and Development Engineering) Clean Energy, Inc. (http://www.ccleanenergy.com/) are also working in the field of MFCs.

Future Trends

MFC technology has developed remarkably recently, and research nimbus of MFC technology is continuously expanding. MFC research is witnessing sustained efforts in almost every aspect of its improvement. This technology holds commitment towards sustainable power generation, and various other applications. It offers direct electricity generation from invaluable and harmful substances like wastewater. MFCs are expected to witness a relatively higher growth rate during the next decade; however, a number of significant challenges viz. cost reduction, scale up, and increment in electricity production must be addressed before MFCs step into the global green energy market.

REFERENCES

Agarry, S.E. 2017. Bioelectricity generation and treatment of petroleum refinery effluent by *Bacillus cereus* and *Clostridium butyricum* using microbial fuel cell technology. Nigerian J. Technol. 36: 543-551.

Allen, R.M. and H.P. Bennetto. 1993. Microbial fuel-cells: electricity production from carbohydrates. Appl. Biochem. Biotechnol. 39: 27-40.

Bean, R.C., Y.H. Inami, P.R. Basford, M.H. Boyer, W.C. Shepherd, E.R. Walwick and R.E. Kay. 1964. Study of the fundamental principles of bio-electrochemistry. NASA Final Technical Report Research Laboratories, Philco Corporation, p 107.

Behera, M., P.S. Jana, T.T. More and M.M. Ghangrekar. 2010. Rice mill wastewater treatment in microbial fuel cells fabricated using proton exchange membrane and earthen pot at different pH. Bioelectrochem. 79: 228-233.

Bennetto, H.P. 1990. Electricity generation by microorganisms. Biotechnol. Edu. 1: 163-168.

Bombelli, P., R.J. Dennis, F. Felder, M.B. Cooper, D.M.R. Iyer, J. Royles, S.T. Harrison, A.G. Smith, C.J. Harrison and C.J. Howe. 2016. Electrical output of bryophyte microbial fuel cell systems is sufficient to power a radio or an environmental sensor. Open Sci. 3. DOI: 10.1098/rsos.160249.

Bond, D.R. and D.R. Lovley. 2003. Electricity production by *Geobacter sulfurreducens* attached to electrodes. Appl. Environ. Microbiol. 69: 1548-1555.

Buitrón, G. and C. Cervantes-Astorga. 2013. Performance evaluation of a low-cost microbial fuel cell using municipal wastewater. Water Air Soil Pollution. 224: 1470-1478.

Canfield, J.H. and B.H. Goldner. 1964. Research on applied bioelectrochemistry. NASA Technical Report Magna Corporation, Anaheim, p 127.

Canfield, J.H., B.H. Goldner and R. Lutwack. 1963. Utilization of human wastes as electrochemical fuels. NASA Technical Report Magna Corporation, Anaheim, p 63.

Cao, Y., H. Yang, X. Ai and L. Xiao. 2003. The mechanism of oxygen reduction on MnO_2-catalyzed air cathode in alkaline solution. J. Electroanal. Chem. 557: 127-134.

Chang, I.S., J.K. Jang, G.C. Gil, M. Kim, H.J. Kim, B.W. Cho and B.H. Kim. 2004. Continuous determination of biochemical oxygen demand using microbial fuel cell type biosensor. Biosen. Bioelectron. 19: 607-613.

Chen, M., Y. Zeng, Y. Zhao, M. Yu, F. Cheng, X. Lu and Y. Tong. 2016. Monolithic three-dimensional graphene frameworks derived from inexpensive graphite paper as advanced anodes for microbial fuel cells. J. Materials Chem. A 4: 6342-6349.

Choi, C. and Y. Cui. 2012. Recovery of silver from wastewater coupled with power generation using a microbial fuel cell. Biores. Technol. 107: 522-525.

Cohen, B. 1931. The bacterial culture as an electrical half-cell. J. Bacteriol. 21: 18–19.

Davila, D., J.P. Esquivel, N. Sabate and J. Mas. 2011. Silicon-based microfabricated microbial fuel cell toxicity sensor. Biosen. Bioelectron. 26: 2426-2430.

Del Duca, M.G. and J.M. Fuscoe. 1964. Thermodynamics and applications of bioelectrochemical energy conversion systems. NASA Technical Report National Aeronautics and Space Administration, Washington, p 63.

Ding, W., S. Cheng, L. Yu and H. Huang. 2017. Effective swine wastewater treatment by combining microbial fuel cells with flocculation. Chemosphere. 182: 567-573.

Ewing, T., P.T. Ha and H. Beyenal. 2017. Evaluation of long-term performance of sediment microbial fuel cells and the role of natural resources. Appl. Energy. 192: 490-497.

Fan, Y., H. Hu and H. Liu. 2007. Enhanced coulombic efficiency and power density of air-cathode microbial fuel cells with an improved cell configuration. J. Power Sour. 171: 348-354.

Fang, Z., H.L. Song, N. Cang and X.N. Li. 2013. Performance of microbial fuel cell coupled constructed wetland system for decolorization of azo dye and bioelectricity generation. Biores. Technol. 144: 165-171.

Fang, Z., X. Cao, X. Li, H. Wang and X. Li. 2017. Electrode and azo dye decolorization performance in microbial-fuel-cell-coupled constructed wetlands with different electrode size during long-term wastewater treatment. Biores. Technol. 238: 450-460.

Fangzhou, D., L. Zhenglong, Y. Shaoqiang, X. Beizhen and L. Hong 2011. Electricity generation directly using human feces wastewater for life support system. Acta Astronautica. 68: 1537-1547.

Ghadge, A.N. and M.M. Ghangrekar. 2015. Performance of low cost scalable air-cathode microbial fuel cell made from clayware separator using multiple electrodes. Biores. Technol. 182: 373-377.

Ghasemi, M., W.R.W. Daud, S.H. Hassan, T. Jafary, M. Rahimnejad, A. Ahmad and M.H. Yazdi. 2016. Carbon nanotube/polypyrrole nanocomposite as a novel cathode catalyst and proper alternative for Pt in microbial fuel cell. Int. J. Hydrogen Energy. 41: 4872-4878.

Gong, D. and G. Qin. 2012. Treatment of oilfield wastewater using a microbial fuel cell integrated with an up-flow anaerobic sludge blanket reactor. Desalination Water Treat. 49: 272-280.

Gregory, K.B., D.R. Bond and D.R. Lovley. 2004. Graphite electrodes as electron donors for anaerobic respiration. Environ. Microbiol. 6: 596-604.

Huang, J., N. Zhu, T. Yang, T. Zhang, P. Wu and Z. Dang. 2015. Nickel oxide and carbon nanotube composite (NiO/CNT) as a novel cathode non-precious metal catalyst in microbial fuel cells. Biosen. Bioelect. 72: 332-339.

Huang, J., P. Yang, Y. Guo and K. Zhang. 2011. Electricity generation during wastewater treatment: an approach using an AFB-MFC for alcohol distillery wastewater. Desalination 276: 373-378.

Huang, L. and B.E. Logan. 2008. Electricity generation and treatment of paper recycling wastewater using a microbial fuel cell. Appl. Microbiol. Biotechnol. 80: 349-355.

Huang, L., T. Li, C. Liu, X. Quan, L. Chen, A. Wang and G. Chen. 2013. Synergetic interactions improve cobalt leaching from lithium cobalt oxide in microbial fuel cells. Biores. Technol. 128: 539-546.

Huang, Y., Z.B. He, J. Kan, A.K. Manohar, K.H. Nealson and F. Mansfeld. 2012. Electricity generation from a floating microbial fuel cell. Biores. Technol. 114: 308-313.

Ichihashia, O., K. Matsuurab, K. Hirookaa and T. Takeguchi. 2016. Application of Zirconium-Based Materials as Catalyst of Air-Cathode in Microbial Fuel Cells. J. Water Env. Technol. 14: 106-113.

Ieropoulos, I., J. Greenman, C. Melhuish and I. Horseld. 2010. EcoBot-III: a robot with guts. pp. 733-740. *In*: H. Fellermann, M. Dorr, M. Hanczyc, L.L. Laursen, S. Maurer, D. Merkle et al. [eds.]. Artificial Life X: Proceedings of the 12th International Conference on the Synthesis and Simulation of Living Systems. Massachusetts, USA: Massachusetts Institute of Technology Press, USA.

Ieropoulos, I., A. Stinchcombe, I. Gajda, S. Forbes, I. Merino-Jimenez, G. Pasternak, D. Sanchez-Herranz and J. Greenman. 2016. Pee power urinal-microbial fuel cell technology field trials in the context of sanitation. Environ. Sci.: Water Res. Technol. 2: 336-343.

Ieropoulos, I., P. Ledezma, A. Stinchcombe, G. Papaharalabos, C. Melhuish and J. Greenman. 2013. Waste to real energy: the first MFC powered mobile phone. Physical Chemistry Chemical Physics 15: 15312-15316.

Jadhav, D.A. 2017. Performance enhancement of microbial fuel cells through electrode modifications along with development of bioelectric toilet, PhD Thesis (unpublished), Indian Institute of Technology, Kharagpur, West Bengal, India.

Jadhav, D.A., A.N. Ghadge, D. Mondal and M.M. Ghangrekar. 2014. Comparison of oxygen and hypochlorite as cathodic electron acceptor in microbial fuel cells. Biores. Technol. 154: 330-335.

Jiang, S.P. and P.K. Shen (eds.). 2013. Nanostructured and advanced materials for fuel cells. CRC Press

Jiang, Y., A.C. Ulrich and Y. Liu. 2013. Coupling bioelectricity generation and oil sands tailings treatment using microbial fuel cells. Biores. Technol. 139: 349-354.

Jiang, Y., P. Liang, P. Liu, X. Yan, Y. Bian and X. Huang. 2017. A cathode-shared microbial fuel cell sensor array for water alert system. Int. J. Hydrogen Energy 42: 4342-4348.

Kaewkannetra, P., W. Chiwes and T.Y. Chiu. 2011. Treatment of cassava mill wastewater and production of electricity through microbial fuel cell technology. Fuel. 90: 2746-2750.

Kamaraj, S.K., A.E. Rivera, S. Murugesa, J. García-Mena, C.F. Reyes and J. Tapia-Ramírez. 2015. Electricity generation from nopal biogas waste biomass using clay cup (cantarito) modified microbial fuel cell. Proc. 6th European Fuel Cell - Piero Lunghi Conference, EFC. Naples; Italy: Code 124051.

Kelly, I., O. Holland, M. Scull and D. McFarland. 1999. Artificial autonomy in the natural world: building a robot predator. Adv. Artificial Life 289-293.

Khaled, F., O. Ondel and B. Allard. 2016. Microbial fuel cells as power supply of a low-power temperature sensor. J. Power Sour. 306: 354-360.

Kharkwal, S., Y.C. Tan, M. Lu and H.Y. Ng. 2017. Development and Long-Term Stability of a Novel Microbial Fuel Cell BOD Sensor with MnO_2 Catalyst. Int. J. Molecular Sci. 18: 276.

Kim, B., I. Chang, M. Hyun, H. Kim and H. Park. 2001. A biofuel cell using wastewater and active sludge for wastewater treatment. International patent: WO0104061.

Kim, B.H., I.S. Chang, G.C. Gil, H.S. Park. and H.J. Kim. 2003a. Novel BOD (biological oxygen demand) sensor using mediator-less microbial fuel cell. Biotechnol. Let. 25: 541-545.

Kim, M., S.M. Youn, S.H. Shin, J.G. Jang, S.H. Han, M.S. Hyun, G.M. Gadd and H.J. Kim. 2003b. Practical field application of a novel BOD monitoring system. J. Environ. Monitoring. 5: 640-643.

Kim, B.H., H.S. Park, H.J. Kim, G.T. Kim, I.S. Chang, J. Lee et al. 2004. Enrichment of microbial community generating electricity using a fuel-cell-type electrochemical cell. Appl. Microbiol. Biotechnol. 63: 672-681.

Kim, H.J., M.S. Hyun, I.S. Chang and B.H. Kim. 1999. A microbial fuel cell type lactate biosensor using a metal-reducing bacterium, *Shewanella putrefaciens*. J. Microbiol. Biotechnol. 9: 365-367.

Kim, J., S.A. Cheng, S. Oh and B.E. Logan. 2007. Power generation using different cation, anion, and ultrafiltration membranes in microbial fuel cells. Env. Sci. Technol. 41: 1004-1009.

Kluger, J. 1997. Uncovering the secrets of Mars. Time: July 14th, 26-36.

Krieg, T., F. Enzmann, D. Sell, J. Schrader and D. Holtmann. 2017. Simulation of the current generation of a microbial fuel cell in a laboratory wastewater treatment plant. Appl. Energy 195: 942-949.

Kumar, R., L. Singh, A.W. Zularisam and F.I. Hai. 2016. Potential of porous Co_3O_4 nanorods as cathode catalyst for oxygen reduction reaction in microbial fuel cells. Biores. Technol. 220: 537-542.

Liu, Z., J. Liu, S. Zhang, X.H. Xing and Z. Su. 2011. Microbial fuel cell based biosensor for in situ monitoring of anaerobic digestion process. Biores. Technol. 102: 10221-10229.

Lovley, D.R. and K. Nevin. 2016. Smart (subsurface microbial activity in real time) technology for real-time monitoring of subsurface microbial metabolism. U.S. Patent Application. 15/131,421.

Ma, M., Y. Dai, J.L. Zou, L. Wang, K. Pan and H.G. Fu. 2014. Synthesis of iron oxide/partly graphitized carbon composites as a high-efficiency and low-cost cathode catalyst for microbial fuel cells. ACS Appl. Mater. Interfaces 6: 13438-13447. DOI: 10.1021/am501844p

Ma, M., S. You, J. Qu and N. Ren. 2016. Natural eggshell membrane as separator for improved coulombic efficiency in air-cathode microbial fuel cells. RSC Adv. 6: 66147-66151.

Mathuriya, A.S. 2016c. A trickling bio-electrochemical filter for waste reduction and electricity generation. (No. 201611008125) Intellectual Property Journal India 20/2016: 18668.

Mathuriya, A.S. and V.N. Sharma. 2009. Bioelectricity production from paper industry waste using a microbial fuel cell by Clostridium species. J. Biochemical Technol. 1: 49-52.

Mathuriya, A.S. and V.N. Sharma. 2010a. Bioelectricity production from various wastewaters through microbial fuel cell technology. J. Biochemical Technol. 2: 133-137.

Mathuriya, A.S. and V.N. Sharma. 2010b. Treatment of brewery wastewater and production of electricity through microbial fuel cell technology. Int. J. Biotechnol. Biochem. 6: 71-80.

Mathuriya, A.S. and J.V. Yakhmi. 2014. Microbial fuel cells–Applications for generation of electrical power and beyond. Crit. Rev. Microbiol. 42: 127-143.

Mathuriya, A.S. 2014. Enhanced tannery wastewater treatment and electricity generation in microbial fuel cell by bacterial strains isolated from tannery waste. Env. Eng. Manage. J. 13: 2945-2954.

Mathuriya, A.S. 2016a. Evaluation of novel microbial fuel cell design for treatment of different wastewaters simultaneously. J. Env. Sci. 42: 105-111.

Mathuriya, A.S. 2016b. Novel microbial auto flow fuel cell (MAFFC) architecture for large scale energy efficient wastewater treatment. (No: 201611016850 A) Intellectual Property Journal India 23/2016: 23155.

Mehta, P., A. Hussain, B. Tartakovsky, V. Neburchilov, V. Raghavan, H. Wang and S.R. Guiot. 2010. Electricity generation from carbon monoxide in a single chamber microbial fuel cell. Enzyme Microb. Technol. 46: 450-455.

Melhuish, C. and M. Kubo. 2007. Collective energy distribution: maintaining the energy balance in distributed autonomous robots using trophallaxis. Distrib. Auton. Robotic Syst. 6: 275-284.

Melhuish, C., I. Ieropoulos, J. Greenman and I. Horsfield. 2006. Energetically autonomous robots: food for thought. Auton. Robots. 21: 187-198.

Mohan, S.V., S.V. Raghavulu and P. Sarma. 2008. Biochemical evaluation of bioelectricity production process from anaerobic wastewater treatment in a single chambered microbial fuel cell (MFC) employing glass wool membrane. Biosens. Bioelectron. 23: 1326-1332.

Monasterio, S., M. Mascia and M. Di Lorenzo. 2017. Electrochemical removal of microalgae with an integrated electrolysis-microbial fuel cell closed-loop system. Sep. Purif. Technol. 183: 373-381.

Niessen, J., F. Harnisch, M. Rosenbaum, U. Schroder and F. Scholz. 2006. Heat treated soil as convenient and versatile source of bacterial communities for microbial electricity generation. Electrochem. Comm. 8: 869-873.

Paitier, A., A. Godain, D. Lyon, N. Haddour, T.M. Vogel and J.M. Monier. 2017. Microbial fuel cell anodic microbial population dynamics during MFC start-up. Biosen. Bioelectr. 92: 357-363.

Pant, D., G. Van Bogaert, L. Diels and K. Vanbroekhoven. 2010. A review of the substrates used in microbial fuel cells (MFCs) for sustainable energy production. Biores. Technol. 101: 1533-1543.

Parameswaran, P., H. Zhang, C.I. Torres, B.E. Rittmann and R. Krajmalnik-Brown. 2010. Microbial community structure in a biofilm anode fed with a fermentable substrate: the significance of hydrogen scavengers. Biotechnol. Bioeng. 105: 69-78.

Potter, M.C. 1911. Electrical effects accompanying the decomposition of organic compounds. Proc. Roy. SOC. London Ser. B 84: 260-276.

Potter, M.C. 1910. On the difference of potential due to the vital activity of microorganisms. Proc. Univ. Durham Philosophical Soc. 3: 245-249.

Rabaey, K. and W. Verstraete. 2005. Microbial fuel cells: novel biotechnology for energy generation. Trends in Biotechnol. 23: 291-298.

Qian, F., Z. He, M.P. Thelen and Y. Li. 2011. A microfluidic microbial fuel cell fabricated by soft lithography. Biores. Technol. 102: 5836-5840.

Qiu, R., B. Zhang, J. Li, Q. Lv, S. Wang and Q. Gu. 2017. Enhanced vanadium (V) reduction and bioelectricity generation in microbial fuel cells with biocathode. J. Power Sour. 359: 379-383.

Ren, L., Y. Ahn and B.E. Logan. 2014. A two-stage microbial fuel cell and anaerobic fluidized bed membrane bioreactor (MFC-AFMBR) system for effective domestic wastewater treatment. Env. Sci. Technol. 48: 4199-4206.

Ryu, E.Y., M.A. Kim and S.J. Lee. 2011. Characterization of microbial fuel cells enriched using Cr (VI)-containing sludge. J. Microbiol. Biotechnol. 21: 187-191.

Shukla, A.K., P. Suresh, S. Berchmans and A. Rajendran. 2004. Biological fuel cells and their applications. Curr. Sci. 87: 455-468.

Sonawane, J.M., S.B. Adeloju and P.C. Ghosh. 2017. Landfill leachate: a promising substrate for microbial fuel cells. Int. J. Hydrogen Energy (available online 11.4.17).

Song, T.S., Y. Jin, J. Bao, D. Kang and J. Xie. 2016. Graphene/biofilm composites for enhancement of hexavalent chromium reduction and electricity production in a biocathode microbial fuel cell. J. Hazard. Mat. 317: 73-80.

Stein, N.E., H.M. Hamelers, G. van Straten and K.J. Keesman. 2012. On-line detection of toxic components using a microbial fuel cell-based biosensor. J. Process Control 22: 1755-1761.

Sun, J., Z. Bi, B. Hou, Y.Q. Cao and Y.Y. Hu. 2011. Further treatment of decolorization liquid of azo dye coupled with increased power production using microbial fuel cell equipped with an aerobic biocathode. Water Res. 45: 283-291.

Sun, M., F. Zhang, Z.H. Tong, G.P. Sheng, Y.Z. Chen, Y. Zhao, Y.P. Chen, S.Y. Zhou, G. Liu, Y.C. Tian and H.Q. Yu. 2010. A gold-sputtered carbon paper as an anode for improved electricity generation from a microbial fuel cell inoculated with *Shewanella oneidensis* MR-1. Biosens. Bioelectron. 26: 338-343.

Tao, X.I.E., Y.M. Gao, Q. Zheng, X.H. Wang, L.I. Yuan, L.U.O. Nan and L.I.U. Rui. 2017. A double-microbial fuel cell heavy metals toxicity sensor. DEStech Trans. Env. Ener. Earth Sci. (icesee), DOI: 10.12783/dteees/icesee2017/7853.

Tender, L.M., S.A. Gray, E. Groveman, D.A. Lowy, P. Kauffman, J. Melhado, R.C. Tyce, D. Flynn, R. Petrecca and J. Dobarro. 2008. The first demonstration of a microbial fuel cell as a viable power supply: powering a meteorological buoy. J. Power Sour. 179: 571-575.

Theodosiou, P., A. Faina, F. Nejatimoharrami, K. Stoy, J. Greenman, C. Melhuish and I. Ieropoulos. 2017. EvoBot: towards a robot-chemostat for culturing and maintaining microbial fuel cells (MFCs). Living Machines 2017: Biomimetic and Biohybrid Systems. Conference on Biomimetic and Biohybrid Systems 26-28 July, Stanford, CA, USA, pp. 453-464.

Thomas, Y.R., M. Picot, A. Carer, O. Berder, O. Sentieys and F. Barriere. 2013. A single sediment-microbial fuel cell powering a wireless telecommunication system. J. Power Source 241: 703-708.

Tommasi, T., A. Chiolerio, M. Crepaldi and D. Demarchi. 2014. A microbial fuel cell powering an all-digital piezoresistive wireless sensor system. Microsystem Technol. 20: 1023-1033.

Trapero, J.R., L. Horcajada, J.J. Linares and J. Lobato. 2017. Is microbial fuel cell technology ready? An economic answer towards industrial commercialization. Appl. Ener. 185: 698-707.

Velasquez-Orta, S.B., I.M. Head, T.P. Curtis and K. Scott. 2011. Factors affecting current production in microbial fuel cells using different industrial wastewaters. Biores. Technol. 102: 5105-5112.

Walter, X.A., A. Stinchcombe, J. Greenman and I. Ieropoulos. 2017. Urine transduction to usable energy: a modular MFC approach for smartphone and remote system charging. Appl. Ener. 192: 575-581.

Wang, X.Q., C.P. Liu, Y. Yuan and F.B. Li. 2014. Arsenite oxidation and removal driven by a bio-electro-Fenton process under neutral pH conditions. J. Hazard. Mat. 275: 200-209.

Wang, Z. and B. Lim. 2017. Mixed cellulose ester filter as a separator for air-diffusion cathode microbial fuel cells. Env. Technol. 38: 979-984.

Watanabe, K. 2008. Recent developments in microbial fuel cell technologies for sustainable bioenergy. J. Biosci. Bioeng. 106: 528-536.

Wilkinson, S. and C.S. Campbell. 1996. Green bug robots—renewable environmental power for miniature robots. *In*: Proc. 4th IASTED Int. Conf. Robot. Manuf. Aug 19–22; Honolulu, Hawaii. Paper 247-113: 275-278.

Wilkinson, S. 1999. Accrued but novel carrot powered 'Gastrobot' for middle or high school demonstrations. *In*: Proc. 7th IASTED Int. Conf. Robot. Appl. Oct 28-30; Santa Barbara, California. Paper 304-024: 347-351.

Wilkinson, S. 2000a. "Gastrobots"—benefits and challenges of microbial fuel cells in food powered robot applications. Autonomous Robots. 9: 99-111.

Wilkinson, S. 2000b. "Gastronome"—A Pioneering Food Powered Mobile Robot. IASTED Int. Conf. Robot. Appl. Paper # 318-037.

Winfield, J., L.D. Chambers, A. Stinchcombe, J. Rossiter and I. Ieropoulos. 2014. The power of glove: soft microbial fuel cell for low-power electronics. J. Power Sour. 249: 327-332.

Winfield, J., I. Ieropoulos, J. Rossiter, J. Greenman and D. Patton. 2013. Biodegradation and proton exchange using natural rubber in microbial fuel cells. Biodegrad. 24: 733-739.

Wu, D., L. Huang, X. Quan and G.L. Puma. 2016. Electricity generation and bivalent copper reduction as a function of operation time and cathode electrode material in microbial fuel cells. J. Power Sour. 307: 705-714.

You, S.J., J.N. Zhang, Y.X. Yuan, N.Q. Ren and X.H. Wang. 2010. Development of microbial fuel cell with anoxic/oxic design for treatment of saline seafood wastewater and biological electricity generation. J. Chem. Technol. Biotechnol. 85: 1077-1083.

Yuan, G.E., Y. Li, J. Lv, G. Zhang and F. Yang. 2017. Integration of microbial fuel cell and catalytic oxidation reactor with iron phthalocyanine catalyst for Congo red degradation. Biochem. Eng. J. 120: 118-124.

Yuan, H., L. Deng, Y. Qi, N. Kobayashi and J. Tang. 2014. Nonactivated and Activated Biochar Derived from Bananas as Alternative Cathode Catalyst in Microbial Fuel Cells. Scientific World J. Article ID 832850, 8 http://dx.doi.org/10.1155/2014/832850

Zhang, L., X. Yin and S.F.Y. Li. 2015. Bio-electrochemical degradation of paracetamol in a microbial fuel cell-Fenton system. Chem. Eng. J. 276: 185-192.

Zhang, X., S. Cheng, X. Huang and B.E. Logan. 2010. The use of nylon and glass fiber filter separators with different pore sizes in air-cathode single-chamber microbial fuel cells. Energy Environ. Sci. 3: 659-664.

Zhao, C.E., W.J. Wang, D. Sun, X. Wang, J.R. Zhang and J.J. Zhu. 2014. Nanostructured graphene/TiO_2 hybrids as high-performance anodes for microbial fuel cells. Chemistry 220: 7091-7097.

Zheng, S., F. Yang, L. Liu, Q. Xiong, T. Yu, F. Zhao, U. Schröder and H. Hou. 2015. Binder-free carbon black/stainless steel mesh composite electrode for high-performance anode in microbial fuel cells. J. Power Sour. 284: 252-257.

Microbial Fuel Cell (MFC) Variants

Sahriah Basri* and Siti Kartom Kamarudin

Universiti Kebangsaan Malaysia Fuel Cell Institute Research Complex,
Bangi, Selangor, Malaysia, 43600

Introduction

Microbial Fuel Cell (MFC) is a device that uses microbes or bacteria as
biocatalysts to produce electricity via chemical reactions. It is also known
as the Bio-Electrochemical System (BES), which is the combination of classic
abiotic electrochemical reactions with biological catalyst redox activity
(Carlo et al 2017). The discovery of energy derived from bacteria began in
1911 with M. Potter, a botanist at the University of Durham, who discovered
the ability of E. coli to produce electricity (Bullen et al. 2006). Through the
study of how microorganisms destroyed organic compounds, he found that
electricity was also produced. With this, he managed to reap new energy
sources, which were able to build a primitive Microbial Fuel Cell. In fact, little
progress was established in the primitive design until the 1980s. M.J. Allen
and H. Peter Bennetto of Kings College in London revolutionised the design
of the original microbial fuel cell. Driven by their desire to provide cheap
and reliable power to third world countries, Allen and Bennetto achieved
progress by combining their understanding of the electron transport chain
and the significant advancement in technology to produce the basic designs
that are still in use as MFCs until today. Yet, the use of MFC in third world
countries is still in its infancy due to the complexity of its design, which
makes it difficult for poor rural farmers to build. Such progress by the King's
College team has provided proof to the scientific community that MFC can
be a useful technology and increased interest in their development should be
promoted (Justin 2010).

MFCs offer many potential advantages over other means of localised
power generation. In general, since fuel cells do not use combustion, their
efficiencies are not limited by the Carnot cycle. The microorganisms in MFCs
can derive energy from many different types of fuels (Bullen et al. 2006),

*Corresponding author: sahriah@ukm.edu.my

making them convenient for situations where refined fuels are not available. While the substrate molecules are oxidised via microbe metabolism as opposed to combustion, there are no harmful, partially oxidised by-products, such as carbon monoxide (Haoran et al. 2016). Although different types of MFCs have been designed for various operating conditions, MFCs are generally operated at room temperature (Youngho and Bruce 2010, Carmen et al. 2018) and neutral pH (Manaswini and Ghangrekar 2009), so they can be utilised in harsh conditions where it is normally thought impractical or undesirable, unlike many other types of fuel cells.

As scientists and researchers around the world began investigating MFCs, the major question to explore was how does an electron get from an electron transport chain to the anode? Nowadays, researchers are working to optimise the electrode material, the type and combination of bacteria, and the transfer of electrons in MFCs. While the idea of using energy generated by bacteria has been around for almost 100 years, researchers have begun to fully understand and appreciate potential MFC and how to enhance its potential for commercialisation.

MFC is very valuable as there are many advantageous applications in reducing pollution and cost of wastewater treatment in a sustainable and environmentally friendly way. Currently, MFCs are used to produce electricity and at the same time as a purification method of wastewater. Besides producing high efficiency hydrogen (Yu et al. 2012), MFC has resolved problems related to power consumption by maintaining and preserving the environment, such as providing cheap and accessible power to remote areas in Africa (Kaewkannetra et al. 2011), where 74% of the population live without electricity. Moreover, it can also be used in water bioremediation containing organic pollutants, such as toluene and benzene, which are compounds found in petrol. One of the major advantages of using MFCs in remote sensing rather than traditional batteries is the production of bacteria, making MFC last longer than traditional batteries. The sensor can be left alone in remote areas for years without maintenance.

With future development, MFC has the potential to produce hydrogen for fuel cells, desalinate seawater, wastewater treatment, and provide sustainable energy sources for remote areas. Although historically MFCs were used only as a novelty in science exhibitions, they are now a growing reality with great potential for improvements in cleaning techniques and power generation processes. Hence, this chapter will focus on the varying types of BES systems, such as Microbial Fuel Cells (MFC), Microbial Electrolysis Cells (MEC), Microbial Desalination Cells (MDC), and Microbial Electrosynthesis Cells (MES).

Microbial Fuel Cell (MFC)

Basically, an MFC system consists of two compartments known as anode and cathode. Both are separated with a membrane called a cation exchange membrane (CEM) or proton exchange membrane (PEM). The transfer of electrons in the compound have been moved by extracellular electron transfer (EET) to the electron acceptor at the anode. After the reaction in the anode, the electrons then pass through an external circuit to produce electricity. Then, the electron continues to combine with protons and a terminal electron acceptor at the cathode, where the process is mediated by microorganisms.

MFCs can be used to produce hydrogen for use as an alternative fuel. When used for hydrogen production, MFCs need to be added by external energy sources to overcome energy barriers to make all organic matter, carbon dioxide, and hydrogen gas. MFC standard is converted to hydrogen production by maintaining both anaerobic columns and adding MFC with 0.25 volts of electricity. Hydrogen foam is formed in the cathode and collected for use as fuel. Although electricity is used instead of being produced as a normal MFC, this hydrogen-producing method is highly effective as the protons and electrons are produced by the bacteria in the anode, and more than 90% become hydrogen gas. Conventional hydrogen production requires 10 times the amount of energy as a customised MFC, making MFC the most effective and environmentally friendly way to produce hydrogen for use as fuel.

Besides, MFCs are able to convert organic substances in wastewaters into usable energy sources by using Electrochemical Active Bacteria (EAB) as biocatalysts. Recently, researchers have expanded the study with numerous microorganisms either in active or passive MFC systems. The research studied potential microorganisms in MFCs which included metal reducing bacteria, such as *Geobacter* spp. and *Shewanella* spp., and phototrophic bacteria, e.g., *Rhodopseudomonas* spp. (Schilir et al. 2016) and *Arthrobacter* sp. Other than that, *Stenotrophomonas* sp. and *Sphingobacterium* sp. bacteria are commonly used on the polarity-inverting electrode. Zhang et al. (2018) further explored *Arthrobacter* sp., which is thought to be proficient in transporting electrons either to electrodes or from electrodes.

In breweries and food manufacturing industries, waste can be treated by MFCs because of their organic-rich wastewater compounds that can serve as food for microorganisms (Mabel et al. 2017). The manufacture of breweries is appropriate for the implementation of MFCs (Edith et al. 2014), since the composition of their wastewater is always the same (Simate 2015). This condition persists, allowing bacteria to adjust and become more efficient. Currently, Fosters, an Australian beer company, has started testing MFCs to clean its sewage for generating electricity and clean water (International Business 2007). The company installed small-scale MFCs for brewing water treatment, where each long tube is a large microbial fuel cell. Twelve MFCs were put in parallel to clean up large volumes of wastewater. Sewage water flowed to the top, cleared by bacteria, and came out from the bottom as purified water. Associates from the University of Queensland, and the Fosters plan to increase the power of MFC cleaners and electrical output and eventually build 660 gallons, 2 kilowatt MFCs that will manage to clean up all of the company's wastewater. The power generated from cleaning the brewing waste is expected to pay the initial cost of the MFC within ten years.

Sewage wastewater can also be exchanged through MFCs to decompose the organic waste contained within it. Research has shown that MFC can reduce the amount of organic matter found in sewage wastes up to 80% (Liu et al. 2004). This process is very similar to the treatment of brewing water, with the only difference being that water should be first prepared to remove toxins and other non-biodegradable substances. This is a challenging step because the sewage water is often changed according to the composition and may require considerable treatment before it can be cleared by MFCs. However, this extensive treatment is permitted by electricity generated during wastewater cleaning. Electricity output from MFC will help offset the high cost of wastewater processing. Wastewater treatment plants may cause less dumping of pollutants into the oceans and rivers if the money saved from electricity bills is placed for further cleaning of wastewater. Besides, MFC is capable of grouping seawater bacteria to reduce nitrate in synthetic water samples and producing electricity by oxidising organic matter (Naga et al. 2018).

A new single cathode chamber and multiple anode chambers of the MFC design (MAC-MFC) were developed by Abhilasha (2016) as shown in Fig. 2.1. At parallel operations with SC-MFC, MAC-MFC showed a more stable and higher performance than any other anode chambers in MAC-MFC, which only performed as backup to others. Thus, MFCs performance is influenced by other factors *viz.* anolyte, microbial activity, cathode electron transfer efficiency, internal resistance, and PEM.

The MFC stacking effect with continuous mode operation was evaluated by Swathi et al. (2017) on the stable power output. Three samples of single-chambered air cathode CMFCs with the difference being Terry cotton ($CMFC_T$), Nafion ($CMFC_N$), and without membrane ($CMFC_{ML}$) were varied and ran in continuous mode, as shown in Fig. 2.2. Results on COD removal

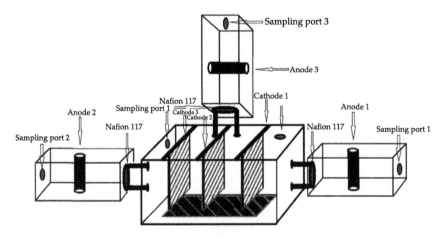

Figure 2.1. Schematic representation of a standard single anode-cathode chamber microbial fuel cell (SC-MFC) and multiple anode chamber and single cathode chamber microbial fuel cell (MAC-MFC) design (Abhilasha 2016).

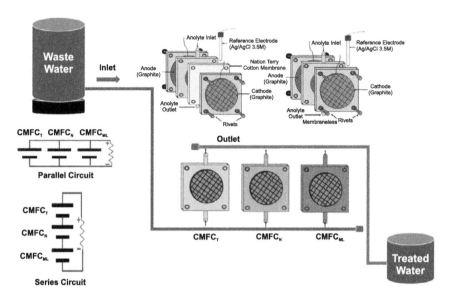

Figure 2.2. MFC stacking design system (Swathi et al. 2017).

efficiency and maximum power density (PD) were obtained for $CMFC_N$ (50%, 0.1 W/m^2), followed by $CMFC_{ML}$ (47%, 0.062 W/m^2), and $CMFC_T$ (39%, 0.025 W/m^2), which were stable throughout the operation.

Rachnarin and Roshan (2017) demonstrated factors affecting the operation of PMFC based on the effect of microbes, effect of plants, and configuration of the design. The study showed that the system performance was dependent on factors of the operating parameters and electrode materials. Soil and plant types were also influenced by the shaping of microbial communities, which in the end affected the power output. Moreover, voltage performance trend was marginally affected by the exudation nature that were utilized by microbes.

The study of MFC is heading towards multi-stacking to improve energy efficiency and for efficiency application purposes. Various studies use wastewater as fuel and catalyst sources. This is the advantage of MFC, where catalysts do not necessarily use expensive metals, such as platinum or aurum. However, the bacteria contained in wastewater are the main elements in ensuring that the reaction occurs. Various structure designs were also studied to optimize the size of the developed MFC. The primary concern of all researchers is to produce cheap MFCs with compact designs. Hence, MFC is a potential source of energy that is very important in the future.

Microbial Electrolysis Cell (MEC)

Microbial electrolysis cell is theoretically related to MFC technology, which was initially introduced in 2005 (Carlo et al. 2017). Whilst MFCs produce an electric current from the microbial decomposition of organic compounds,

MECs partially reverse the process to generate hydrogen or methane from organic materials by applying an electric current. MEC used microbial communities as biocatalysts of the bio-electrochemical processes, where they expressively affect the production of biohydrogen in their overall performance (Masoud et al. 2015). Review on the microbial community obviously stated that the microbial behavior control has unavoidable influence on the MEC efficiency improvement.

Figure 2.3 shows the half biological (HB) MFC, HB-MEC, and full biological (FB) design system by Tahereh et al. (2018) to investigate the hydrogen production from a full biological MEC. Results showed that FB-MFC overall performance had more promising results compared to the FB-MEC systems. However, FB-MEC system was slightly higher in cathodic hydrogen recovery which had operated under recirculation batch mode (65% vs 56%). Moreover, the hydrogen production rate was also higher compared to the FB-MEC under the mode of operation (0.45 m³ H₂/(m³.d) vs. 0.17 m³ H₂/(m³.d)). Nonetheless, the two-stage bioethanol refinery (Sugnaux et al. 2016) was also one of the choices in the MEC structural design.

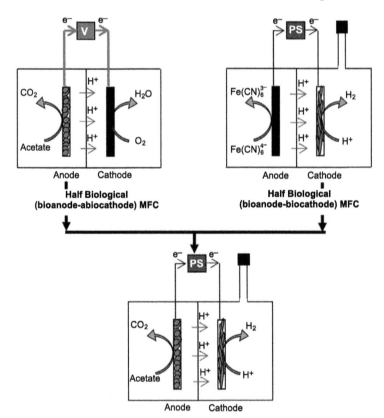

Figure 2.3. Overview of the HB-MFC and HB-MEC setups and assembling the FB-MEC setup by transferring the previously enriched MFC bioanode and MEC biocathode to the two-chamber experimental setup of the FB-MEC (Tahereh et al. 2018).

The study on anode performance of MECs with urine samples was carried out by Sónia et al. (2018) by using different anodes, C-Tex (cellulose-based), Keynol (phenolic-based), and PAN (polyacrylonitrile-based). From the results, anode biofilm developed on C-Tex with proteobacteria phyla was identified in the higher current generation, which can be signified with the bacteria assigned to it. Besides, Laura et al. (2016) reportedly produced high current density and hydrogen production in alkaline (pH=9.3) exoelectrogenesis condition. They recorded better performance than neutral conditions with 50 mA·m^{-2} and 2.6 L$_{H_2}$·L$^{-1}_{REACTOR}$·d^{-1}. The coulombic efficiency (CE) for alkaline MFC recognised a much higher pH neutrality for MFC, which is 60% and 43%, respectively. The anodic exoelectrogenic biofilm was primarily enriched with alkaliphilic bacteria, such as *Geoalkalibacter* sp., with exoelectrogenic activity noticed in previous studies, and *Alkalibacter* sp., which showcased a primary stage with an extremely abundant microorganism in alkaline MECs.

Recently, there have not been any other studies involving MECs combined with an anaerobic digester (AD) to treat food wastes except from that of Amro et al. (2017). Previous studies did not compare the energy production difference when the MEC is located within an AD for only part of the digestion period, which differs from results of an AD-MEC operated during the entire digestion period. This study serves to address these research gaps. A combined AD and MEC was tested by incorporating an MEC into an AD reactor for treating food waste. Figure 2.4 shows the experimental setup and the results of the research. Due to instability and constant low production rate of AD, the application of an MEC to AD (MEC-AD) potential helps to alter the AD microbial community by enriching exoelectrogens and methanogens. This also speeds up the degradation of a substrate (including recalcitrant compounds), thus enabling an increased biogas production (Yu et al. 2018).

MECs are also capable of hydrogen production using a bioelectrochemical pathway known as "bioelectrohydrogenesis", which was recently

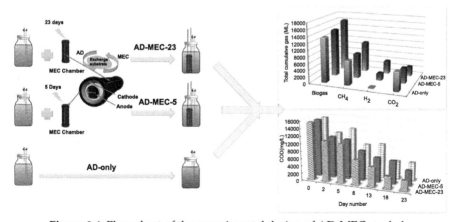

Figure 2.4. Flow chart of the experimental design of AD-MEC study by Amro et al. (2017).

discovered (Shaoan and Bruce 2011, Ruixia et al. 2018, Rengasamy et al. 2017). Abudukeremu et al. (2016) produced hydrogen from the combination of protons and electrons at the cathode. MEC reactors are superficially known to provide voltage (≥ 0.2 V) under a biological condition of P = 1 atm (1.01×10^5 Pa), T = 30°C and pH = 7. It is completed by the voltage input via a power supply. Nevertheless, MECs need comparatively low energy input of 0.2–0.8 V compared to typical water electrolysis (1.23–1.8 V). Reducing electrode spacing experimentally (Shaoan and Bruce 2011) makes it capable of increasing the rates of hydrogen production from 0 to 0.6 V applied voltages. The results prove that altering the design and reducing electrode spacing helps to enable the increased rate of hydrogen production. However, high surface area anodes are needed and the closest electrode spacing does not permit the increased rate of hydrogen production.

Besides hydrogen, MEC is a promising technology for methane generation (Moreno et al. 2016). The study of the effects of applied voltage on activities of microorganisms in MEC by Aqiang et al. (2016) found that 0.8 V was the optimal applied voltage for wastewater treatment. It was reported that the design reactors are useful in maintaining industrial practice. Moreno et al. (2016) suggested that hydrogenotrophic methanogenesis decreased the homoacetogenic activity incidence, thus improving the performance of MEC.

Microbial Desalination Cell (MDC)

Microbial desalination cell is the new, green, and environmentally-friendly technology that desalinates seawater and produces bioelectricity (Abdullah et al. 2018, Bahareh et al. 2018a). Recently, due to its potential for use in biodiesel production and oxygen generation, microalgae had received great attention for bio-electrochemical systems application. The MDC is a newly-developed technology which integrates with electrodialysis and the MFC process for the production of renewable energy, water desalination, and wastewater treatment. Besides, MDC also promotes water desalination methods to generate electric current by bacteria, which is based on the removal of ionic species from water in proportions. Unfortunately, the insufficient deionisation and low current generation have created challenges to improve the process of this technology.

Ashwaniy and Perumalsamy (2017) had studied integrated desalination with wastewater treatment and bioelectricity production in MDC by utilising microalgae as the bio-cathode (shown in Fig. 2.5) and the results were compared to the chemical cathode. The study utilised petroleum wastewater in the anode chamber and microalgae *Scenedesmus abundans* in the cathode chamber. The results highlighted that microalgae bio-cathodes performed better than chemical cathodes while increasing COD removal and producing a considerable amount of bioelectricity.

Abdolmajid et al. (2017) studied MDC by using ozone competency as a new cathodic electron acceptor. They examined the standard reduction potentials and cell half-reactions of oxygen and ozone in the bio-

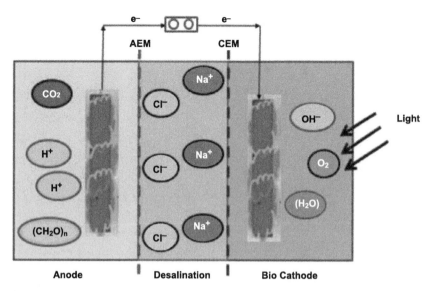

Figure 2.5. Microbial desalination cell design (Ashwaniy and Perumalsamy 2017).

electrochemical systems. Furthermore, the data was compared between salinity removal and electricity generation with O_3-MDC through O_2-MDCs data. The results proved that faster desalination occurred under maximum current generation. Nevertheless, once ozone was used as an electron acceptor, the desalination performance enhanced the reactor and had lower internal resistance, and the current density and power density were several times higher than those obtained in O_2-MDC.

Morvarid et al. (2017) studied algal cathode MDC performance by using dairy wastewater, as shown in Fig. 2.6. They compared algal MFC and two different NaCl concentrations, which resulted in saline water concentration by using the middle chamber and desalination mechanism of algal MDC performance. The anolyte contained lactose used in dairy wastewater, lipid, protein and casein (that are treated biologically in the anode), which in turn improved bioelectricity production (Mostafa et al. 2015). The wastewater was able to promote desalination in the MDC due to its low electrical conductivity compared to the synthetic wastewater. Recent research efforts were done to study generations of bioelectricity and salt water simultaneous desalination in an MDC using the microalgae *Chlorella vulgaris* as a biocathode.

A simultaneous removal was achieved by Yan et al. (2017) by using a combined system of microbial desalination cell-microbial electrolysis cell (MDC-MEC) with saline in seawater, metal in industrial wastewater, and nitrogen in municipal wastewater. From the study, the fluctuation of pH problems was successfully solved by combining desalination with nitrogen removal in the MDC system. This is due to the neutralisation of protons during the nitrogen removal reactions, as denitrification occurred in the anode and the nitrification process took place in the cathode. At the same time, the MDC power output was generated by using ESC to support MEC

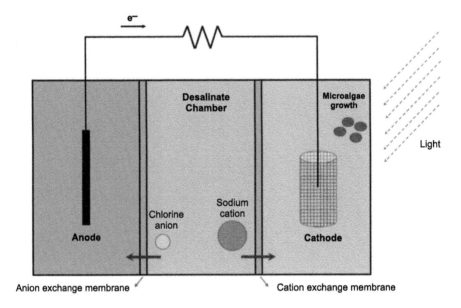

Figure 2.6. Schematic of microalgae biocathode in microbial desalination cell
(Morvarid et al. 2017).

for lead (II) reduction. Moreover, the advantages of this integrated MDC-MEC system are also to use multiple types of saline water and wastewater to be cured concurrently.

Conventionally, MDCs are used to extract organically harvested energy from wastewater for *in situ* saline water desalination. However, it cannot be treated like industrial wastewater as it contains high-level organics that may cause blocking in the thin membrane stack. Thus, in this study, a multi-stage MDC (M-MDC) was invented and operated for diluted industrial desalination while simultaneously treating wastewater.

Moreover, Photosynthetic Microbial Desalination Cells (PMDCs) were fabricated by Bahareh et al. (2018b), which also resulted in high removal rates of nutrient compounds and organic carbon PMDCs which had been initiated by 2015 (Baherah and Veera 2015). It was discovered that PMDCs can be constructed to maximise the harvested energy recovered through either bioelectricity or biomass production. Moreover, the analysis of microbial composition with different samples of biosolids in the PMDCs also discovered a diverse microbial community group.

Microbial Electrosynthesis (MES)

Microbial electrosynthesis cell is a process that can produce biocommodities from the reduction of substrates with microbial catalysts and an external electron supply (Hyo et al. 2017). This process is expected to become a new application of a cell factory for novel chemical production, wastewater treatment, and carbon capture and utilisation. However, MES is still subject to several problems that need to be overcome for commercialisation; therefore continuous development in the field of metabolic engineering is essential. The development of MES can open up new opportunities for sustainable biocommodity production platforms. Generally, MES offers exclusive environmental and social benefits: organic and inorganic resources, remediation, and recovery of elements (Jhuma 2017, Ali et al. 2017).

However, MES still has some challenges: 1) CO_2 reduction in MESs is relatively slow and the growth of the cathodic microbial communities tends to be slower than that of anodic microbial communities, and 2) uncertainty about how microorganisms accept electrons and how they use these electrons to synthesise. Thus, to solve these problems, many researchers have placed more effort in MES methods—be it electrochemically, economically, and biologically. Moreover, the review by Frauke et al. (2018) pointed out that the transport chains of engineering electrons were difficult due to inadequate information and data about the actual mechanisms of the electron transfer, the membrane assembly, and the complication of extracellular electron transfer (EET); which is essential to complete the electron transfer chain (ETC). This ETC is generated from the extracellular electrode surface in the intracellular metabolism of electron transfer. However, recently, researchers have started to study development of tools in strain construction for MES application, which have already begun designs, such as CRISPR/Cas systems.

Based on the biological study, Nabin et al. (2016) developed a 3-D-graphene functionalised carbon touched on composite cathode which enabled the accelerated electron transfer of a microbial catalyst *Sporomusa ovata* in an MES reactor. The alteration with a network of 3-D-graphene had increased the conversion of CO_2 to acetate where the electrosynthesis rate was 6.8 folds. Moreover, it also expressively improved the current consumption and biofilm density. Yinbo et al. (2017) also studied the sulfate-reducing effects on a mixed-culture performance with *Acetobacterium* and *Desulfovibrionaceae* bacteria. Both studies used different catalysts to improve the MES performance. In another article, Yinbo et al. (2017) attempted to solve the physical structure by fabricating an effective microbial electrosynthesis of a bioanode system, which converts acetate from CO_2 by using wastewater in a bioenergy system. The MESs of bioanode were fabricated using bipolar membrane (BPM) and proton exchange membrane (PEM) as separators which operated with a range of voltages.

Moreover, three-dimensional rGO/biofilm was constructed by Song et al. (2017), enabling highly efficient electron transfer rates within biofilms, and between biofilms and electrodes, demonstrating that the development of 3-D electroactive biofilms, with higher extracellular electron transfer rates, is an effective way of increasing MES efficiency.

In a different way, Raúl et al. (2018) pursued the influence of bacterial communities in biocathodes during start-up by electrode and potential inocula diversity. Competitions and interactions of possible syntrophies between eubacteria and archaea were discussed together with their potential roles in the current production and product formation. Through this study, it was proven that reductive potentials are due to the procedure of an inconsistent start-up and it was unrelated with the inoculum used. However, this may prove beneficial by imposing oxidative potentials to develop with prepared reductive potentials or biocathodic operation from electroactive biofilm.

Xiaohu et al. (2017) developed an economical and innovative method of H_2O_2 production by using a microbial reverse-electrodialysis electrolysis cell (MREC) system. In the MREC system, the generated electrical potential by salinity-gradient and the exoelectrogens between fresh water and salt were exploited to initiate the high-rate of H_2O_2 production.

Conclusions

Microbial fuel cell has become an exciting and promising research area. Many MFC applications will help to reduce the use of fossil fuels and generate energy from waste. This chapter discussed four different types of BES, which include Microbial Fuel Cell, Microbial Electrolysis Cell, Microbial Desalination Cell, and Microbial Electrosynthesis Cell. Along with the understanding of the MFC concept, many MFC-based applications have emerged, such as wastewater treatment, microbial electrolysis cells, sediment MFCs, and bioremediation. Several MFC applications were explained in this paper. Among those MFC-based technologies, the most immediate and useful one is currently used as a wastewater treatment method, hydrogen production, and others. Thus, the variants of MFC, MEC, MDC, and MES are new types of bioreactors that use exoelectrogenic biofilms for electrochemical energy production. In recent years, a large number of studies have been conducted to explore Microbial Fuel Cells in many aspects, such as electron transfer mechanisms, enhancing power outputs, and reactor developments and applications. Although MFCs are a promising technology for renewable energy production, they are faced with several challenges. For instance, in-depth studies are required to identify the types of bacteria or catalyst, material, components, and parameters of MFC needed to focus on producing eco-friendly MFCs as a fuel source.

Acknowledgment

This research was financially supported by the Research University Grant of Geran Galakan Penyelidik Muda (Project No: GGPM-2017-029) and Gerak Galakan Penyelidikan – Komuniti (Project No: GGPK-2016-002) from Universiti Kebangsaan Malaysia (UKM).

REFERENCES

Abdolmajid, G., A.E. Ali, H.S. Mohammad, H. Mohammad and Ehrampoush. 2017.

Ozone cathode microbial desalination cell: an innovative option to bioelectricity generation and water desalination. Chemosphere 188: 470-477.

Abdullah, A.M., A. Waqar, S.B. Mahad, K. Mohammad and R.D. Bipro. 2018. A review of microbial desalination cell technology: configurations, optimization and applications. Journal of Cleaner Production 183: 458-480.

Abhilasha, S.M. 2016. Novel microbial fuel cell design to operate with different wastewaters simultaneously. Journal of Environmental Sciences 42: 105-111.

Abudukeremu, K., S. Yibadatihan, A. Peyman, F.A. Nadia, K. Chandrasekhare and S.K. Mohd. 2016. A comprehensive review of microbial electrolysis cells (MEC) reactor designs and configurations for sustainable hydrogen gas production. Alexandria Engineering Journal 55: 427-443.

Ali, K.M., D. Patrick, K.B. Satinder, D.T. Rajeshwar, L.B. Yann and B. Gerardo. 2017. Microbial electrosynthesis of solvents and alcoholic biofuels from nutrient waste: a review. Journal of Environ. Chem. Eng. 5: 945-954.

Amro, H., W. Freddy, H.G. Xiao, Y. Liang, L. Stephanie and Q. Ling. 2017. Next generation digestion: complementing anaerobic digestion (AD) with a novel microbial electrolysis cell (MEC) design. International Journal of Hydrogen Energy 42: 28681-28689.

Aqiang, D., Y. Yu, S. Guodong and W. Donglei. 2016. Impact of applied voltage on methane generation and microbial activities in an anaerobic microbial electrolysis cell (MEC). Chemical Engineering Journal 283: 260-265.

Ashwaniy, V.R.V. and M. Perumalsamy. 2017. Reduction of organic compounds in petro-chemical industry effluent and desalination using scenedesmus abundans algal microbial desalination cell. Journal of Environmental Chemical Engineering 5: 5961-5967.

Bahareh, K. and G.G. Veera. 2015. Sustainable photosynthetic biocathode in microbial desalination cells. Chemical Engineering Journal 262: 958-965.

Bahareh, K., S. Renotta, P.B. John and G.G. Veera 2018a. Bioelectricity production from wastewater in photosynthetic microbial desalination process under different flow configurations. Journal of Industrial and Engineering Chemistry 58: 131-139.

Bahareh, K., G.G. Veera, S. Renotta and P.B. John 2018b. Evaluation of anammox biocathode in microbial desalination and wastewater treatment. Chemical Engineering Journal 342: 410-419.

Bullen, R.A., T.C. Arnot, J.B. Lakeman and F.C. Walsch. 2006. Biofuel cells and their development. Biosens. Bioelectron 21: 2015-2045.

Carlo, S., A. Catia, E. Benjamin and I. Ioannis. 2017. Microbial fuel cells: from fundamentals to applications. A review. Journal of Power Sources 356: 225-244.

Carmen, M., M. Fernández, A. Yeray, F.L. Luis, V. José and A.R. Manuel. 2018. Thermally-treated algal suspensions as fuel for microbial fuel cells. Journal of Electroanalytical Chemistry 814: 77-82.

Edith, M., F. Ronen and L. Janice. 2014. Application of carbon black and iron phthalocyanine composites in bioelectricity production at a brewery wastewater fed microbial fuel cell. Electrochimica Acta 128: 311-317.

Frauke, K., L. Bin, Y. Shiqin and O.K. Jens. 2018. Balancing cellular redox metabolism in microbial electrosynthesis and electro fermentation—a chance for metabolic engineering. Metabolic Engineering 45: 109-120.

Haoran, Y., D. Lifang, C. Xixi, Z. Tao, Z. Shungui, C. Yong and Y. Yong. 2016. Recycling electroplating sludge to produce sustainable electrocatalysts for the efficient conversion of carbon dioxide in a microbial electrolysis cell. Electrochimica Acta 222: 177-184.

Hyo, J.S., A.J. Kyung, W.N. Chul and M.P. Jong. 2017. A genetic approach for microbial electrosynthesis system as biocommodities production platform. Bioresource Technology 45: 1421-1429.

International Business. 2007. Microbial 'fuel cell' squeezes energy from brewery wastewater. 2 May 2017.

Jhuma, S. 2017. Microbial electrosynthesis. Encyclopedia of Sustainable Technologies: 455-468.

Justin, M. 2010. Microbial Fuel Cells: Generating Power from Waste. 4 May 2010.

Kaewkannetra, P., W. Chiwes and T.Y. Chiu. 2011. Treatment of cassava mill wastewater and production of electricity through microbial fuel cell technology. Fuel 90: 2746-2750.

Laura, R., A.B. Juan and G. Albert. 2016. Increased performance of hydrogen production in microbial electrolysis cells under alkaline conditions. Bioelectrochemistry 109: 57-62.

Liu, H., R. Ramnarayanan and B.E. Logan. 2004. Production of electricity during wastewater treatment using a single chamber microbial fuel cell. Environmental Science & Technology 38: 2281-2285.

Mabel, K.A., J.A. Helton, S. Rodrigo and A.S. Edson. 2017. Treatment of brewery wastewater and its use for biological production of methane and hydrogen. International Journal of Hydrogen Energy 42: 26243-26256.

Manaswini, B. and M.M. Ghangrekar. 2009. Performance of microbial fuel cell in response to change in sludge loading rate at different anodic feed pH. Bioresource Technology 100: 5114-5121.

Masoud, H., M.M. Mohammad and Y. Soheila. 2015. Biocatalysts in microbial electrolysis cells: a review. International Journal of Hydrogen Energy 41: 1477-1493.

Moreno, R., M.I. San-Martín, A. Escapa and A. Morán. 2016. Domestic wastewater treatment in parallel with methane production in a microbial electrolysis cell. Renewable Energy 93: 442-448.

Morvarid, K.Z., K. Hamid-Reza and V. Manouchehr. 2017. Electricity generation, desalination and microalgae cultivation in a biocathode-microbial desalination cell. Journal of Environmental Chemical Engineering 5: 843-848.

Mostafa, R., A. Arash, D. Soheil, Z. Alireza and O. Sang-Eun. 2015. Microbial fuel cell as new technology for bioelectricity generation: a review. Alexandaria Engineering Journal 54: 745-756.

Nabin, A., A. Halder, P.L. Tremblay, Q. Chi and T. Zhang. 2016. Enhanced microbial electrosynthesis with three-dimensional graphene functionalized cathodes fabricated via solvothermal synthesis. Electrochimica Acta 217: 117-122.

Naga, S.M.V.V., R.K. Kesava, R. Bernardo and T. Tonia. 2018. Denitrification of water in a microbial fuel cell (MFC) using seawater bacteria. Journal of Cleaner Production 178: 449-456.

Rachnarin, N. and R. Roshan. 2017. Plant microbial fuel cells: a promising biosystems engineering. Renewable and Sustainable Energy Reviews 76: 81-89.

Raúl, M., S. Ana, M.A. Raúl, E. Adrián and M. Antonio. 2018. Impact of the start-up process on the microbial communities in biocathodes for electrosynthesis. Bioelectrochemistry 121: 27-37.

Rengasamy, K., Y.C. Ka, S. Ammaiyappan, B. Arpita and W.C.W. Jonathan. 2017. Bioelectrohydrogenesis and inhibition of methanogenic activity in microbial electrolysis cells – a review. Biotechnology Advances 35: 758-771.

Ruixia, S., J. Yong, G. Zheng, L. Jianwen, Z. Yuanhui, L. Zhidan and J.R. Zhiyong 2018. Microbial electrolysis treatment of post-hydrothermal liquefaction wastewater with hydrogen generation. Applied Energy 212: 509-515.

Simate, G.S. 2015. Water treatment and reuse in breweries. Brewing Microbiology 425-456. DOI https://doi.org/10.1016/C2014-0-03102-4

Schilir, T., T. Tommasi, C. Armato, D. Hidalgo, D. Traversi, S. Bocchini, G. Gill and C.F. Pirri. 2016. The study of electrochemically active planktonic microbes in microbial fuel cells in relation to different carbon-based anode materials. Energy 106: 277-284.

Shaoan, C. and E.L. Bruce. 2011. High hydrogen production rate of microbial electrolysis cell (MEC) with reduced electrode spacing. Bioresource Technology 102: 3571-3574.

Sónia, G.B., P. Luciana, S.G.P.S. Olívia, F.R.P. Manuel, T.H. Annemiek, K. Philipp, M.A. Maria and A.P. Maria. 2018. Influence of carbon anode properties on performance and microbiome of Microbial Electrolysis Cells operated on urine. Electrochimica Acta 267: 122-132.

Song, T.S., Z. Hongkun, L. Haixia, Z. Dalu, W. Haoqi, Y. Yang, Y. Hao and X. Jingjing. 2017. High efficiency microbial electrosynthesis of acetate from carbon dioxide by a self-assembled electroactive biofilm. Bioresource Technology 243: 573-582.

Sugnaux, M., M. Happe, P.C. Christian, G. Olivier, H. Gérald, B. Maxime and F. Fabian. 2016. Two-stage bioethanol refining with multi litre stacked microbial fuel cell and microbial electrolysis cell. Bioresource Technology 221: 61-69.

Swathi, K., S. Omprakash, K.B. Sai, G. Velvizhi1, M.S. Venkata. 2017. Stacking of microbial fuel cells with continuous mode operation for higher bioelectrogenic activity. Bioresource Technology 257: 210-216.

Tahereh, J., R.W.D. Wan, G. Mostafa, H.A.B. Mimi, S. Mehdi, H.K. Byung, A.C.M. Alessandro, M.J. Jamaliah and I. Manal. 2018. Clean hydrogen production in a full biological microbial electrolysis cell. International Journal of Hydrogen Energy. (In press)

Xiaohu, L., A. Irini and Z. Yifeng. 2017. Salinity-gradient energy driven microbial electrosynthesis of hydrogen peroxide. Journal of Power Sources 341: 357-365.

Yan, L., S. Jordyn, H.Z.X. Yuankai, M. Jeffrey and L. Baikun. 2017. Energy-positive wastewater treatment and desalination in an integrated microbial desalination cell (MDC)-microbial electrolysis cell (MEC). Journal of Power Sources 356: 529-538.

Yinbo, X., L. Guangli, Z. Renduo, L. Yaobin and L. Haiping. 2017a. Acetate production and electron utilization facilitated by sulfate-reducing bacteria in a microbial electrosynthesis system. Bioresource Technology 241: 821-829.

Yinbo X., L. Guangli, Z. Renduo, L. Yaobin and L. Haiping. 2017b. High-efficient acetate production from carbon dioxide using a bioanode microbial electrosynthesis system with bipolar membrane. Bioresource Technology 233: 227-235.

Youngho, A. and E.L. Bruce. 2010. Effectiveness of domestic wastewater treatment using microbial fuel cells at ambient and mesophilic temperatures. Bioresource Technology 101: 469-475.

Yu, H.J., H.R. Jae, H.K. Cho, K.L. Woo, L.L. Hyo, H.Z. Rui and H.A. Dae. 2012. Enhancing hydrogen production efficiency in microbial electrolysis cell with membrane electrode assembly cathode. Journal of Industrial and Engineering Chemistry 18: 715-719.

Yu, Z., L. Xiaoyun, Z. Shuai, J. Jing, Z. Tuoyu, K. Aman, K. Apurva, L. Pu and L. Xiangkai. 2018. A review on the applications of microbial electrolysis cells in anaerobic digestion. Bioresource Technology 255: 340-348.

Zhang, G., Z. Hanmin, Y. Fenglin, Z. Rong and W. Junlei. 2018. Sequencing polarity-inverting microbial fuel cell for wastewater treatment. Biochemical Engineering Journal 133: 106-112.

CHAPTER

3

Fundamentals of Photosynthetic Microbial Fuel Cell

Rashmi Chandra[1]* and Garima Vishal[2]

[1] Tecnologico de Monterrey, School of Engineering and Science, Campus Monterrey, Ave. Eugenio Garza Sada 2501, Monterrey, N.L., CP 64849, Mexico
[2] Department of Chemical Engineering, Indian Institute of Technology, New Delhi, India

Introduction

The main and most primitive energy source for life on planet earth is the sunlight. Solar energy is inserted to the earth atmosphere through photosynthesis. This is a physico-chemical process used by plants, algae, and certain bacteria convert energy of photons to chemical energy from organic matter (Georgianna and Mayfield 2012, Strik et al. 2011). In this regard, PhFCs are a recently established research area where solar energy is used to produce bioelectricity. Its sustainable and renewable nature expanded important interest in basic and applied research (Chandra et al. 2012, Venkata Mohan et al. 2014).

PhFCs integrate photosynthetic and electrochemically active microbes (He et al. 2009) or plants (Strik et al. 2008) for the *in situ* generation of electricity or chemicals, for example H_2, CH_4, ethanol, and H_2O_2 (Strik et al. 2011). Additional benefits of PhFCs are the removal of atmospheric CO_2 and their wastewater treatment application (Rosenbaum et al. 2010). Most advances reported in the recent years were demonstrated using short-scale units operating under monitored environment in the laboratory. However, in many cases the scale up feasibility, life cycle analysis (LCA), and techno-economic analysis (TEA) have not been completed (Bensaid et al. 2015, Dekker et al. 2009, Rosenbaum et al. 2010, Santoro et al. 2017). This chapter focuses on renewable power generation approaches based on the coupling of photosynthetic bioreactors.

---

*Corresponding author: rashmichandrabhu@gmail.com; rashmichandra@itesm.mx

In PhFCs, oxidation of organic compounds results in electrons production by photosynthetic bacteria. These electrons or photosynthates are transported to the anode via conductive materials (Bensaid et al. 2015). These conductive materials consist of membrane bound proteins and "nanowires" (Logan et al. 2006). In the majority of PhFCs, electrons reach the cathode from the anode associate with protons and terminal electron acceptors (TAE) at cathode (Bensaid et al. 2015). PhFC research has allowed the generation of a wide variety of different system configurations, the use of photosynthetic mechanism, and respective organisms. One important distinction between each system is the source of energy, carbon, and electron that will lead to power generation.

Mechanism of Electron Generation in PhFCs

Oxygenic Systems

Electrical circuit can be set up across light energy (excitation of electron), e^- source from the metabolism of substrate, and the e^- sink (O_2) through introducing non-catalyzed electrodes along with membrane which facilitates the potential development of bioelectricity. In recent literature, it has been found that various genera of cyanobacteria generate electrogenic activity. In counterpart to heterotrophs organisms, cyanobacteria is light-dependent to produce the electrogenic activity by providing electrons to the acceptor located outside of the membrane under light conditions (Pisciotta et al. 2011). It is still unknown whether cytochrome bd quinol oxidase transports electrons directly to acceptor located outside of the membrane, or whether extra downstream compounds are implicated.

Microalgae and cyanobacteria

Electric energy can be obtained from solar energy by forming phototrophic biofilm on the anode of PhFC. Oxygenic photosynthetic species commence charge separation results in delivering electrons and protons in a photophosphorylation cascade (Gajda et al. 2015). Hence, oxygenic photosynthesis involves a 3-membrane bound protein complexes called photosystem I (PSI), photosystem II (PSII), and a cytochrome bf complex. These e^- are transported among these huge protein systems by movable compounds called plastoquinone (PQ) and plastocyanin (PC), as presented in Fig. 3.1. PQ, a lipophilic carrier molecule, seems to take the major part in the electrogenic pathway of cyanobacteria (Roberts et al. 2004). These compounds transport electrons over fairly extensive lengths and perform a distinctive part in photosynthetic energy conversion. Studies by Pisciotta et al. showed (Pisciotta et al. 2011) that the photosynthetic electron transfer chain (ETC) and, mainly water photolysis by PSII are the origins of e^- released extracellularly by cyanobacteria under light conditions. Further evidence showed that plastoquinone (PQ) are crucial transporters of electrons to the extracellular medium, and support the whole process where the H_2O is the

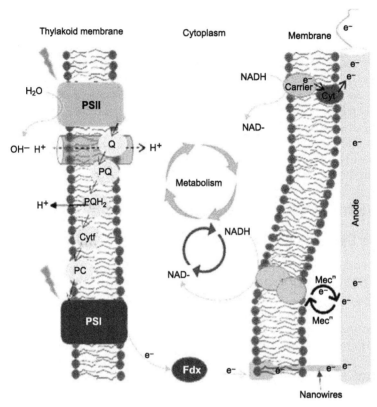

Figure 3.1. Schematic diagram of conversion of light energy to electron by Z scheme and transfer of electron from microbe to anode. PSII: Photosystem II; Q: Quinone; PQ: Plastoquinone; cyt f: Cytochrome; PC: Plastocyanin; PSI: Photosystem I; Fdx: ferredoxin.

origin of electrons released to the surrounding, rather than organic matter. Then, electrons are transferred to PSI where chemical energy is obtained by the transformation of light energy, i.e., when the discharged reducing power is transformed into electricity in existence of the electrode-membrane arrangement. This type of PhFC, called photosynthetic microbial fuel cell with oxygenic photosynthesis (pMFCOX), can intercept electrons directly at cellular level from the electron transport chain (ETC) through the use of outer electrode (anode) where electrons are counterbalanced at the cathode via external resistance (Chandra et al. 2012, Ivashin et al. 1998).

Terrestrial and aquatic plants systems

An emerging technology to produce sustainable energy are plant-MFCs (Wetser et al. 2015), which integrate the rhizome of live plants inside the anode compartment, as represented in Fig. 3.2. Plant rhizodeposits consist of root exudates, plant residues, and few gases. These rhizodeposits can be utilized as substrates by electrochemically active microbes, which donate

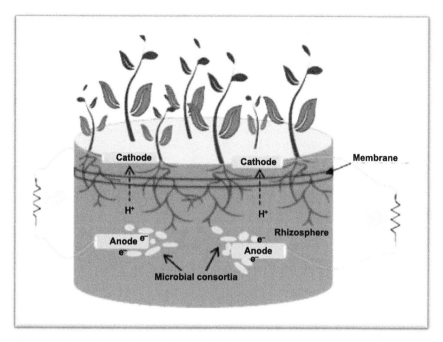

Figure 3.2. Plant-based microbial fuel cell showing a synergy between rhizosphere and microbes.

their e⁻ to anode (Strik et al. 2008). The photosynthates produced due to CO_2 fixation are liberated as root exudates by the plants. It can be observed that anode has been used as electron depositor by the microorganism, and due to potential difference, the electron flows in the circuit. The remaining protons are carried across the membrane onto the cathode, where O_2 has been brought down to produce water (Villaseñor et al. 2013). Although this energy generation technology relies on plant photosynthesis, it is not directly solar light dependent since plants excrete organic matter continuously (day and night).

A list of photosynthetic microbial fuel cells has been represented in Table 3.1. It's a kind of Bioelectrochemical system (BES) which can convert chemical energy to electricity. Electrodes within organic compounds formed in the rhizosphere area of plants produce electricity due to electron flow through the system. At cathode, microbes of rhizosphere, which convert solar energy to electricity, utilize the electrons which are in contact with hydrogen molecule and removes the toxicity by oxidation. This technology can use organic waste to generate energy. Although this technology is at its initial stage but the research is growing very fast.

Performance of PMFC can be affected by light intensity and photoperiodism. Research suggests that optimal amount of light is required for the growth of microbes and for operating the system. Rhizodeposition and photosynthetic pathways play a major role in PMFCs. Few plants are

Table 3.1. Photosynthetic microbial fuel cell and their characteristics

Type of photosynthetic mechanism	OCV	Current	Type of electrode	Power	Reference
Oxygenic photosynthesis at cathode	795 mV	1.73 mA	Graphite	57.0 mW/m^2	Venkata Mohan et al. 2014
Aquatic plants systems (Oxygenic photosynthesis)	803 mV	4.03 mA	Graphite	115.50 mW/m^2	Chiranjeevi et al. 2013
Terrestrial plants (Oxygenic photosynthesis)	1004 mV	4.52 mA	Graphite	163 mW/m^2	Chiranjeevi et al. 2013
Oxygenic photosynthesis at anode	46 mV	0.6 mA	Graphite	3.55 µW/m^2	Venkata Subhash et al. 2013
Anoxygenic photosynthesis (single chambered)	505 mV	3.1 mA	Graphite	72.96 mW/m^2	Chandra et al. 2012
Anodic anoxic photosynthesis and cathodic oxygenic photosynthesis	605 mV	2.14 mA	Graphite	112.20 mW/m^2	Chandra et al. 2017

more efficient for power generation, which can be classified as C3, C4, and CAM. C4 is the most efficient in all of these categories. Few C3 plants are also efficient. It has also been seen that C4 never gets saturated with light, and gives photosynthetic efficiency of 6%, as compared to C3 plants, which give 4.6%. The theoretical maximum electricity output of a plant-MFC for a plant growth area is 03.20 W/m^2 (Strik et al. 2011), while a long-term output of only 0.155 W/m^2 has been achieved (Helder et al. 2012). Beneath open-air circumstances, it was possible to reach a power density of 88.00 mW/m^2, which is under maximum obtained at laboratory environment controlled (440 mW/m^2). Cathode potential was dependent on solar radiation, allowing the power output susceptible of the diurnal cycle. The temperature could affect the anode potential of plant-MFCs, driving to a reduction in electrical current generation throughout low temperature and no electrical energy generation during frost episodes (Helder et al. 2013). Plant-MFCs have been installed and studied in rice paddy fields, where they have shown the generation of electrical energy by rhizosphere microbial population through oxidation of organic matter in the rhizospheres (Kaku et al. 2008). Analogous evidence of concept was again confirmed through use of Reed mannagrass (*Glyceria maxima*), reaching 67 mW/m^2 as highest power output (Strik et al. 2008).

Microbes donate electron to the anode in rhizosphere zone by two methods, direct or mediated electron transfer method. Better the adaptation of environment by microbes, higher would be the efficiency. The action of microbes is controlled by the class of the plant, with which they are in contact with. *Geobactor* are the best performing bacteria. However, the microbial community was found to be influenced by the soil, as it affects the power output generated. The soil-root consortium is the place where microbes are present in contact with the plant. Soil plays an important role in the performance of PMFC. The two soil types used in PMFC were agricultural soil and forest soil. The agricultural soil was 17 times more efficient than forest soil. Design of PMFC should consider cost effectiveness and efficient system which can work for long time. In order to improve efficiency, we should consider the substrate (fuel), type of microbial fuel cell, parameters to decrease internal and external resistance, ionic strength, electrode material, and spacing.

Photosynthetic algae or cyanobacteria at anode with artificial mediator

Photosynthetic microbial fuel cell research started with cyanobacterial species of *Synechocystis* and *Anabaena* by documenting the increment in power output in light and decrement in power during the dark. In such class of PhFC intracellular carbon oxidized and electron were recovered through electrode integrated PhFC. Such arrangement works under oxygenic photosynthesis and resulting discharge of reducing equivalents (such as H^+ and e^-), as well as with O_2 through the photolysis of H_2O. Liberation of oxygen, electrons, and H^+ are concurrent mechanisms, where O_2 neutralizes H^+ and electrons, owing to its elevated electronegativity. Dissolved oxygen behaves like a terminal electron acceptor, which traps all protons and electrons produced through cyanobacterial metabolism deriving in low level of electrogenesis. Reducing equivalents produced have a positive impact upon electrogenic activity; however, at the same time the generation of O_2 has a negative impact upon overall performance. Studies are reported with 2-hydroxy-1,4-naphtoquinone as a non-natural intermediary to exchange the electron against microbes to the electrode, specially anode (mediated PhFC). The utilization of non-stainable and non-ecofriendly redox mediators prevents sustainability and real field application of the PhFC research.

Anoxygenic systems

Photosynthetic organisms use pigments, such as chlorophyll and bacteriochlorophyll (*Bchl*) molecules to harness light from sun as energy source and organic/inorganic carbon material as electron source to lead reaction center of the photosystem. Solar radiations were absorbed by the pigment and channeled to the reaction center. They get promoted from basal state and triggers chain photochemical reactions, provoking a separation of negative and positive charge across the membrane. In anoxygenic photosynthetic bacteria, the Bchl-a shows the energy efficiency of 95 to 99%

accessory pigments linked to the reaction center (RC). RC-LH complex has the capability to transform light energy to proton motive force with the role of quinone pool and second membrane-attached e⁻ transporter protein (cytochrome (cyt) bc1 complex) (Chandra et al. 2017). At first, the excited e⁻ from the *Bchl-a* goes to quinone, followed by cytochrome-complex before exiting across microbial membrane. In this kind of photochemical reactions sunlight offers energy and acetate (organic residue) as the e⁻. The negative Gibbs free energy of the photo electrochemical reaction and positive emf involves that electrical current can be generated by using organic carbon as the substrate. Separation of charge initiate a series of e⁻ transfer, which are joined to the translocation of protons through the membrane and establish an electrochemical proton gradient that leads to a development of bio potential, and finally to electro-genesis. On the other hand, oxygenic photosynthesis needs the joining of three membrane protein complexes (PSI, PSII, and cytochrome bf complex) working in sequence for the electron transfer from H_2O to NADP+ to produce O_2 (Venkata Subhash et al. 2013). Plastoquinone and plastocyanin, which are small movable substances, transport electrons between those protein complexes. These small molecules transport electrons above relatively long distances and play a sole role in photosynthetic energy conversion.

MFC device holding both oxygenic and anoxygenic phototrophs generate a mixotrophic environment. Bioelectricity production was observed to be better under light periods in comparison with dark in spite of higher dissolved O_2 with light, probably due to the fact that anoxygenic phototrophs were more efficient in electron production (Chandra et al. 2012). To gather the benefits of keeping very low hydrogen concentration, linking the photosynthetic hydrogen generation accompanied *in situ* hydrogen oxidation via an electrocatalytic-conversion phase offers a hopeful technology. As a result, H_2/H^+ works as a natural electron mediator between the anode and the microbial metabolism. In addition, generating hydrogen in a photofermentation process from organic matter by anoxygenic photosynthetic purple bacterium has been researched in joining with electrocatalytic electrodes for instantaneous hydrogen elimination (Cho et al. 2008, Rosenbaum et al. 2005). Cho et al. (2008) reported the immediate connection of photosynthetic activity and power generation using *Rhodobacter sphaeroides* in the light (3 W/m³) against dark (0.008 W/m³) (Rosenbaum et al. 2010).

Anoxygenic photosynthesis at anode

Anoxygenic photosynthetic bacteria use Bchl to capture the energy from sunlight and utilize organic residues/CO_2 as electron at reaction center of the photo-system. Solar energy trapped by accessory pigment directs to the Bchl reaction center and gets promoted from ground state to excited state and starts a sequence of photochemical reactions, which generate a potential difference across the membrane without O_2 production as represented in Fig.

3.3 (Chandra et al. 2017, 2015). At first, the excited e- from the Bchl-a gets into quinon pool and flows to cytochrome complex, before passing across the cell membrane. Photosynthetic bacteria utilize volatile fatty acids as electron donor for bioelectricity generation. The key step for PhFC is a stable, cost-beneficial, proton-exchange membrane.

Mixotrophic conditions provide the practicality of syntrophic synergy across anoxygenic and oxygenic photosynthetic environment in a sole system. In such an environment, it is possible that the heterotrophic O_2 intake is produced by autotrophs pending bacterial respiration, which allows positive effect on the fuel cell power generation. Additionally, to electricity produced by organisms, the mixotrophic function allows the neutralization of atmospheric CO_2, which is generated as a result of respiration pending PhFC function, and uses the carbon in organic form found in the anolyte (wastewater), allowing its processing. The practicality of using mixotrophic operation way to effective operational mode of PhFC owing practicability of anoxygenic microenvironment by syntrophic cooperative of photosynthetic bacteria and algae has been reported. PhFC process reported practical microenvironment for the development of bacteria with photosynthetic activity regarding algae. Photosynthesis under anoxygenic conditions allows to keep lower dissolved oxygen (DO) which raises the power generation. Furthermore, PhFC exhibited successful enrichment of acid-rich effluent

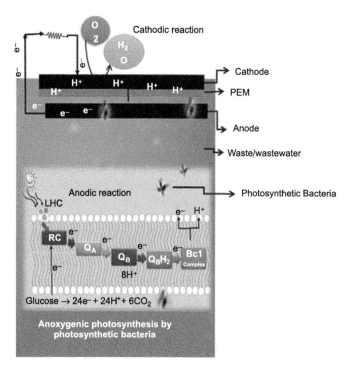

Figure 3.3. Anoxygenic photosynthesis at anode; Reaction center: RC; QuinoneA: Q_A; QuinoneB: Q_B; PQ: Plastoquinone; cyt f: Cytochrome f; PC: Plastocyanin.

produced by acidogenic hydrogen generation reactions as substrate for energy production. Photo-mixotrophic operation condition allows the interaction among oxygenic and anoxygenic photosynthesis mechanism, and acts in favor for photobioelectrocatalytic fuel cell process. A complete system setup has been represented in Fig. 3.4.

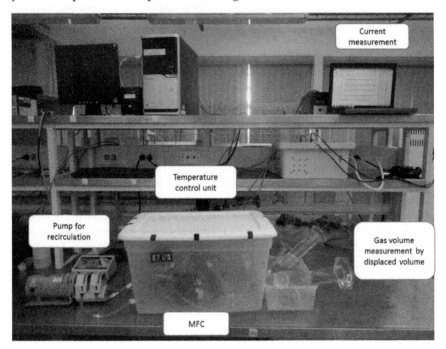

Figure 3.4. A complete experimental setup of microbial fuel cell operation.

Electron Transfer

Some research proposes that in natural microbial communities living in the microbial fuel cell, extracellular molecules are continuously intricate into transport of e⁻ between bacterial cell and external electrode (Watanabe et al. 2009, Shi et al. 2016). Thus, it is crucial to comprehend the characteristics of the electron intermediaries located in biocathode microbial fuel cell, such as chemical properties, stability, water solubility, and toxicity, since it is possible to optimize cathodic electrode materials to enhance yields using this information.

Intracellular and extra cellular electron transfer

In *Rhodoferax ferrireducens* has been reported the efficient electron transport via membrane-associated mediating complexes from bacteria linked to the electrode. The production of sterically accessible membrane proteins raises the electron-transfer flow from bacteria to the non-soluble electron acceptor located outside the cell (Rabaey and Boon 2005).

Recent research related to extracellular electron transport have concluded the value of outer membrane cytochrome proteins, appendages with properties to conduct electricity, soluble intermediaries, secondary metabolites, and chemotaxis in the metal oxide respiring or electrode respiring mechanism. The nano structures with properties to conduct electricity consist of natural protein with electronic activity called bacterial nanofilaments (nanowires) that can aim the microbe–mineral interface (MMI) and facilitate charge movement among distances during the MMI mechanism. However, nowadays many researchers have studied only a few bacterial nanofilaments, principally from strains of the *Geobacteraceae, Merismopediaceae, Shewanellaceae* individually and in syntrophy. The induction of filaments with electrical activity in a metabolically adaptable strain of *Rhodopseudomonas palustris* RP2 in energetically engineered environments has been reported (Venkidusamy et al. 2015). *Shewanella putrefaciens* has a versatile anodic extracellular electron transport mechanism, including excreted flavins, menaquinone-related redox mediators, and outer membrane cytochromes (Fredrickson et al. 2008, Newman and Kolter 2000), and can use an outer membrane-bound redox substance for electron transport in a microbial cathodic oxygen reduction. Studies with *Shewanella oneidensis*, an important electrode reducer, suggest that soluble electron shuttles are the mediators for most of its electron transport to the anode (Lovley 2008). *S. oneidensis* MR-1 also releases extracellular flavins facilitating electron transfer from bacteria to minerals and electrodes. Electrons are transported from the cytoplasmic membrane protein menaquinol oxidase (CymA) to outer-membrane cytochromes (MtrC/OmcA) by the important cytochrome:porin pair of MtrA (soluble periplasmic cytochrome) and MtrB (outer-membrane porin) (McCormick et al. 2015). Although the mechanism of electron movement across the outer membrane remains unknown, the Mtr pathway of *S. oneidensis* MR-1 is the best-characterized microbial extracellular electron transfer pathway (Shi et al. 2016).

Direct electron transport

Direct electron transport is carried out by the physical interaction that occurs between the electrode surface and the microbial membrane. Different outer membrane redox proteins as cytochromes directly capture electrons from the electrode surface. The necessity of a direct physical interaction between microbial membranes and insoluble electron acceptors which are the support for comprehension of the mechanism of extracellular electron exchange to insoluble electron acceptors in the natural environment has been reported (Eaktasang et al. 2016). Very few reports are available for the electron transport mechanism of PhFC. This electron transfer mechanism has been reported for *Geobacter* sp. and mixed cultures using CO_2, O_2, Cr(VI), or U(VI), fumarate, nitrate, tetrachloroethene, as electron acceptor (Gregory et al. 2004). While most electrochemically active bacteria in biocathodes are Gram negative, transitory electrochemical analyses have shown that some Gram[+] bacteria, including *Micrococcus luteus, Bacillus subtilis,* and

Staphylococcus carnosus perform direct electron transfer too, indicating a potentially widespread bacteria capability (Cournet et al. 2010). It has not been yet determined whether straightforward electron transport through photosynthetic microorganisms and the electrode is conceivable without artificial redox mediators, electrocatalytic electrodes, or heterotrophic bacteria (Rosenbaum et al. 2010).

Mediated electron transport

Electrons released by oxidation-reduction reactions inside the cell are scavenged by mediator chemicals entering the microbial cell in their oxidized state and exiting in their reduced state. When electrons are delivered into the culture broth, the mediator molecule spontaneously suffers an oxidation reaction, releasing electrons at the anode, thus entering the electrical circuit (Powell et al. 2011). Several requirements must be met by soluble mediators, including non-toxicity and adequate bacterial uptake and excretion rate being the most important ones (Rabaey and Boon 2005). Some mediators are potassium ferricyanide, anthraquinone 2,6-disulfonic acid, cobalt sepulchrate, thionine, neutral red, and azure A (Bond and Lovley 2003). It has been proposed that several exogenous mediators, such as neutral red and bromocresol green may attack any NADH regeneration step in the intracellular electron-transport pathways and transfer electrons to the extracellular anode (Mao and Verwoerd 2013a). Several toxic and unstable mediators, such as neutral red, methylene blue, and thionic and phenolic molecules are used as electron transporters between the microbial membrane and electrodes. Nevertheless, the active forms of such compounds are not suitable for long-term usage (Fu et al. 2010). The detection of soluble electron mediators could help understand electron transport mechanisms achieved by the phototrophic communities, and the connection between the generation of the electron mediators and phototrophic activities allows a deeper research (Xiao and He 2014). Finally, adding exogenous mediators could be advantageous for the electron transport between the cathode and microbes, as it augments costs and most of them are toxic and short-lived (Huang and Angelidaki 2008).

Indirect electron transport

Indirect electron transfer has emerged as a secondary way to carry out electrons to the anode in a PhFC. Whereas few microbes can achieve a direct electron transport, other microorganisms excrete redox-active molecules to accomplish an indirect electron transport with electrodes. This process is carried out without physical contact between the bacterial cell membrane and the cathode electrode surface. Exoelectrogenic bacteria, capable of transferring electrons extracellularly, have been well studied due to their biotechnological relevance and are ubiquitous in anoxic sediments and other anaerobic environmental systems (Logan 2009). The most prominent of these are *Geobacter sulfurreducens* and *Shewanella oneidensis*, which are used as model systems for electron transfer, and are recognized for their ability to conduct electrons through specialized appendages, such as microbial nanowires and c-type cytochromes (Eaktasang et al. 2016, Li et al. 2016).

Pili

Nanowires are extensions of the outer membrane containing the c-type cytochrome MtrC, which is involved in Fe(III) and Mn(IV) and the outer membrane cytochrome (OmcA) to make physical connections with neighboring cells (Shi et al. 2016). Such nanostructures are composed by the outer membrane Mtr proteins MtrC and OmcA in the entire surface. This observation agrees with the anteriorly proposed multi-step redox hopping mechanism of electron transport along the nanostructure extension (McCormick et al. 2015). Although many microorganisms can donate electrons to the anode, only a few iron reducers, such as *Geobacter sulfurreducens* and *Shewanella oneidensis* MR-1 have been reported to generate nanowires than can be used as mediators for direct electron transport the electrode (Eaktasang et al. 2016). The conductive pili of *G. sulfurreducens* PCA are assemblies of the pilin protein PilA (Cologgi et al. 2011). The conductivity of *Geobacter* spp. nanowires increases when temperature or pH decreases. This type of temperature-dependent and pH-dependent conductivity is a property that is shared with some conductive polymers where *Geobacter* spp. nanowires are proposed to transfer electrons by a metallic-like electron transfer mechanism (Shi et al. 2016). Despite outer membrane c-type cytochromes not being critical for conductivity along the pilus, they are probably intricated in some electron transport steps. For instance, the hexa-heme OmcS seem to be involved in the electron transport from nanowires to Fe(III) oxides, and could allow electron flow between nanofilaments (McCormick et al. 2015).

Electron Losses

Potential losses both at cathode and anode and electrons discharged among other elements are some of the principal concern which will carry to low electron transfer productivity (Srikanth and Venkata Mohan 2012). A primary PhFC bottleneck is the electron transmission from bacteria to the anode. The process where some compounds are oxidized at the surface of anode and reducing one at the bacterial surface or in the bacterial interior needs an activation energy, defined as overpotentials, which is crucial in the lower range of power generation, resulting in a transfer resistance, and hence in potential electron losses (Srikanth and Venkata Mohan 2012, Villaseñor et al. 2013). In this context, technological and bacterial adaptations can alleviate high overpotentials. Among these improvements include the increasing specific surface of the electrode to enhance its contact with bacteria. This increment is limited to a certain range because bacteria create biofilms, clogging small pores impeding the free feed flow along the electrode environment. Additionally, the substitution or omission of the proton exchange membrane can significantly raise the power yield (Rabaey and Boon 2005). Some ways to reduce activation losses in bacteria are by developing their extracellularly oriented mediation capability. Three different procedures have been developed: electron transfer through mobile redox shuttle, direct membrane complex mediated electron transfer, and electron flux through conductive pili, called nanowires (Logan et al. 2006). Bacterial optimization embraces the expression of soluble mobile redox mediators or of membrane-coupled intermediary. A redox mediator is a molecule that helps to shuttle electrons between bacteria and an electron acceptor. It can be oxidized faster at the electrode decreasing overpotential losses (Rabaey and Boon 2005). Moreover, the addition of avcatalyst, such as platinum in the cathode chamber for electron transfer from cathode to oxygen helps to reduce the overpotential activation (Schröder et al. 2003, Srikanth and Venkata Mohan 2012). Another voltage loss is due to ohmic loss that is triggered by electrical resistances of membrane, electrolyte, and electrodes (Rabaey and Boon 2005, Rismani-Yazdi et al. 2008), and can be easily got over by using mixed consortia and wastewater (as a result of the great conductivity of electrolyte) (Srikanth and Venkata Mohan 2012). As a result of the huge oxidative strength of the anode chamber, concentration loss is carried out in the space where the compound is being oxidized quicker at the anode, provoking the generation of huge amounts of electrons than that can be carried to the anode surface and to the cathode (Srikanth and Venkata Mohan 2012, Villaseñor et al. 2013a).

Factors Affecting Power Generation

Biotic

Several studies have projected the limit of PhFC power yield as the potential difference that occurs with the NADH and the cathode reactions if microorganisms are employed as a biotic catalyst. In that regard, NADH can be targeted as the electron source and liberate the maximum achievable power from a microorganism (Mao and Verwoerd 2013b). The consortium can produce soluble intermediaries to support the electricity generation (Cao et al. 2008). Self-excreted intermediaries perform crucial role in the electricity production; the increase of the intermediary molecules over time intensify the electron transfer rate. Anthraquinone-2,6-disulfonate (AQDS) has been confirmed as self-excreted mediator (Martinez et al. 2017, Wang et al. 2016). Besides, the indole group is usually found in known mediators, such as a pyocyanin and ACNQ (Cao et al. 2008).

Abiotic

In recent years, there has been an increasing concern in the use of phototrophic (mainly photoautotrophic) microorganisms in the cathode chamber of an MFC because of several advantages, such as oxygen supplementation, carbon dioxide reduction, generating profitable biomass, and/or treating wastewater (Cao et al. 2008, Xiao and He 2014).

Substrate

A complex substrate helps to establish diversified microbial community that are electrochemically active, whereas simpler substrates are easier to metabolize and enhance the bioelectricity of the fuel cell (Ewing et al. 2017, Kurokawa et al. 2016, Pandey et al. 2016, Pant et al. 2010). Nowadays, there exists an increasing interest in the use of organic wastes as substrate in biological reactions to produce bioenergy. However, the use of such sources depends directly on the economic viability and efficiency, both depending on the chemical composition and concentrations of the components of the waste material. The impact of the substrate is not only ligated to the structuring of the microbial biofilm at the anode, but also the PhFC performance comprising Coulombic efficiency (CE) and power density (PD) (Pant et al. 2010).

Carbon organic

In the majority of the MFC research so far, the electricity generation has been done using acetate as substrate (Pant et al. 2010). Since the acetate is a non-complex substrate, it has been used largely as a carbon source to promote electroactive bacteria (Bond and Lovley 2003) and as a substrate because it does not react in diverged microbial reactions (fermentations and methanogenesis) at room temperature (Aelterman et al. 2006). Additionally, acetate is the final output of several metabolic routes related to sugars and lipids (Biffinger et al. 2008). A work developed by Chae et al. (2009) found that the highest CE (72.3%) and power output was acetate when used as substrate and compared it to several substrates. Nevertheless, the protein-rich wastewater provided a more diverse microbial film due to the complex substrate composition.

Inorganic carbon (CO_2)

Photosynthetic organisms use carbon dioxide (CO_2) as carbon source, and photosynthetic reactions involve the reduction of this molecule. Actually, to increase microalgal production, carbon dioxide is added to microalgal, and it is seen as a legitimate technique. In this context, PhFC integrating anaerobic digestion (oxidation of organic compounds) at the anode and photosynthetic microorganisms (microalgae) in the cathode chamber represent an appropriate focus to surpass carbon limitation conditions in systems using algal without the need for external CO_2 supply. In that regard, the CO_2 produced at the anode of a PhFC could be redirected to the cathode chamber to improve microalgae development (Uggetti and Puigagut 2016).

Light energy

Illumination is a salient feature for growth of photosynthetic microorganisms. Several studies found that it is possible to significantly increase the electricity generation through the increment of illumination intensity, probably through augmenting the generation of dissolved oxygen (Gouveia et al.

2014, He et al. 2014, Wu et al. 2014). Nonetheless, some studies have found that more electricity can be generated under lower light intensity (Juang et al. 2012). Since illumination is a limiting factor, prolonged periods of light could increase oxygen generation, and thus power generation. However, an extended light period was found to decrease electricity generation, such findings support the evidence that dark period is fundamental to control healthy consortia of cyanobacteria and regarding organisms (Wu et al. 2013, Xiao et al. 2012).

Electrode

The effect of distance between electrodes in voltage output and current densities, different spaces between anodes and cathodes were tested under both dark and light conditions. The findings suggest that reducing the distance between electrodes could improve the voltage and current density. Such results are in concordance with other researches suggesting that narrow space between anode and cathode, besides internal resistances in the electrolyte, are smaller and output measurements can increase (Mao and Verwoerd 2014).

The electrode material is one of the most important aspects in the design of a microbial fuel cell. Extensive materials and shapes have been used in different studies in microbial fuel cells. These materials have been modified in their roughness to improve the bacterial union and electron transfer to the electrode surface as well (Hemalatha et al. 2017, Mashkour and Rahimnejad 2015, Wei et al. 2011). The carbonaceous material electrode-based are the most commonly used as electrodes in bioelectrochemical systems due to the specific characteristics that they possess, such as low cost, good biocompatibility, good chemical stability, and high conductivity. Currently, several electrodes of different material, such as platinum or titanium have been synthetized, and have shown good conductivity (Guo et al. 2016, Hemalatha et al. 2017, Shukla and Kumar 2018).

pH and temperature

The best performance of PhFC has been found in acidic conditions. Such results could be explained due to the carbonic acid augments ionic force and decreased internal resistance. High ionic force indicates a high proton concentration (H^+), which results in a high electrolyte conductivity and subsequently, a high fuel cell performance (Mao and Verwoerd 2014).

Temperature influences to a high degree the anode potential of the PhFC; driving to a decrement in electric current generation when the microorganisms are located in cold environments and inhibited during frost periods (Helder et al. 2013). Mao and Verwoerd et al. (2014) observed OCV increased with the rising of temperature, and the maximal OCV, 0.39 V, was obtained at 40°C. The best power density achieved was 6.5 mWm^{-2}. In that regard, the highest PhFC performance has been found at a higher temperature because higher temperature increases the reaction rate as well the electron transport.

It has been suggested that a high temperature increments the activity of the substrates and diminishes the activity of products, and these behaviors have an important impact on PhFC performance (Mao and Verwoerd 2014).

Terminal electron acceptor at cathode

Oxygen generation in photosynthetic reactions represents an interesting characteristic in the cathode process, because mechanic aeration requires vast quantities of energy. The oxygen reduction rate in cathodes is another important limiting factor of PhFCs. One common method is the implementation of a catalyst at the cathode to upgrade the oxygen reduction rate (Wang et al. 2014). Recently, several researches in the fields of PhFC have taken advantage of the photosynthetic activity of algae at the cathode compartment in relation to their capacity to discard CO_2 from the cathode as well as provide high concentrations of oxygen (Gajda et al. 2013).

Photosynthetic MFC Implementation in Practical Applications

PhFCs have the potential to be used in a large range of applications (Xiao and He 2014). Anaerobic digestion represents an interesting strategy to energetically utilize the biomass generated in the processing of domestic wastewater with algal-based PhFC (Gonzalez-Fernandez et al. 2015). Regardless of promising results due to low microalgal generation rates and biomass concentration, that is strongly dependent on the availability of nutrients and carbon concentration, the high costs of energy valorization of algal biomass via anaerobic digestion limit their use. Biomass generated by photosynthetic activity can be recycled as an energy source as well as biofuels and value-added compounds generation through further transformation (Ward et al. 2014). Equally, algal biomass could represent an important substrate for current electrical generation in PhFCs (Xiao and He 2014).

Wastewater treatment

MFC-based wastewater treatment technologies have studied and developed in recent years (Villaseñor et al. 2013a). Nowadays, MFCs can be used for degrading several compounds from domestic wastewaters and industrial sewage. They have been used in various volume scales, as well in long-term studies (Xiao and He 2014). Currently, wastewater represents an inexpensive source of nutrients to improve microalgal growth rate and concentration, as well as allow the contaminant discard (Uggetti and Puigagut 2016). However, domestic effluents lack the sufficient inorganic carbon to optimize algal generation (Park et al. 2011). The commercialization of algal bioreactors has suffered several challenges, mainly due to their large water footprint (Olguín 2012). The main drawback of feeding wastewater into a cathode chamber is the generation of heterotrophic bacteria, which combined with the electron donor competition of the organic compounds, disturb the correct PhFC functionality.

Wetland microbial fuel cells

Horizontal subsurface flow constructed wetland (HSSF-CW) is a system where water is circulated through a porous bed of gravel on which plants grow (García et al. 2010). An HSSF-CWs treating domestic wastewater can act simultaneously as an MFC producing electric power. However, when a large concentration of organic compounds is loaded, the matter cannot be completely oxidized in the anode, thus traveling to the cathode. In there, the organic material enters anaerobic conditions and the MFC stops functioning. On the other hand, when low organic concentrations are loaded, the wastewater organic material is completely oxidized in the anode, allowing the electrical current generation (Villaseñor et al. 2013a). A limiting drawback is the voltage fluctuation caused by the light/darkness changes due to the photosynthetic activity of the macrophytes, which alters the conditions in the cathodic compartment (Villaseñor et al. 2013a).

There are various applications of MFC in hydrogen production and development of biosensors. MFC and PhFC provide a sustainable source of hydrogen production. Commonly, such systems are named bio-electrochemically assisted microbial reactors (BEAMR) (Liu et al. 2005). PhFC and MFC can be used for pollutant analysis. In that sense, MFCs represent an attractive alternative for developing electrochemical sensors and small remote telemetry.

Advances in PhFCs

Micro-fabricated devices have shown to be capable platforms for elucidating biological phenomena. Microfluidic devices have been recognized as powerful tools for research in microbial systems, mainly due to their micrometer dimensions. Currently, such devices can efficiently handle and control cells and molecules in a high spatial and temporal precision (Han et al. 2013). Recently, the advantages of microfluidic devices have been used to solve several problems in microbial areas, such as the behavior understanding of heterogeneous populations in microbial fuel cells (Han et al. 2013). The advantages of micro-MFCs originate from their unique structural features and scales. In order to surpass the limited voltage yield of miniaturized MFCs, the connection of multiple MFCs has been proposed (Oh and Logan 2007). Nevertheless, the voltage reversal in these connected systems is still an unsolved problem which needs to be solved for long-term implementation of MFC-based sensor networks and/or environmental toxin monitoring systems (Han et al. 2013). Multiple product recovery also give good option for a biorefinery platform (Chandra et al. 2018, Olguín 2012).

Final Thoughts

The use of PhFC has gained the attention of scientists in the past years due to the diverse potential applications and the involvement of different disciplines. More studies regarding the use of PhFC in microfluidic devices are expected due to the ease of change parameters and different microsystems are obtained that will help to understand and know better the mechanisms of producing electricity and some chemicals. Also, the different changes that could be done in the configuration, electrode material, and the several environmental parameters can produce a different response in the production of electricity by these devices. A deep understanding on the transport of electrons needs to be studied to make the process more efficient. A large-scale process is the next step in this technology, so it can be applied and used with more frequency in the industry and/or to be standardized for domestic purposes. It is important to highlight the importance of the well-defined photosynthetic mechanism i.e., oxygenic and anoxygenic of the photosynthetic microbes and their behavior in different reactor configuration in order to understand the mechanisms of their operation. By now, plant microbial fuel cell is promising towards bioelectricity generation without disturbing the ecosystem, specially for greenhouses, and most important is that these technologies are very useful for the development of photo-sensing devices.

Acknowledgements

We acknowledge the School of Engineering and Sciences and the Emerging Technologies and Molecular Nutrition Research Group for their support. We acknowledge the financial support from CONACYT (Mexico) as SNI-C fellowship to RC. The author also wants to thank Aavesh green sustainability solutions S. De R. L. De. C. V. for their support.

REFERENCES

Aelterman, P., K. Rabaey, P. Clauwaert and W. Verstraete. 2006. Microbial fuel cells for wastewater treatment. Water Sci. Technol. 54: 9.

Bensaid, S., B. Ruggeri and G. Saracco. 2015. Development of a photosynthetic microbial electrochemical cell (PMEC) reactor coupled with dark fermentation of organic wastes: medium term perspectives. Energies 8: 399-429.

Biffinger, J.C., J.N. Byrd, B.L. Dudley and B.R. Ringeisen. 2008. Oxygen exposure promotes fuel diversity for Shewanella oneidensis microbial fuel cells. Biosens. Bioelectron. 23: 820-826.

Bond, D.R. and D.R. Lovley. 2003. Electricity production by geobacter sulfur reducens attached to electrodes. Appl. Environ. Microbiol. 69: 1548-1555.

Cao, X., X. Huang, N. Boon, P. Liang and M. Fan. 2008. Electricity generation by an enriched phototrophic consortium in a microbial fuel cell. Electrochem. Commun. 10: 1392-1395.

Chae, K.J., M.J. Choi, J.W. Lee, K.Y. Kim and I.S. Kim. 2009. Effect of different substrates on the performance, bacterial diversity, and bacterial viability in microbial fuel cells. Bioresour. Technol. 100: 3518-3525.

Chandra, R., G. Venkata Subhash and S. Venkata Mohan. 2012. Mixotrophic operation of photo-bioelectrocatalytic fuel cell under anoxygenic microenvironment enhances the light dependent bioelectrogenic activity. Bioresour. Technol. 109: 46-56.

Chandra, R., J. Annie Modestra and S. Venkata Mohan. 2015. Biophotovoltaic cell to harness bioelectricity from acidogenic wastewater associated with microbial community profiling. Fuel 160: 502-512.

Chandra, R., J.S. Sravan, M. Hemalatha, S. Kishore Butti and S. Venkata Mohan. 2017. Photosynthetic synergism for sustained power production with microalgae and photobacteria in a biophotovoltaic cell. Energy & Fuels.

Chandra, R., C. Castillo-Zacarias, P. Delgado and R. Parra-Saldívar. 2018. A biorefinery approach for dairy wastewater treatment and product recovery towards establishing a biorefinery complexity index. J. Clean. Prod. 183: 1184-1196.

Chiranjeevi, P., R. Chandra and S. Venkata Mohan. 2013. Ecologically engineered submerged and emergent macrophyte based system: an integrated eco-electrogenic design for harnessing power with simultaneous wastewater treatment. Ecological Engineering 51: 181-190.

Cho, Y.K., T.J. Donohue, I. Tejedor, M.A. Anderson, K.D. McMahon and D.R. Noguera. 2008. Development of a solar-powered microbial fuel cell. J. Appl. Microbiol. 104: 640-650.

Cologgi, D.L., S. Lampa-Pastirk, A.M. Speers, S.D. Kelly and G. Reguera. 2011. Extracellular reduction of uranium via Geobacter conductive pili as a protective cellular mechanism. Proc. Natl. Acad. Sci. U.S.A. 108: 15248-15252.

Cournet, A., M.L. Délia, A. Bergel, C. Roques and M. Bergé. 2010. Electrochemical reduction of oxygen catalyzed by a wide range of bacteria including Gram-positive. Electrochem. Commun. 12: 505-508.

Dekker, A., A. Ter Heijne, M. Saakes, H.V.M. Hamelers and C.J.N. Buisman. 2009. Analysis and improvement of a scaled-up and stacked microbial fuel cell. Environ. Sci. Technol. 43: 9038-9042.

Eaktasang, N., C.S. Kang, H. Lim, O.S. Kwean, S. Cho, Y. Kim and H.S. Kim. 2016. Production of electrically-conductive nanoscale filaments by sulfate-reducing bacteria in the microbial fuel cell. Bioresour. Technol. 210: 61-67.

Ewing, T., P.T. Ha and H. Beyenal. 2017. Evaluation of long-term performance of sediment microbial fuel cells and the role of natural resources. Appl. Energy 192: 490-497.

Fredrickson, J.K., M.F. Romine, A.S. Beliaev, J.M. Auchtung, M.E. Driscoll, T.S. Gardner, K.H. Nealson, A.L. Osterman, G. Pinchuk, J.L. Reed, D.A. Rodionov, J.L. Rodrigues, D.A. Saffarini, M.H. Serres, A.M. Spormann, I.B. Zhulin and J.M. Tiedje. 2008. Towards environmental systems biology of Shewanella. Nat Rev Microbiol 6: 592-603.

Fu, C.C., T.C. Hung, W.T. Wu, T.C. Wen and C.H. Su. 2010. Current and voltage responses in instant photosynthetic microbial cells with Spirulina platensis. Biochem. Eng. J. 52: 175-180.

Gajda, I., J. Greenman, C. Melhuish and I. Ieropoulos. 2015. Self-sustainable electricity production from algae grown in a microbial fuel cell system. Biomass and Bioenergy 82: 87-93.

Gajda, I., J. Greenman, C. Melhuish and I. Ieropoulos. 2013. Photosynthetic cathodes for microbial fuel cells. Int. J. Hydrogen Energy 38: 11559-11564.

García, J., D.P.L. Rousseau, J. Morató, E. Lesage, V. Matamoros and J.M. Bayona. 2010. Contaminant removal processes in subsurface-flow constructed wetlands: a review. Crit. Rev. Environ. Sci. Technol. 40: 561-661.

Georgianna, D.R. and S.P. Mayfield. 2012. Exploiting diversity and synthetic biology for the production of algal biofuels. Nature 488: 329-335.

Gonzalez-Fernandez, C., B. Sialve and B. Molinuevo-Salces. 2015. Anaerobic digestion of microalgal biomass: challenges, opportunities and research needs. Bioresour. Technol. 198: 896-906.

Gouveia, L., C. Neves, D. Sebastião, B.P. Nobre and C.T. Matos. 2014. Effect of light on the production of bioelectricity and added-value microalgae biomass in a photosynthetic alga microbial fuel cell. Bioresour. Technol. 154: 171-177.

Gregory, K.B., D.R. Bond and D.R. Lovley. 2004. Graphite electrodes as electron donors for anaerobic respiration. Environ. Microbiol. 6: 596-604.

Guo, S., J. Yi, Y. Sun and H. Zhou. 2016. Recent advances in titanium-based electrode materials for stationary sodium-ion batteries. Energy Environ. Sci. Energy Environ. Sci 9: 2978-3006.

Han, A., H. Hou, L. Li, H.S. Kim and P. de Figueiredo. 2013. Microfabricated devices in microbial bioenergy sciences. Trends Biotechnol. 31: 225-232.

He, H., M. Zhou, J. Yang, Y. Hu and Y. Zhao. 2014. Simultaneous wastewater treatment, electricity generation and biomass production by an immobilized photosynthetic algal microbial fuel cell. Bioprocess Biosyst. Eng. 37: 873-880.

He, Z., J. Kan, F. Mansfeld, L.T. Angenent and K.H. Nealson. 2009. Self-sustained phototrophic microbial fuel cells based on the synergistic cooperation between photosynthetic microorganisms and heterotrophic bacteria self-sustained phototrophic microbial fuel cells based on the synergistic cooperation between photosynth. Environ. Sci. Technol. 43: 1648-1654.

Helder, M., D.P.B.T.B. Strik, H.V.M. Hamelers, R.C.P. Kuijken and C.J.N. Buisman. 2012. Bioresource Technology New plant-growth medium for increased power output of the Plant-Microbial Fuel Cell. Bioresour. Technol. 104: 417-423.

Helder, M., D.P.B.T.B. Strik, R.A. Timmers, S.M.T. Raes, H.V.M. Hamelers and C.J.N. Buisman. 2013. Resilience of roof-top plant-microbial fuel cells during Dutch winter. Biomass and Bioenergy 51: 1-7.

Hemalatha, M., J.S. Sravan, D.K. Yeruva and S. Venkata Mohan. 2017. Integrated ecotechnology approach towards treatment of complex wastewater with simultaneous bioenergy production. Bioresour. Technol. 242: 60-67

Huang, L. and I. Angelidaki. 2008. Effect of humic acids on electricity generation integrated with xylose degradation in microbial fuel cells. Biotechnol. Bioeng. 100: 413-422.

Ivashin, N., B. Källebring, S. Larsson and Ö. Hansson. 1998. Charge separation in photosynthetic reaction centers. J. Phys. Chem. B 102: 5017-5022.

Juang, D.F., C.H. Lee and S.C. Hsueh. 2012. Comparison of electrogenic capabilities of microbial fuel cell with different light power on algae grown cathode. Bioresour. Technol. 123: 23-29.

Kaku, N., N. Yonezawa and Y. Kodama. 2008. Plant/microbe cooperation for electricity generation in a rice paddy field. Appl. Microbiol Biotechnol. 79(1): 43-49. doi: 10.1007/s00253-008-1410-9.

Kurokawa, M., P.M. King, X. Wu, E.M. Joyce, T.J. Mason and K. Yamamoto. 2016. Effect of sonication frequency on the disruption of algae. Ultrason. Sonochem. 31: 157-162.

Li, C., K.L. Lesnik, Y. Fan and H. Liu. 2016. Millimeter scale electron conduction through exoelectrogenic mixed species biofilms. FEMS Microbiology Letters, 363(15): fnw153, https://doi.org/10.1093/femsle/fnw153

Liu, H., S. Cheng and B.E. Logan. 2005. Production of electricity from proteins using a microbial fuel cell. Environ. Sci. Technol. 39: 658-662.

Logan, B.E. 2009. Exoelectrogenic bacteria that power microbial fuel cells. Nat. Rev. Microbiol. 7: 375-381.

Logan, B.E., B. Hamelers, R. Rozendal, U. Schröder, J. Keller, S. Freguia, P. Aelterman, W. Verstraete and K. Rabaey. 2006. Microbial fuel cells: methodology and technology. Environ. Sci. Technol. 40: 5181-5192.

Lovley, D.R., 2008. The microbe electric: conversion of organic matter to electricity. 19(6): 564-571.

Mao, L. and W.S. Verwoerd. 2013. Model-driven elucidation of the inherent capacity of Geobacter sulfurreducens for electricity generation. J. Biol. Eng. 7: 14.

Mao, L. and W.S. Verwoerd. 2014. Computational comparison of mediated current generation capacity of Chlamydomonas reinhardtii in photosynthetic and respiratory growth modes. J. Biosci. Bioeng. 118: 565-574.

Martinez, C.M., X. Zhu and B.E. Logan. 2017. AQDS immobilized solid-phase redox mediators and their role during bioelectricity generation and RR2 decolorization in air-cathode single-chamber microbial fuel cells. Bioelectrochemistry 118: 123-130.

Mashkour, M. and M. Rahimnejad. 2015. Effect of various carbon-based cathode electrodes on the performance of microbial fuel cell. Biofuel Res. J. 2: 296-300.

McCormick, A.J., P. Bombelli, R.W. Bradley, R. Thorne, T. Wenzel and C.J. Howe. 2015. Biophotovoltaics: oxygenic photosynthetic organisms in the world of bioelectrochemical systems. Energy Environ. Sci. 8: 1092-1109.

Newman, D.K. and R. Kolter. 2000. A role for excreted quinones in extracellular electron transfer. Nature 405: 94-97.

Oh, S.E. and B.E. Logan. 2007. Voltage reversal during microbial fuel cell stack operation. J. Power Sources 167: 11-17.

Olguín, E.J. 2012. Dual purpose microalgae-bacteria-based systems that treat wastewater and produce biodiesel and chemical products within a biorefinery. Biotechnol. Adv. 30: 1031-1046.

Pandey, P., V.N. Shinde, R.L. Deopurkar, S.P. Kale, S.A. Patil and D. Pant. 2016. Recent advances in the use of different substrates in microbial fuel cells toward wastewater treatment and simultaneous energy recovery. Appl. Energy 168: 706-723.

Pant, D., G. Van Bogaert, L. Diels and K. Vanbroekhoven. 2010. A review of the substrates used in microbial fuel cells (MFCs) for sustainable energy production. Bioresour. Technol. 101: 1533-1543.

Park, J.B.K., R.J. Craggs and A.N. Shilton. 2011. Recycling algae to improve species control and harvest efficiency from a high rate algal pond. Water Res. 45: 6637-6649.

Pisciotta, J.M., Y. Zou and I.V. Baskakov. 2011. Role of the photosynthetic electron transfer chain in electrogenic activity of cyanobacteria. Appl. Microbiol. Biotechnol. 91: 377-385.

Powell, E.E., R.W. Evitts, G.A. Hill and J.C. Bolster. 2011. A microbial fuel cell with a photosynthetic microalgae cathodic half cell coupled to a yeast anodic half cell. Energy Sources, Part A Recover. Util. Environ. Eff. 33: 440-448.

Rabaey, K. and N. Boon. 2005. Microbial Phenazine Production Enhances Electron Transfer in Biofuel Cells. Environ. Sci. Technol. 39: 3401-3408.

Rahimnejad, M., A. Adhami, S. Darvari, A. Zirepour and S.-E. Oh. 2015. Microbial fuel cell as new technology for bioelectricity generation: a review. Alexandria Eng. J. 745-756.

Rismani-Yazdi, H., S.M. Carver, A.D. Christy and O.H. Tuovinen. 2008. Cathodic limitations in microbial fuel cells: an overview. J. Power Sources 180: 683-694.

Roberts, A.G., M.K. Bowman and D.M. Kramer. 2004. The inhibitor DBMIB provides insight into the functional architecture of the Qo site in the cytochrome b6f complex. Biochemistry 43(24): 7707-7716.

Rosenbaum, M., Z. He and L.T. Angenent. 2010. Light energy to bioelectricity: photosynthetic microbial fuel cells. Curr. Opin. Biotechnol. 21: 259-264.

Rosenbaum, M., U. Schröder and F. Scholz. 2005. In situ electrooxidation of photobiological hydrogen in a photobioelectrochemical fuel cell based on rhodobacter sphaeroides. Environ. Sci. Technol. 39: 6328-6333.

Santoro, C., C. Arbizzani, B. Erable and I. Ieropoulos. 2017. Microbial fuel cells: from fundamentals to applications. A review. J. Power Sources 356: 225-244.

Schröder, U., J. Nießen and F. Scholz. 2003. A generation of microbial fuel cells with current outputs boosted by more than one order of magnitude. Angew. Chemie - Int. Ed. 42: 2880-2883.

Shi, L., H. Dong, G. Reguera, H. Beyenal, A. Lu, J. Liu, H. Yu and J.K. Fredrickson. 2016. Extracellular electron transfer. Nat. Rev. Microbiol. 14: 651-662.

Shukla, M. and S. Kumar. 2018. Algal growth in photosynthetic algal microbial fuel cell and its subsequent utilization for biofuels. Renew. Sustain. Energy Rev. 82: 402-414.

Srikanth, S. and S. Venkata Mohan. 2012. Change in electrogenic activity of the microbial fuel cell (MFC) with the function of biocathode microenvironment as terminal electron accepting condition: influence on overpotentials and bio-electro kinetics. Bioresour. Technol. 119: 241-251.

Strik, D.P.B.T.B., H. Terlouw, H.V.M. Hamelers and C.J.N. Buisman. 2008. Renewable sustainable biocatalyzed electricity production in a photosynthetic algal microbial fuel cell (PAMFC). Appl. Microbiol. Biotechnol. 81: 659-668.

Strik, D.P.B.T.B., R.A. Timmers, M. Helder, K.J.J. Steinbusch, H.V.M. Hamelers and C.J.N. Buisman. 2011. Microbial solar cells: applying photosynthetic and electrochemically active organisms. Trends Biotechnol. 29: 41-49.

Uggetti, E. and J. Puigagut. 2016. Photosynthetic membrane-less microbial fuel cells enhance microalgal biomass concentration. Bioresour. Technol. 218: 1016-1020.

Venkata Mohan, S., S. Srikanth, P. Chiranjeevi, S. Arora and R. Chandra. 2014. Algal biocathode for in situ terminal electron acceptor (TEA) production: synergetic association of bacteria-microalgae metabolism for the functioning of biofuel cell. Bioresour. Technol. 166: 566-574.

Venkata Subhash, G., R. Chandra and S. Venkata Mohan. 2013. Microalgae mediated bio-electrocatalytic fuel cell facilitates bioelectricity generation through oxygenic photomixotrophic mechanism. Bioresour. Technol. 136: 644-653.

Venkidusamy, K., M. Megharaj, U. Schröder, F. Karouta, S.V. Mohan and R. Naidu. 2015. Electron transport through electrically conductive nanofilaments in Rhodopseudomonas palustris strain RP2. RSC Adv. 5: 100790-100798.

Villaseñor, J., P. Capilla, M.A. Rodrigo, P. Cañizares and F.J. Fernández. 2013. Operation of a horizontal subsurface flow constructed wetland - Microbial fuel cell treating wastewater under different organic loading rates. Water Res. 47: 6731-6738.

Wang, D., T. Song, T. Guo and Q. Zeng. 2014. Electricity generation from sediment microbial fuel cells with algae-assisted cathodes. Int. J. Hydrogen Energy 39: 13224-13230.

Wang, W., S. You, X. Gong, D. Qi, B.K. Chandran, L. Bi, F. Cui and X. Chen. 2016. Bioinspired nanosucker array for enhancing bioelectricity generation in microbial fuel cells. Adv. Mater. 28: 270-275.

Ward, A.J., D.M. Lewis and F.B. Green. 2014. Anaerobic digestion of algae biomass: a review. Algal Res. 5: 204-214.

Watanabe, K., M. Manefield, M. Lee and A. Kouzuma. 2009. Electron shuttles in biotechnology. Curr. Opin. Biotechnol. 20: 633-641. doi:10.1016/j.copbio.2009.09.006

Wei, J., P. Liang and X. Huang. 2011. Recent progress in electrodes for microbial fuel cells. Bioresour. Technol. 102: 9335-9344.

Wetser, K., E. Sudirjo, C.J.N. Buisman and D.P.B.T.B. Strik. 2015. Electricity generation by a plant microbial fuel cell with an integrated oxygen reducing biocathode. Appl. Energy 137: 151-157.

Wu, X., T. Song, X. Zhu, P. Wei and C.C. Zhou. 2013. Construction and operation of microbial fuel cell with Chlorella vulgaris biocathode for electricity generation. Appl. Biochem. Biotechnol. 171: 2082-2092.

Wu, Y. Cheng, Z. Jie Wang, Y. Zheng, Y. Xiao, Z. Hui Yang and F. Zhao. 2014. Light intensity affects the performance of photo microbial fuel cells with *Desmodesmus* sp. A8 as cathodic microorganism. Appl. Energy 116: 86-90.

Xiao, L. and Z. He. 2014. Applications and perspectives of phototrophic microorganisms for electricity generation from organic compounds in microbial fuel cells. Renew. Sustain. Energy Rev. 37: 550-559.

Xiao, L., E.B. Young, J.A. Berges and Z. He. 2012. Integrated photo-bioelectrochemical system for contaminants removal and bioenergy production. Environ. Sci. Technol. 46: 11459-11466.

Bio-based Products in Fuel Cells

Beenish Saba[1,2]*, Ann D. Christy[1], Kiran Abrar[2] and Tariq Mahmood[2]

[1] Department of Food, Agricultural and Biological Engineering, Ohio State University, Coffee Road, Columbus, Ohio 43210

[2] Department of Environmental Sciences, PMAS Arid Agriculture University Rawalpindi, Pakistan

Introduction

Utilization of renewable resources to produce fuels and chemicals is increasing. Organic waste materials are renewable and can be used as an alternative resource for biofuel production. As interest has reached its apex in bio-electrochemistry, how bacteria transfer electrons to solid state electrodes and what benefits can be obtained from this bioelectrochemical system (BES) are the contemporary achievements of researchers to discover this technology, whose initial target was the conversion of both inorganic and organic types of waste into different energy products (Logan et al. 2006). Traditionally, BES is composed of an anode and a cathode that are separated by an ion exchange membrane (Du et al. 2008). Currently, BES applications have become evident, which include microbial desalination cells (MDCs), microbial fuel cells (MFCs), microbial solar cells (MSCs), and microbial electrosynthesis cells (MECs) (Bajracharya et al. 2016). Recently, researchers have concentrated on the microbial electrosynthesis cell's (MEC's) ability to diminish substrates into utilizable chemicals (Mohanakrishna et al. 2016). In MECs, without the utilization of land and chemicals, desired fuels can be manufactured in a direct and sustainable way. MECs would attenuate not only the social demand on fossil fuels, but also reduce greenhouse effects (Rabaey et al. 2011). Over the past years, efforts have been focused for developing MECs into multiple platforms that can synthesize diverse fuels, such as acetate, butanol and butyrate (Zhang and Angelidaki 2014).

Microbial Electrosynthesis Cells (MECs)

Microbial electrosynthesis (MEC) is a new emerging platform where

*Corresponding author: beenishsaba@uaar.edu.pk

synthesis of organic chemicals takes place at a cathode by applying electrical energy, and microorganisms are utilized as a catalyst for the electrochemical minimizing of low-grade compounds (Rabaey and Rozendal 2010). MECs hold the multifaceted potential for capturing of CO_2, serving as a waste biorefinery, removal of pollutants, and biosynthesis of renewable and clean electro-fuels or other precious commodities. The development of MECs from a concept to a technology has greatly been uplifted due to viable advances in process mechanisms and design (Zhen et al. 2016). The efficacy of MECs varies with different substrates used under operational conditions, as it can evidently convert any biodegradable waste into biofuel, H_2, and other products Lu and Ren 2016. In the scientific community, MECs have obtained interest as a new platform technology for removal of different organic and inorganic contaminants, as well as production of highly valuable and flexible chemicals, such as formate, acetate, hydrogen, methane, alcohols, and hydrogen peroxide (Albo et al. 2015).

Microbial electrolysis cells (MECs) is one of the more favorable technologies used to produce chemicals from wastes, and over the past few years it has improved a lot. Numerous challenges will be faced by this technology in having practical applications of wastewater treatment, including degrading complicated recalcitrant compounds and control of microbial reactions in the bioreactors. Ellis (2013) accomplished a study of *Clostridium saccha-roperbutylacetonicum N1-4* used algal biomass in the wastewater for production of butanol, acetone, and ethanol (ABE). Ten percent algae were used as a feedstock for carrying out batch fermentation process. The MECs have many merits, such as high energy efficiency, low cost, and good environmental sustainability of the products (Lovley and Nevin 2013). Moreover, since the applied energy can be provided by renewable sources (wind and solar), and even the wastewater, the MECs also provide new methods for storage of renewable energy as electrical energy in chemical bonds (Xiang et al. 2017). Microbial electrosynthesis cell (MEC) using biocatalyst refers to the generation of multi-carbon compounds, such as CH_4, H_2, oxo-butyrate, acetate, bio-alcohols, and bio-plastics under current conditions (Bajracharya et al. 2017).

Different types of products including fermentable organics (e.g., glycerol, glucose, xylose, biodiesel by-products, galactose, mannose, etc.), non-fermentable organics (e.g., lactic acid, butyric acid), and acetate, can be produced. For the biorefinery wastes, such as lignocellulosic biomass and by-products, fermentation effluent, high strength wastewater (e.g., industrial and food processing wastewaters) can be used as feedstocks for the production of bio-based products (Lu et al. 2016). CO_2 is both the most inexpensive and most extensive carbon resource. MECs offer merits by using CO_2 as a substrate. CO_2 is available in the atmosphere, oceans, and soils as an abundant source. Therefore, a new manufacturing paradigm could be created which might provide a sustainable energy and material supply by conversion of CO_2 into fuels and other value-added chemicals (Ragauskas et al. 2006). Among the existing technologies, microbial electrosynthesis cell

(MEC) is a new strategy in which CO_2 is converted to chemicals and fuels by a microbial reduction process, while using electricity as the energy source (Zhang et al. 2013).

The key parameters that have an influence on performance of MECs include microbial physiology, the type of electrolytes and substrates type, redox potential, and electrode materials (Sharma et al. 2014). Electrons can be accepted by endoelectrogens (current consumption), on the other hand, electrons are transferred to anode by exoelectrogens (current generation) for the reduction of substrate from cathode (Choi and Sang 2016). In previous years, electrochemical reactions which include biocatalysts (microorganisms) have been developed, and microbial reduction reactions produce chemicals at the cathode (Rosenbaum and Henrich 2014). The performance of MECs is generally enhanced by using exoelectrogenic bacteria, such as *Acidobacteria, Actinobacteria, α-Proteobacteria, δ-Proteobacteria* and *Firmi-cutes* (Kadier et al. 2016a). Exo-electrogenic bacteria in MECs oxidize the organic substrate to produce protons (H^+), electrons (e^-), and carbon dioxide (CO_2). The protons reach the cathode and electrons move towards the anode, and at cathode reduction of protons to hydrogen takes place under the applied potential (Modestra et al. 2015a).

There is a special group of bacteria which has the ability of sequestering the uptake of CO_2 chemolithoautotrophically and anaerobically, and thereby synthesizing valuable chemicals/products by utilization of hydrogen as an electron donor (Mohan et al. 2014). Enrichment of this specific group of bacteria is therefore of high importance in microbial electrosynthesis for utilization of CO_2 as a substrate for high yield of desired products. Homoacetogenic bacteria have the ability to fix CO_2 which makes them notable bacteria in variety of MECs experiments (Modestra et al. 2015b). Modestra and Mohan (2017) designed a double-chambered biochemical system, and used acetogenic (homoacetogenic) bacteria as inoculum source at cathode chamber with an input voltage of –0.8 V (Ag/AgCl). As a result, synthesis of carboxylic acid with a major proportion of acetic acid takes place. MEC was fabricated consisting of two chambers separated by an ion exchange membrane. Nutrient waste was used as a substrate. At a set voltage of –1.12 V, *C. pasteurianum* was the source of inoculum. A variety of different products, mainly butanol and ethanol, were synthesized (Mostafazadeh et al. 2017). A two-chambered MEC was erected utilizing raw granular sludge taken from beer brewery as a biocatalyst. Different cathodic potentials (–0.5, –0.6, –0.7) were used along with sulfate reducing bacteria as a catholyte. At –0.7 V, maximum acetate production was attained (Xiang et al. 2017). Lu et al. (2016) constructed a two-chambered MEC and set the applied voltage at –0.129 V. At anode, exoelectrogenic microbes were utilized, which converted the biorefinery waste into current and protons. Electrons move towards the cathode and reduce protons to H_2. MFC made of glass material was built consisting of two chambers. At an applied voltage of 200 to 954 mV, culture of microalga *Chlorella vulgaris* and bacterial communities along with effluent water of a chocolate factory were inoculated at the cathode chamber.

Bioelectricity was generated with the flow of electrons from anolyte to catholyte due to degradation of organic waste (Huarachi-Olivera et al. 2018).

Aryal et al. (2017) manufactured a H-type MEC reactor. The cathode chamber was inoculated with different *Sporomusa* cultures at the cathodic potential of –690 mV. Acetate production was faster by *Sporomusa ovata* as compared to other strains. Tubular BES consisting of two chambers was configured and enriched culture of *Megasphaera sueciensis* was the source of inoculum at cathodic potential of –0.8 V. Maximum butyrate formation was attained along the synthesis of other products (acetate, ethanol, and butanol) (Batle-Vilanova et al. 2017). Li et al. (2017) fabricated a microbial reverse-electro dialysis electrolysis cell (MREC). Exoelectrogens and salinity gradient between fresh and salt water were used to generate the electric potential for the synthesis of H_2O_2 at the cathodic potential of -0.485 ± 0.025 V.

Mohanakrishna et al. (2016) constructed a H-type dual-chambered MEC reactor by following two-stage operation. Gaseous CO_2 was used as a substrate source. In the first stage, optimization of two cathodic potential –600 mV to –800 mV was done utilizing bicarbonates as a carbon source. Enriched acetogenic bacteria were used as a biocatalyst in the second stage. At –800 mV, acetate generation was maximum. A large scale biochemical system was created by using *Clostridium ljungdahlii* as an inoculum source. Gaseous CO_2 was used as a substrate and resulted in a variety of bio-based products, e.g., formic, acetic, propionic acids; methanol, and ethanol (Christodouloua et al. 2017). A detailed summary of substrates and microbes used in MECs is presented in Table 4.1.

Types of Microbial Electrosynthesis Cells (MECs)

In MECs rigorous research efforts have been developed which brought advances in not only new platform designs but also understanding the bioelectrochemical principles. A variety of electrochemical systems have been projected for different purposes with different configurations are shown in Fig. 4.1.

Microbial electrolysis cell (MEC)

Initially MEC was focused on electrohydrogenesis (H_2 evolution) through reduction of proton at the cathode using the waste stream treatment. A variety of waste substrates with different biodegradability from model carbon sources can be used in MEC, such as methanol (Montpart et al. 2014), starch, glucose and glycerol (Montpart et al. 2015), liquid fractions of municipal solid waste (LPW) (Zhen et al. 2016), human urine (Tice and Kim 2014), fermentation (Nam et al. 2014), and domestic wastewater (Ditzig et al. 2007).

MEC has various advantages for H_2 evolution, as it requires less energy input (0.2-1.0 V) in comparison to traditional hydrogen fermentation (Santos et al. 2013). MEC can also biocatalyze the electrochemical CO_2 conversion to various low carbon electro-fuels, including methane (Cheng et al. 2009),

Table 4.1. Summary of substrates and microbes in 2017 MEC literature.

Sr. No	MECs Type	Substrate	Microbes	Energy Input	Products	Reference
1.	Double chambered bio electrochemical system(BES)	CO_2	Acetogenic, (homoacetogenic) bacteria	–0.8 V	Carboxylic acid	Modestra and Mohan, 2017
2.	Double- chamber microbial electro synthesis system (MES)	Nutrient water	C. pasteurianum	-1.12 V	Solvents, acids, mainly butanol, etha nol	Mostafazad eh et al.,2017
3.	Two-chamber MES	raw granular sludge	*sulfate-reducing* bacteria	-0.7V	acetate	Xiang et al.,2017
4.	Two-chamber MES	Bio refinery waste	exoelectrogenic	-0.129 V	H2, ethanol	Lu et al.,2016
5.	Two chamber MFC	effluent waters of a chocolate factory	*microalga* Chlorella vulgaris and bacterial communities	200 - 954 mV	Biofilms,	Huarachi- Olivera et al.,2018
6.	H-cell two- chamber MES	CO_2	Sporomusa ovata	-690 mV	acetate	Aryal et al.,2017
7.	Two-chambered tubular BES	CO2 and acetate	Megasphaera sueciensis	-0.8V	butyrate	Batle- Vilanova et al.,2017
8.	Microbial reverse-electro dialysis electrolysis cell (MREC).	salinity- gradient salt and fresh water	exoelectrogens	-0.485 ± 0.025 V	Hydrogen peroxide (H2O2)	Li et al.,2013
9.	H-type MES reactors	Bicarbonat es and CO_2	*acetogenic* bacteria	-800mV	Acetate, ethanol	Mohanakris hna et al.,2016
10.	Large scale bio electrochemical systems	Gaseous CO_2+	*Clostridium ljungdahlii*	-0.393V	(formic, acetic, propionic acids; methanol and ethanol	Christodoul oua, et al.,2017

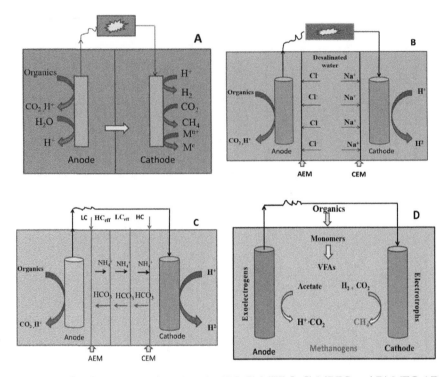

Figure 4.1. Configurations of MECs: A) MEC, B) MEDC, C) MREC and D) MEC-AD.

methanol (Albo et al. 2015), ethylene, and alcohol (Li and Liao 2013) at the cathode surface. The electrosynthesis of an important industrial chemical, i.e., hydrogen peroxide (H_2O_2), can be utilized for bioremediation and disinfection by using the cathodic reaction with an input of external voltage (0.5 V).

Microbial desalination cell (MDC)

A method newly developed for desalination of seawater by using the electric current produced by exoelectrogenic bacteria at the anode and reduction of oxygen at cathode to synthesize pure water is called microbial desalination cell (MDC) (Davis et al. 2013). MDC is an adaptable technology (Sevd et al. 2015), but it is not economically competitive due to inconsistent power generation, which negatively affects the desalination performance.

Microbial electrodialysis cell (MEDC)

The MDC modification can be done by bubbling oxygen near the cathode and imposing a small voltage between electrodes. This method is known as microbial electrodialysis cell (MEDC) (Mehanna et al. 2010). At a set voltage of 0.55 V, the use of MEDC was investigated at two different NaCl concentrations. In the desalination chamber, H_2 production increased and conductivity was reduced. This approach improved desalination efficiency by controlling the potentials between the electrodes, and the energy demand for desalination was offset by using H_2 production, thus making the process self-sustaining. Some of the technical issues, such as high internal resistance due to the addition of middle chamber installation, pH imbalances, and variable analyte concentration are limiting MEDCs performance (Davis et al. 2013).

Microbial reverse electrodialysis electrolysis cell (MREC)

To succeed the thermodynamic obstruction to cathodic bio-electrosynthesis in the typical MEC system, additional energy input must be provided. No method might have been able to do this without an external voltage supplied until the development of MREC.

A distinctive method for spontaneous H_2 production by coordinating a small reverse electrodialysis (RED) stack into an MEC was recommended by Kim and Logan (2011). MREC is a hybrid system in which the energy used for driving the evolution of H_2 comes from organic matter oxidation by the anode exoelectrogens. The requirement for external power sources in the RED stack is removed by using the salinity gradient between the sea water and fresh water. Though this coupled system effectively requires zero energy input, still many challenges are faced by this real application, such as the high costs associated with the use of several pairs of alternatively stacked cation and anion exchange membrane, which caused fouling by biofilm growth.

Microbial electrosynthesis cell-anaerobic digestion (MEC-AD) system

Another kind of biochemical system that relies on syntropic interactions between fermentative products and exoelectrogens is the MEC-AD system. The conversion of numerous substrates to metabolites by fermenters which are easily approachable to electro active microorganisms make the overall fermentation process convenient for the recovery of electrons (Cheng and Logan 2007).

Energy-rich substrates, for example activated sludge (Sun et al. 2015), trash slurry (Sasak et al. 2011), corn silage (Koch et al. 2015), and cattle manure (Samani et al. 2016), have been analyzed for this framework. This framework indicates great pledge and many favorable circumstances for bioethanol generation in examination with absolute anaerobic digestion.

Bio-based Products

Bio-based products synthesized from organic waste can help to boost economy and can lower the dependance on fossil fuels consumption. Bio-based products are fully or mostly determined from the composition of the materials of biological origin. The higher efficiency of the process can be obtained by using fermentation and bio catalysis instead of traditional chemical synthesis, which result in the reduction of toxic waste by products generation, as well as decreased energy and water consumption. As they are inferred from renewable crude materials, for example, plants, an expanded dependence ahead for bio-based items might lessen assistance on climatic CO_2.

Acetate

One of the fascinating applications of MEC is the generation of precious multi-carbon chemicals through bio-electrochemical CO_2 depletion. MEC of acetate acid derivation starting with CO_2 has been consummated through utilizing homoacetogenic bacteria as the biocatalyst (Nevin et al. 2011). As stated by Wood-Ljungdahl (WL) pathway homoacetogens, such as *Clostridium* spp. and *Sporomusa* spp. have proven ability to decrease CO_2 biologically to acetate acid derivation (Schumann and Muller 2014), and an amount of *Clostridium* spp. is shown for generation of ethanol from a mixture of CO, CO_2, and H_2 (Liew et al. 2010). In MES, the electricity-driven methodology of CO_2 diminishment by using homoacetogens could provide maintainable biofuels and feedstock of chemicals (Rabaey and Rozendal 2010).

Butanol

As bio-butanol is prepared from natural plant matter, its processing needs an unbiased CO_2 equalization. Bio-butanol can be generated from cellulose and wood residues with no effect on the food supply (Swana et al. 2011). Similarly contrasted with CO_2, utilization of organic matter (e.g., wastewater) offers profits, giving work to less demanding biocatalyst growth and lowering electron consumption, and prevalent infrastructure can be used for bio-butanol production (Rabaey et al. 2011). The bio-products' firmness and quality used for synthesis are closely affiliated to the type of biomass. In MES processes, carbon dioxide reduction reactions and catalyzation of organic molecules (at the cathode) by bacteria are related to the respiratory metabolism of the bacteria.

Formate

Formate is one of the valuable chemicals used directly in the formation of de-icing agents, animal feed, and leather treatment chemicals. Formate has been suggested as an electron carrier in microbial electrosynthesis (Du et al. 2017). *Methanococcus maripaludis* requires electron uptake for the formation of formate or H_2 by undefined, cell derived enzyme, by using methanogen, at the cathode. A standout amongst those electrocatalytic enzyme i.e., hetero disulphide reductase supercomplex, empower electromethanogenesis and catalyze the formate generation in the electrochemical reactor. Enzymatic electrosynthesis is an optimistic technology towards the production of formate (Lienemann et al. 2018).

Methane

Methane fuel is no doubt generally utilized all over the world. Different

MEC investigations using methane as a target product have been freshly accounted with successful outcome. In a two-chambered MEC, a biocathode attached with methanogen was set for production of methane by using CO_2 at a cathodic potential less than 0.7 V (Cheng et al. 2009). Methane yield was attained by shifting the mode of electrical exposure or the type of flushed substrates (CO_2/N_2) at 0.9 V. Similarly, the cathode served as an electron donor for methanogenesis by using *methanobacteriaceae* (particularly methano bacterium) at the electrode surface of microbe, thus supporting the undeviating electron transfer process between electrodes of the cell (Zhen et al. 2015).

Factors Affecting MECs Performance

Several parameters influence MEC performance, for example, inocula source, membrane, compartment separators, as well as cathode material. Therefore, scaling of MEC towards the commercial requisitions will require further understanding and optimizations of working parameters.

Inoculum source

The inoculum source as well as acclimation and enrichment methods seem to be of considerable significance for process firmness and start up. Both mixed and pure cultures can be used as inoculum. The *Shewanella* species and *Geobacter* species are most commonly used as pure cultures for extracellular electron transfer mechanism directly and indirectly (Choi and Sang 2016). The pure culture use shows a great potential for the degradation of selected pollutants (Hasany et al. 2016) or bio-electrosynthesis of given product (Bajracharya et al. 2015)

The oxidation of numerous complex organic matter present in wastewater is difficult for pure cultures without syntrophy of other fermenters (Selembo et al. 2009). The bioconversion of macromolecules to readily biodegradable substrates would be preferred by a syntrophic consortium between fermenter partners and exoelectrogens. Only *Rhodoferax ferrireducens* shows potential in the complete mineralization of glucose (Chaudhary and Lovely 2003). In most cases, the more successful as well as cost-effective inoculation in MECs includes a variety of microorganisms within a mixed culture.

Membrane and separators

The MEC utilizes an ion exchange membrane for separation of the cathode and the anode chambers. Many potential benefits in hydrogen producing MECs are provided by the manipulation of the membrane unit. Membrane insertion decreases H_2 loss, and prevents release of gaseous by-products (CO_2, H_2S, etc.) (Rozendal et al. 2007). Use of membrane seems to be necessary for liquid phase electro-fuels, such as acetate, which is used as the target products in an MEC system, as it inhibits the diffusion of produced acetate towards anode chamber. The danger of short circuiting of diffusion of products between anode and cathode is reduced by using a membrane (Kadier et al. 2016b). Use of membrane also provides a substantial environment for purification of high products and bacterial growth.

Electrode material

As the MEC stage center component, properties of the electrodes influence microbial attachment, transfer of electron at an electrode surface, and dynamics of the cathode reaction. Carbon-based materials are the earliest and most regularly utilized type of electrodes due to good conductivity, large surface area, manufacturing ease, high chemical stability, and low capital cost (Liu et al. 2014). Carbon-based materials can be used both as the anode or cathode, and include plain graphite plate, graphite rod (Zhen et al. 2015), graphite fiber brush (Call and Logan 2008), carbon paper (Liu et al. 2014), and carbon cloth (Fu et al. 2015).

Graphite rods have a low cost, are stable chemically and have a good electrical conductivity, but low surface area and porosity. Carbon paper is thin and has low strength, hence it is difficult to connect with wires (Zhou et al. 2011). Carbon cloth is more porous and flexible, providing a high surface for growth of microbes (Wei et al. 2011). Hence, the use of these materials present challenges for commercial development.

Conclusion

Microbial electrosynthesis is a new and promising field in bioelectrochemical systems. This technology can accelerate production of bio-based products and biofuels from reduction of organic substrate. Main challenge in the commercialization of the technology is manipulation of microorganisms. The efficiency of this system can be increased by enhancement of microbial interaction with electrodes and direct extracellular electron transfer. Types of waste substrate used, and operation conditions greatly affect the type and quantity of bio product.

REFERENCES

Albo, J., M.A. Guerra, P. Castae and A. Irabien. 2015. Towards the electrochemical conversion of carbon dioxide into methanol. Green Chem. 17: 2304-2324.

Aryal, N., P.L. Tremblay, D.M. Lizak and T. Zhang. 2017. Performance of different Sporomusa species for the microbial electrosynthesis of acetate from carbon dioxide. Bioresour. Technol. 233: 184-190.

Bajrachary, S., A. Heijne, X.D. Benetton, K. Vanbroekhoven, C.J.N. Buisman and D.P.B.T.B. Strik. 2015. Carbon dioxide reduction by mixed and pure cultures in microbia₁ electrosynthesis using an assembly of graphite felt and stainless steel as a cathode. Bioresour. Technol. 195: 14-24.

Bajracharya, S., M. Sharma, G. Mohanakrishna, X.D. Benneton, D.P.B.T.B. Strik, P.M. Sarma and D. Pant. 2016. An overview on emerging bioelectrochemical systems

(BESs): technology for sustainable electricity, waste remediation, resource recovery and chemical production. Renew. Energ. 98: 153-170.

Bajracharya, S., R. Yuliasni, K. Vanbroekhoven, C.J.N. Buisman, D.P.B.T.B. Strik and D. Pant. 2017. Long-term operation of microbial electrosynthesis cell reducing $CO2$ to multi-carbon chemicals with a mixed culture avoiding methanogenesis. Bioelectrochem. 113: 26-34.

Batle-Vilanova, P., R. Ganigue, S. Ramio-Pujol, L. Baneras, G. Jimenez, M. Hidalgo, M.D. Balaguer, J. Colprim and S. Puig. 2017. Microbial electrosynthesis of butyrate from carbon dioxide: Production and extraction. Bioelectrochemistry 117: 57-64.

Call, D. and B.E. Logan. 2008. Hydrogen production in a single chamber microbial electrolysis cell lacking a membrane. Environ. Sci. Technol. 42: 3401-3406.

Chaudhuri, S.K. and D.R. Lovley. 2003. Electricity generation by direct oxidation of glucose in mediatorless microbial fuel cells. Biotechnol. 21: 1229–1232.

Cheng, S., D. Xing, D.F. Call and B.E. Logan. 2009. Direct biological conversion of electrical current into methane by electromethanogenesis. Environ. Sci. Technol. 43: 3953-3958.

Cheng, S. and B.E. Logan. 2007. Sustainable and efficient biohydrogen production via electrohydrogenesis. PNAS 104: 18871-18873.

Choi, O. and B. Sang. 2016. Extracellular electron transfer from cathode to microbes: application for biofuel production. Biotechnol. Biofuels. 9(11): 1-14.

Christodoulou, X., T. Okoroafor, S. Parry and S.B. Velasquez-Orta. 2017. The use of carbon dioxide in microbial electrosynthesis: advancements, sustainability and economic feasibility. J. CO_2 Util. 18: 390-399.

Davis, R.J., Y. Kim and B.E. Logan. 2013. Increasing desalination by mitigating anolyte pH imbalance using catholyte effluent addition in a multi-anode bench scale microbial desalination cell. Sustain. Chem. Eng. 1: 1200-1206.

Dernotte, J., C.M. Rousselle, F. Halter and P. Seers. 2010. Evaluation of butanol-Gasoline blends in a port fuel-injection, spark-ignition engine. Gas. Sci. Technol. Rev. 65(2): 345-351.

Ditzig, J., H. Liu and B.E. Logan. 2007. Production of hydrogen from domestic wastewater using a bioelectrochemically assisted microbial reactor (BEAMR). Int. J. Hydrog. Ener. 32: 2296-2304.

Du, D., R. Lan, J. Humphreys and S. Tao. 2017. Progress in inorganic cathode catalysts for electrochemical conversion of carbon dioxide into formate or formic acid. J. Appl. Electrochem. 47: 661-678.

Du, Z., Q. Li, M. Tong, S. Li and H. Li. 2008. Electricity generation using membrane-less microbial fuel cell during wastewater treatment. Chin. J. Chem. Eng. 16 (5): 772–777.

Ellis, J.T. 2013. Utilizing Municipal and Industrial Wastes for the Production of Bioproducts: from Metagenomics to Bioproducts. All Graduate Theses and Dissertations. Paper 1702.

Fu, Q., Y. Kuramochi, N. Fukushima, H. Maeda, K. Sato and H. Kobayashi. 2015. Bioelectro-chemical analyses of the development of a thermophilic biocathode catalyzing electromethanogenesis. Environ. Sci. Technol. 49: 1225-1232.

Hasany, M., M.M. Mardanpour and S. Yaghmaei. 2016. Biocatalysts in microbial electrolysis cells: a review. Int. J. Hydrog. Ener. 41: 1477-1493.

Huarachi-Olivera, R., A. Duenas-Gonza, U. Yapo-Pari, P. Vega, M. Romero-Ugarte, J. Tapia, L. Molina, A. Lazarte-Rivera, D.G. Pacheco-Salazar and M. Esparza. 2018. Bioelectrogenesis with microbial fuel cells (MFCs) using the Chlorella vulgaris and bacteria communities. Electron. J. Biotechn. 31: 34-43.

Kadier, A., M.S. Kalil, P. Abdeshahian, K. Chandrasekhar, A. Mohamed, N.F. Azman, W. Logrono, Y. Simayi and A.A. Hamid. 2016a. Recent advances and emerging challenges in microbial electrolysis cells (MECs) for microbial production of hydrogen and value-added chemicals. Renew. Sustain. Ener. Rev. 61: 501-525.

Kadier, A., M.S. Kalil, P. Abdeshahian, K. Chandrasekhar, A. Mohamed and N.F. Azman. 2016b. Recent advances and emerging challenges in microbial electrolysis cells (MECs) for microbial production of hydrogen and value-added chemicals. Renew. Sust. Ener. Rev. 61: 501-525.

Kim, Y. and B.E. Logan. 2011. Hydrogen production from inexhaustible supplies of fresh and salt water using microbial reverse-electrodialysis electrolysis cells. PNAS 108: 16176–16181.

Koch, C., A. Kuchenbuch, J. Kretzschmar, H. Wedwitschka, J. Liebetrau and M.E. Ullera. 2015. Coupling electric energy and biogas production in anaerobic digesters impacts on the microbiome. RSC. Adv. 5: 31329-31340.

Li, H. and J.C. Liao. 2013. Biological conversion of carbon dioxide to photosynthetic fuels and electrofuels. Energ. Environ. Sci. 6: 2892-2899.

Li, X., I. Angelidaki and Y. Zhang. 2017. Salinity-gradient energy driven microbial electrosynthesis of hydrogen peroxide. J. Powr. Sourc. 341: 357-365.

Lienemanna, M., J.S. Deutzmannb, R.D. Miltonb, M. Sahinc and A.M. Spormannb. 2018. Mediator-free enzymatic electrosynthesis of formate by the Methanococcus maripaludis heterodisulfide reductase supercomplex. Bioresour. Technol. 254: 278-283.

Liew, F.M., M. Köpke, S.D. Simpson, K.P. Nevin, T.L. Woodard, A.E. Franks, Z.M. Summers and D.R. Lovley. 2010. Gas Fermentation for Commercial Biofuels Microbial electrosynthesis: feeding microbes electricity to convert carbon dioxide and water to multicarbon extracellular organic. M. Bio. 1: 103-110.

Liu, H., S. Grot and B.E. Logan. 2005. Electrochemically assisted microbial production of hydrogen from acetate. Environ. Sci. Technol. 39: 4317-4320.

Liu, X.W., W.W. Li and H.Q. Yu. 2014. Cathodic catalysts in biochemical systems for energy recovery from wastewater. Chem. Soc. Rev. 43: 7718-7745.

Logan, B.E., B. Hamelers, R. Rozendal, U. Schröder, J. Keller, S. Freguia, P. Aelterman, W. Verstraete and K. Rabaey. 2006. Microbial fuel cells: methodology and technology. Environ. Sci. Technol. 40(17): 5181-5192.

Lovley, D.R. and K. P. Nevin. 2013. Electrobiocommodities: powering microbial production of fuels and commodity chemicals from carbon dioxide with electricity. Curr. Opin. Biotechnol. 24(3): 385-390.

Lu, L. and Z.J. Ren. 2016. Microbial electrolysis cells for waste biorefinery: a state of the art review. Bioresour. Technol., 215: 254-264.

Mehanna, M., T. Saito, J.L. Yan, M. Hickner, X.X. Cao and X. Huang. 2010. Using microbial desalination cells to reduce water salinity prior to reverse osmosis. Ener. Environ. Sci. 3: 1114-1120.

Modestra, J.A., M.L. Babu and S.V. Mohan. 2015a. Electro-fermentation of real-field acidogenic spent wash effluents for additional biohydrogen production with simultaneous treatment in a microbial electrolysis cell. Sep. Purif. Technol. 150: 308-315.

Modestra, J.A., B. Navaneeth and S.V. Mohan. 2015b. Bio-electrocatalytic reduction of CO_2: enrichment of homoacetogens and pH optimization towards enhancement of carboxylic acids biosynthesis. J. CO_2 Util. 10: 78-87.

Modestra, J.A. and S.V. Mohan. 2017. Microbial electrosynthesis of carboxylic acids through CO₂ reduction with selectively enriched biocatalyst: microbial dynamics. J. CO₂ Util. 20: 190-199.

Mohan, S.V., G. Velvizhi, J.A. Modestra and S. Srikanth. 2014. Microbial fuel cell: critical factors regulating bio-catalyzed electrochemical process and recent advancements. Sustain. Energy. Rev. 40: 779-797.

Mohanakrishna, G., K. Vanbroekhoven and D. Pant. 2016. Imperative role of applied potential and inorganic carbon source on acetate production through microbial electrosynthesis. J. CO₂ Util. 15: 57-64.

Montpart, N., E.R. Loobet, V.K. Garlapati, L. Rago, J.A. Baeza and A. Guisasola. 2014. Methanol opportunities for electricity and hydrogen production in bioelectrochemical systems. Int. J. Hydro. Ener. 39: 770-777.

Montpart, N., L. Rago, J.A. Baeza and A. Guisasola. 2015. Hydrogen production in single chamber microbial electrolysis cells with different complex substrates. Water Res. 68: 601-615.

Mostafazadeh, A.K., P. Drogui, S.K. Brar, R.D. Tyagi, Y.L. Bihan and G. Buelna. 2017. Microbial electrosynthesis of solvents and alcoholic biofuels from nutrient waste. A review. J. Environ. Chem. Eng. 5: 940-954.

Nam, J.Y., M.D. Yates, Z. Zaybak and B.E. Logan. 2014. Examination of protein degradation in continuous flow, microbial electrolysis cells treating fermentation wastewater. Bioresour. Technol. 171: 182-186.

Nevin, K.P., S.A. Hensley, A.E. Franks, Z.M. Summers and J.O.T.L. Woodard. 2011. Electrosynthesis of organic compounds from carbon dioxide is catalyzed by a diversity of acetogenic microorganisms. Appl. Environ. Microbiol. 77: 2882-2886.

Rabaey, K., P. Girguis and L.K. Nielsen. 2011. Metabolic and practical considerations on microbial electrosynthesis. Curr. Opin. Biotechnol. 22: 1-7.

Rabaey, K. and R.A. Rozendal. 2010. Microbial electrosynthesis—revisiting the electrical route for microbial production. Nat. Rev. Microbiol. 8: 706-716.

Ragauskas, A.J., C.K. Williams, B.H. Davison, G. Britovsek, J. Cairney, C.A. Eckert, W.J. Frederick, J.P. Hallett, D.J. Leak, C.L. Liotta, J.R. Mielenz, R. Murphy, R. Templer and T. Tschaplinski. 2006. The path forward for biofuels and biomaterials. Science 311(5760): 484-489.

Rosenbaum, M.A. and A.W. Henrich. 2014. Engineering microbial electrocatalysis for chemical and fuel production. Curr. Opin. Biotechnol. 29: 93-98.

Rozendal, R.A., H.V.M. Hamelers, R.J. Molenkmp and J.N. Buisman. 2007. Performance of single chamber biocatalyzed electrolysis with different types of ion exchange membranes. Water Res. 41: 1984-1994.

Samani, S., M.A. Abdoli, A. Karbassi and M.M. Amin. 2016. Stimulation of the hydrolytic stage for biogas production from cattle manure in an electrochemical bioreactor. Water. Sci. Technol., 74: 606–615.

Santos, D.M.F., C.A.C. Sequeira and J.L. Figueiredo. 2013. Hydrogen production by alkaline water electrolysis. Quim. Nova. 36: 1176-1193.

Sasak, K., M. Morita, D. Sasaki, S. Hirano, N. Matsumoto and A. Watanabe. 2011. A bio chemical reactor containing carbon fiber textiles enables efficient methane fermentation from garbage slurry. Bioresour. Technol. 102: 6837-6842.

Schuchmann, K. and V. Müller. 2014. Autotrophy at the thermodynamic limit of life: a model for energy conservation in acetogenic bacteria. Nat. Rev. Microbiol. 12: 809-821.

Selembo, P.A., J.M. Perez, W.A. Lloyd and B.E. Logan. 2009. High hydrogen production from glycerol or glucose by electrohydrogenesis using microbial electrolysis cells. Int. J. Hydrog. Ener. 34: 5373-5381.

Sevd, S., H.Y. Yuan, Z. He and I.M.A. Reesh. 2015. Microbial desalination cells as a versatile technology: functions, optimization and prospective. Desalina. 371: 9-17.

Sharma, M., S. Bajrachary, S. Gildemyn, S.A. Patil, Y. Alvarez-Gallego, D. Pant, K. Rabaey and X. Dominguez-Benetton. 2014. A critical revisit of the key parameters used to describe microbial electrochemical systems. Electrochim. Acta. 140.

Sun, R., A. Zhou, J. Jia, Q. Liang, Q. Liu and D. Xin. 2015. Characterization of methane production and microbial community shifts during waste activated sludge degradation in microbial electrolysis cells. Bioresour. Technol. 175: 68-74.

Swana, J., Y. Yang, M. Behnam and R. Thompson. 2011. An analysis of net energy production and feedstock availability for biobutanol and bioethanol. Bioresour. Technol. 102(2): 2112-2117.

Tice, R.C. and Y. Kim. 2014. Energy efficient reconcentration of diluted human urine using ion exchange membranes in bioelectrochemical systems. Water Res. 64: 61-72.

Wei, J.C., P. Liang and X. Huang. 2011. Recent progress in electrodes for microbial fuel cells. Bioresour. Technol. 102: 9335-9344.

Xiang, Y., G. Liu, R. Zhang, Y. Lu and H. Luo. 2017. High-efficient acetate production from carbon dioxide using a bioanode microbial electrosynthesis system with bipolar membrane. Bioresour. Technol. 233: 227-235.

Zhang, T., H.R. Nie, T.S. Bain, H.Y. Lu, M.M. Cui, O.L. Snoeyenbos-West, A.E. Franks, K.P. Nevin, T.P. Russell and D.R. Lovley. 2013. Improved cathode materials for microbial electrosynthesis. Ener. Environ. Sci. 6(1): 217-224.

Zhang, Y.F. and I. Angelidaki. 2014. Microbial electrolysis cells turning to be versatile technology: recent advances and future challenges. Water Res. 56: 11-25.

Zhen, G., K. Takuro, X. Lu, G. Kumar, Y. Hu and P. Bakonyi. 2016. Recovery of biohydrogen in a single- chamber microbial electrohydrogenesis cell using liquid fraction of pressed municipal solid waste (LPW) as substrate. Int. J. Hydrog. Ener. 41: 17896-17906.

Zhen, G., T. Kobayashi, X. Lu and K. Xu. 2015. Understanding methane bioelectrosynthesis from carbon dioxide in a two-chamber microbial electrolysis cells (MECs) containing a carbon biocathode. Bioresour. Technol. 186: 141-148.

Zhou, M.H., M.L. Chi, J.M. Luo, H.H. He and T. Jin. 2011. An overview of electrode materials in microbial fuel cells. J. Powr. Sourc. 196: 4427-4435.

Electrotroph as an Emerging Biocommodity Producer in a Biocatalyzed Bioelectrochemical System

Supriyanka Rana[1], Lakhveer Singh[2] and Zularisam bin ab Wahid[1]*

[1] Faculty of Engineering Technology, Universiti Malaysia Pahang Lebuhraya Tun Razak Gambang, Kuantan, Malaysia, 26300
[2] Biological and Ecological Engineering, 116 Gilmore Hall, Oregon State University, Corvallis, OR 97331, USA

Introduction

This chapter highlights the role of biocathodic microorganisms i.e., electrotroph in biocommodities production through bioelectrochemical systems, such as microbial fuel cells (MFCs) and microbial electrolysis cells (MECs). Microbial electrosynthesis, a biocathodic-technology, is driven by cathodic microorganisms, which usually deposit in the form of a biofilm on cathode, thereby assisting in catalysing the reduction of electron donors, such as CO_2 or O_2 by consuming electrons on the cathode surface (Kumar et al. 2015). The biocatalysts responsible for such biocatalysis have been named "electrotrophs" by Lovley, i.e., the cathode respiring microbe (Lovley 2008). These microbes can convert chemical energy of organic substrate or waste into useful electrical power and other useful products, in the presence of a variety of electron donors. Electrotrophs can accept the electrons from biocathodic electrode, either directly or indirectly by using the unknown mechanisms that are yet to be clearly understood (He and Angenent 2006). So, microbe cultivation platform, such as 'biofilm' allows the researchers to cultivate the stable microbial consortia in a defined bioelectrochemical system (a cathode colonized with a biofilm is called biocathode) in order to study the intricate microbial processes, such as redox reactions, mechanisms of extracellular

---

*Corresponding author: zularisam@gmail.com

electron transfer, electrode microbe interactions, and also to decipher the components involved in electron transport chain using electrochemistry. The few microbiome-based studies that have been carried out till date have given us a glance of possibilities contained hidden inside electrode microbiome in terms of their ability to produce several novel bioenergy compounds (Eddie et al. 2017). Cathode colonized with a biofilm (called biocathode) is reportedly more efficient in producing electricity in comparison to the abiotic cathode (without biofilm), and this correlation of their relative activity with current production has already been validated by many researchers (Eddie et al. 2017, Rago et al. 2017). However, electrotrophic biofilms on cathode which are capable of reducing CO_2 (anoxic reduction) are more difficult (slower) to grow, and consequently make it harder to study than the exoelectrogenic biofilms on anodes (Morita et al. 2011). Cathodic microbiome is the least studied and understood area in comparison to the anodic microbiome. Another limitation that hinders the growth of this area is the high cathodic activation over-potentials which adversely affects the bioelectrochemical cell's overall performance (Clauwaert et al. 2007a). Despite these drawbacks, the enormous applicability of biocathodes (such as bio-remediation of the polluted land, reducing the pollutants like nitrates or chloro-organics, and treating wastewater) over-weighs its limitations (He and Angenent 2006, Lovley 2006, Ter-Heijne et al. 2007).

Different modes of electron transfer are said to be responsible for such an increase in bioelectricity production in biocathodic MFCs (Choi and Sang 2016, Bergel et al. 2005), which involves direct electron transfer (microbes directly accept electrons from cathode surface) using direct physical contact between the bacterial cell membrane and cathode surface (Huang et al. 2011) using the outer membrane redox macromolecules, such as cytochromes, e.g., in *Geobacter* spp. or mixed cultures (Cao et al. 2009, Tandukar et al. 2009), and flavin or menaquinone e.g., *Shewanella putrefaciens* (Kim et al. 1999, Newman and Kolter 2000, Fredrickson et al. 2008), or by bacterial nanowires e.g., iron-reducing bacteria (Zhao et al. 2018), or by indirect/or mediated mode of electron transfer (microbes indirectly accept electrons by using shuttles (or mediators) e.g., *Acinetobacter calcoaceticus*) (Rabaey et al., 2008), *Leptothrix discophora* (a manganese-oxidizing bacterium) (Rhoads et al. 2005) and *Acidithiobacillus ferrooxidans* (an iron oxidizing bacterium) (Ter-Heijne et al. 2007) (Fig. 5.1). Various terminal electron acceptors, such as oxygen, nitrate, sulfate, iron, manganese, arsenate, fumarate, and perchlorate are used by electrotrophs to allow the complete reduction of substrate on biocathodes (Fig. 5.1). Biocathodic microbe's unique ability to utilize such a diverse range of electron acceptors help it treat a diverse range of wastewaters (Sevda and Sreekrishnan 2013, Blazquez et al. 2016) to reduce associated operation costs, enhance the bioelectricity production (Tran et al. 2009), and helps in broadening the scope of its applicability on an industrial scale (Ucar et al. 2017). Over the past few years, researchers have tried to harness their potential by utilizing them to produce either sustainable biofuels, such as

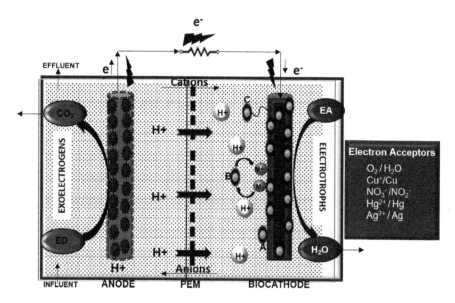

Figure 5.1. Different modes of electron transfer in biocathodic microbial fuel cell. A) Direct microbial catalysis, B) Mediated microbial catalysis and C) Nanowire assisted microbial catalysis. (ED – Electron Donors, EA – Electron Acceptors) (based on the information from He and Angenent 2006, Lu et al. 2012, He et al. 2015, Kanani 2017)

biohydrogen and biomethane (Sun et al. 2008, Singh et al. 2013, Singh and Wahid 2015), and/or other biocommodities, such as ethanol and butanol via processes, such as microbial electrosynthesis (Steinbusch et al. 2010, Marshall et al. 2012).

Electrotrophs: An Overview

Bioelectrochemical systems are electrode-based systems inhabited by specialized microbes with the capacity to eat and breath electricity. Biocathode-based systems are driven by microorganisms, which act as the chief catalytic force to drive redox reductive reactions. Microorganisms usually deposit on both electrodes in the form of electroactive biofilm—anodic microbes called exoelctrogens (Kumar et al. 2016), and cathodic microbes called electrotrophs (Lovley 2011) (Fig. 5.2). Although electrotrophic-biocathodic biofilm are slower to grow on cathode in comparison to the exoelectrogenic biofilms on anode (He and Angenent 2006, Nevin et al. 2008), but solution-like electrical inversion (where bioanode has switched to produce hydrogen-evolving biocathode) is presently used to counter this issue (Rozendal et al. 2008, Pisciotta et al. 2012). Also, a large number of researches have been reported by the researchers targeting the bioanodic microbiome, i.e., exoelectrogens (electron donors), but the biocathodic microbiome, i.e., electrotrophs (electron acceptor) remains one of the least explored territories (Ki et al. 2008, Michaelidou et al. 2011, Jiang et al. 2016, Zhang et al. 2017). The reducing

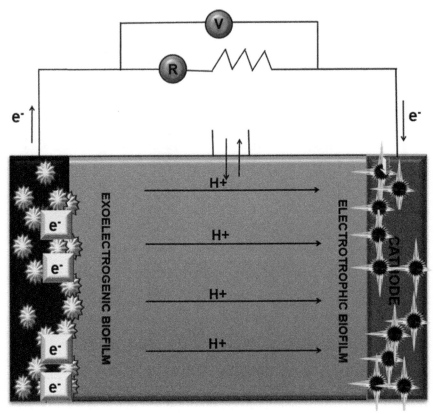

Figure 5.2. Single-chamber membrane-less microbial fuel cell (Adopted from Rismani-Yazdi et al. 2008).

nature of the cathodic electrotrophs strengthen them with the capability to synthesize the valuable multicarbon compounds (Logan and Rabaey 2012). Biocathodes can either be aerobic (when oxygen acts as an electron acceptor), or anaerobic (when nitrates and sulfates act as electron acceptors). Different types of electron acceptor and electrode-oxidizing microorganisms in biocathodes have been documented by Huang and his team (Huang et al. 2011). In biocathodes, microbes accept electrons from an anode and could reduce the harmful contaminants, such as nitrate, perchlorate, and so on, and thereby can be used as biosensor (or bioremediator) in bioremediation process (Cucu et al. 2016, Butler et al. 2010).

Electrotroph-Cathode Interactions: Insights into Microbial Electrochemistry

Study of the interaction between the MFCs and electrodes materials and its scope in industrial applications is known as microbial electrochemistry (Schroder et al. 2015). The significance of the interspecies microbial interactions within MFCs biofilm communities have been recently proposed and studied in detail by only a few researchers, which are insufficient to predict their electrosynthesis mechanisms under changing environmental conditions (Du et al. 2014). Till date, only a few electrotrophs have been known so far in their pure culture forms, which include *Sporomusa ovata, Sporomusa sphaeroides, Morella thermoacetica, Clostridium ljungdahlii, Clostridium aceticum, Geobacter metallireducens,* and *Geobacter sulfurreducens* (Lovley and Nevin 2013). Systematic integration and comparison of electrochemical and biological data help in gaining the in-depth understanding of the microbe-electrode interactions (Dolch et al. 2014, Das and Dash 2017).

Electrotrophic mechanisms in cathodic electron transfer

Biocathodic electrotrophs transfer electrons using different types of extracellular electron transfer mechanisms (Fig. 5.3), such as direct electron transfer using membrane protein, such as c-type cytochromes or hydrogenases (Rosenbaum et al. 2011) (Fig. 5.4) or mediated electron transfer (Choi and Sang 2016) or through physical contact using cellular structures, called nanowires (Zhao et al. 2018). Out of these three electron transfer modes, direct electron transfer is considered as an ideal mode for extracellular electron transfer from a cathode to microbes. Electron transfers occur in two directions, i.e., from microbe to anode and cathode to microbe. Electrons usually flow from low redox potential (electron donor) to a higher redox potential (electron acceptor). Reportedly, the same electron transport chain can be used for either of the directions, i.e., reversible electron transfer might be possible within cytochrome c complex channels (Choi and Sang, 2016). On the other hand, intermediate soluble redox compounds (called mediators), such as phenanzines and flavins are produced during mediated electron transfer which assist the microorganisms in interacting with the electrode surface more easily (Rabaey et al. 2005), but these mechanisms are energetically more demanding than the direct one (Stams and Plugge 2009, Lovley 2011).

Existence of the interspecies syntropy among biocathodic electrotrophs is only due to these complex electron transfer mechanisms, the exploration

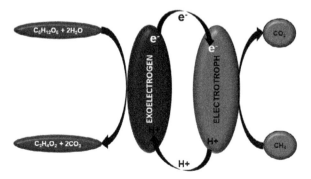

Figure 5.3. Electron transfer from anodic exoelectrogens to cathodic electrotrophs (Adopted from Logan and Rabaey 2012).

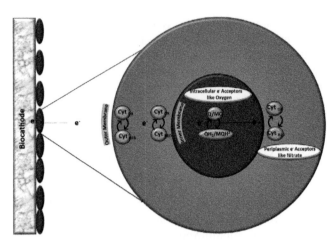

Figure 5.4. Direct mode of extracellular electron transfer mechanisms on biocathode (Adopted from Rosenbaum et al. 2011).

of whose versatility could lead us to discover novel mechanisms or pathways with the potential to yield several novel bioelectrochemical/or biocommodities (Rotaru et al. 2014). Advanced research in their exploration can bring more insights into mechanisms responsible to establish the intricate bond within electrotrophic communities and with electrode (Lovley 2012), that could further help researchers to engineer better electrotrophic strains with enhanced capacity of biocommodities production (Rosenbaum et al. 2011).

Electrotroph as an Electron Acceptor in Bioelectrochemical System

A diverse microbial population constitutes the biofilm on MFCs biocathode, where they synergistically work together to deliver functions, such as electron transfer, power production, and product biosynthesis. Biocathodic electrotrophs are known to convert complex organic substrates into electricity, hydrogen, or other useful biochemical compounds. In bioelectrochemical cells, such as MFCs, these specialized microbes transfer electron by electron transport chain from anodic surface to cathodic surface, and produce power with the simultaneous production of a proton motive force for ATP generation. Despite the lack of an outer membrane, gram-positive bacteria manage to sustain its growth while generating electrical energy in MFCs, but their transfer mechanisms are yet to be understood that could give some insights into how they manage to do that (Wrighton and Coates 2009). In 2006, Logan and Regan have elaborately discussed microbial communities associated with MFCs and their relevance in making this potential bioenergy technology next level in the coming future (Logan and Regan 2006). Gaining better understanding of the diversity, extracellular electron transfer processes, their mechanisms and biofilm ecology by using high throughput sequencing approaches can help us engineer the metabolic pathways of these microbes to attain their maximum yielding potential and to manipulate them to achieve the bioproduction of the desired products. Microbial analysis of biocathodic communities in bioelectrochemical system have been conducted by a few researchers till date and reported communities are enlisted in Table 5.1.

Microbial species in MFCs are profoundly affected by our choice of material that we use as a cathode or anode, because each electrode material type differs in terms of its surface structure and conductivity, which ultimately influence microbial adherence pattern (Rabaey et al. 2004, Zhou et al. 2011). Microbial community analysis of biocathode in MFCs has been done using different material types, such as granular activated carbon, carbon felt cube, granular graphite, and granular semi-coke, and microbes *Comamonas* of *Betaproteobacteria* family are reportedly found to be the dominant species in first three material types which were found to affect the power output positively, while another microbe *Acidovorax* was found in the last one, which was found to affect electron transfer. However, the *Betaproteobacteria* and *Bacteroidetes* were the most dominant families in all material types followed by differing occurrence rate of families, such as *Actinbacteria*, *Verrucomicrobia*, *Acidobacteria*, *Gammaproteobacteria*, and *Alphaproteobaceria* (Sun et al. 2012). Similarly, *Comamonas testosteroni* has been found to be the dominant anodic species under alkaline conditions, suggesting that cathodes and anode share many of its reducing and oxidizing populations, thus indicating their capacity to perform bidirectional electron transfer, i.e., to and from electrodes (e.g., *Shewanella putrefaciens* and *Geobacter sulfurreducens*) (Juang et al. 2011). Furthermore, *Acinetobacter calcoaceticus*, *Sphingobacterium multivorum*, *Micrococcus luteus*, and *Bacillus subtilis* have been found to be responsible to catalyse oxygen reduction reaction in biocathodic MFCs (Rabaey et al. 2008, Cournet et al. 2010). In the last decade, few microbiome-based researches have been reported primarily focusing on microbe assisted biocommodities production, which are enlisted in Table 5.2.

Electrotroph as a Biocommodity Producer
Worldwide, sustainable bioenergy fuels (next generation fuels), such as bioelectricity, bioethanol and biohydrogen are taking over the market share of unsustainable fossil fuels. Electrotrophs are known to produce many valuable chemicals, such as methane, ethanol, and formic acid (Rabaey and Rozendal 2010), and are best known to be used as biosensors to detect biological oxygen demand (BOD) of water to ensure its purity level (Kim et al. 2003). In biocathodic BES, autotrophic microbial communities have been reported to the simultaneous production of three different biocommodities, i.e., methane, acetate, and hydrogen (Marshall et al. 2012). Similarly, cathode associated microbiomes have resulted in the simultaneous production of various biocommodities, where glycerol was the major product, followed by propionate, valerate, 1,3-propanediol, ethanol, propanol, and butyrate.

Table 5.1. Observed microbial dynamics on biocathodic biofilm of MFCs biocathodes

Cathodic microbial species	Reference
Gammaproteobacteria, Rhodobacter sphaeroids	Rosenbaum et al. 2005
Pseudomonas, Rhodobacteraceae, Sphingomonadaceae	Clauwaert et al. 2007b
Sphingobacterium, Acinetobacter, Acidovorax sp.	Rabaey et al. 2008
Proteobacteria, Bacteroidetes, Alphaproteobacterial, Chlorobi, Deltaproteobacteria, Actinobacteria, Gammaproteobacteria	Chen et al. 2008
Firmicutes, Blphaproteobateria, Betaproteobacteria, Bammaproteobacteria, Bacteroidetes	Aldrovandi et al. 2009
Acinetobacter calcoaceticus, Sphingobacterium multivorum, Micrococcus luteus, Bacillus subtilis	Cournet et al. 2010
Xanthomonadaceae, Xanthomonas, Bacteroidetes, Azovibrio, Proteobacteria, Bacteroidetes, Actinobacteria, Planctomycetes, Firmicutes	Chen et al. 2010; Chung et al. 2011
Alphaproteobacteria, Gammaproteobacteria, Firmicutes, Acitobacteria, Flavobacteriaceae	Erable et al. 2010
Pseudomonas, Ralsronia, Gammaproteobacteria, Cyanobacteria	De-Schamphelaire et al. 2010
Shewanella putrefaciens, Geobacter sulfurreducens	Juang et al. 2011
Actinobacteria, Firmicutes, Bacteroidetes, Alphaproteobacteria, Betaproteobacteria, Gammaproteobacteria	Lyautey et al. 2011
Nitrobacter sp., *Achromobacter* sp., *Acinetobacter* sp., *Bacteroidetes*	Zhang et al. 2011
Betaproteobacteria, Bacteroidetes, Actinbacteria, Verrucomicrobia, Acidobacteria, Gammaproteobacteria, Alphaproteobaceria	Sun et al. 2012
Desulfovibrio paquesii	Aulenta et al. 2012
Gammaproteobacteria, Deinococcus-Thermus, Betaproteobacteria, Agrobacterium, Achromobacter	Rimboud et al. 2015, Milner et al. 2016
Sporomusa ovata, Sporomusa sphaeroides, Morella thermoacetica, Clostridium ljungdahlii, Clostridium aceticum, Geobacter metallireducens, Geobacter sulfurreducens	Lovley and Nevin 2013
Citrobacter population, *Pectinatus* sp., *Clostridium* sp.	Dennis et al. 2013
Trichococcus palustris DSM 9172, *Oscillibacter, Clostridium* sp. (*Clostridium propionicum; Clostridium celerecrescens*), *Desulfotomaculum* sp., *Tissierella* sp.	Zaybak et al. 2013
Desulfovibrio putealis, Hydrogenophaga caeni, Methylocystis sp.	Van Eerten-Jansen et al. 2013
Rhizobiales, Phycisphaerales, Planctomycetales, Sphingobacteriales	Wang et al. 2013

(Contd.)

Chloroflexus	Blanchet et al. 2014
Nitrospira, Nitrosomonas, Nitrobacter, Alkalilimncola	Du et al. 2014
Pseudoalteromonas sp., *Marinobacter* sp., *Roseobacter* sp., *Bacillus* sp.	Debuy et al. 2015
Methanobacterium sp.	Batlle-Vilanova et al. 2015
Halomonas, Pseudomonas, Desulphomonas, Alkalimonas, Rodhobacteraceae, Rodhospirellaceae	Reimers et al. 2006, Rago et al. 2017

Table 5.2. Microbiome assisted biocommodities production using bioelectrochemical system

Reviews/Research article	Biocommodity	Reference
Application of biocathode in microbial fuel cells: Cell performance and microbial community	Bioelectricity	Chen et al. 2008
Microbial community analysis of a methane-producing biocathode in a bioelectrochemical system	Methane	Van Eerten-Jansen et al. 2013
Electricity generation by microbial fuel cell using microorganisms as catalyst in cathode	Bioelectricity	Jang et al. 2013
Enhancement of biobutanol production by electromicrobial glucose conversion in a dual chamber fermentation cell using *C. pasteurianum*	Biobutanol	Mostafazadeh et al. 2016
A review on bio-electrochemical systems (BESs) for the syngas and value added biochemicals production	Syngas	Kumar et al. 2017
On the edge of research and technological application: a critical review of electromethanogenesis	Methane	Blasco-Gomez et al. 2017

The positive correlation has been found between: the production of 1,3-propanediol, propanol, and ethanol with the *Citrobacter* population; the production of propionate with the *Pectinatus* population and the production of valerate with the *Clostridium* population (Dennis et al. 2013). Also, facultatively autotrophic electrotrophs (such as *Trichococcus palustris* DSM 9172, *Oscillibacter* sp., and *Clostridium* sp. (*Clostridium propionicum; Clostridium celerecrescens*), *Desulfotomaculum* sp., *Tissierella* sp.) have led to the synthesis of butanol, ethanol, acetate, propionate, butyrate, and hydrogen gas (Zaybak et al. 2013).

Bioelectricity

Microbial fuel cells (MFCs) are best known for their ability to turn organic waste to electricity. Use of unsustainable expensive metals, such as platinum and other non-platinum chemical catalysts as cathodes have been discontinued due to their cost prohibitive nature and long degradation time, and have been replaced with a sustainable and cost-effective alternative, such as biocathodic MFCs for improved oxygen reduction reaction (ORR) and enhanced electricity production (Kumar et al. 2018). For instance, high-performing aerobic biocathode has showed similar results as of platinum coated cathodes (70 μWcm^{-2}, in terms of enhanced power production; nine-fold increase in power production when compared with the unmodified cathode (7 to 62 μWcm^{-2})) (Milner et al. 2016). In this study, the most dominant species was *Gammaproteobacteria*, which proposed to be responsible for the high activity of biologically catalysed ORR in cathodes. In another study, metabolic activity of an electroactive microbe, *Rhodobacter sphaeroids* was used for *in situ* oxidation of photo-biological hydrogen for the generation of electricity using MFCs (Rosenbaum et al. 2005). Furthermore, biocathodic MFCs have been constructed using saline tolerant bacterial strains of oxygen-reducing monospecies—*Pseudoalteromonas* sp., *Marinobacter* sp., *Roseobacter* sp., *Bacillus* sp., which lead to an appreciable increase in bioelectricity (40 mA m^{-2}) (Debuy et al. 2015). Moreover, the cathodic microbial communities were validated to catalyse the cathodic oxygen reduction and the attached biofilm has primarily shown the dominance of *Sphingobacterium*, *Acinetobacter*, and *Acidovorax* sp. Few pure culture isolates from these communities have shown increase in power output (Rabaey et al. 2008). Variation in type of substrates have a significant impact on the performance of MFCs biocathodes, which further steer the growth of different types of microbial communities on cathodic biofilm, thereby ultimately resulting in different ranges of power producing capacities (Kiely et al. 2011).

Biomethane

Biocathode, a major biofuel, can be harnessed by electrotrophic microbial communities through BES by utilizing CO_2 as an electron acceptor. The effect of the cathode potential and other operational parameters on biocommodities' production and the mechanism and cathodic microbes associated with their production has been elaborately discussed in a few research studies (Villano et al. 2010, Van Eerten-Jansen et al. 2015). A methane-producing biocathode has been reportedly producing methane at a maximum rate of 5.1 L CH_4/m^2 projected cathode per day (1.6 A/m^2) at –0.7 V, producing hydrogen as an intermediate, and biocathodic biofilm was primarily dominated by bacterial species closely related to *Desulfovibrio putealis* (anaerobic microorganism), *Hydrogenophaga caeni* (aerobic microorganism), and *Methylocystis* sp. (facultative aerobic microorganism) (Van Eerten-Jansen et al. 2013). In addition, the biogas upgradation to methane has been reported for the first time within a mixed culture biocathodic BES, where *Methanobacterium* sp. was the dominant species responsible for hydrogenotrophic methanogenesis on biocathode (Batlle-Vilanova et al. 2015). The relationship of methane and electricity production through MFCs (double chamber) has been established using sewage sludge as substrate, but only anodic microbial communities have been studied (Xiao et al. 2014). However, the electrosynthesis of such multi-commodities can be adversely affected by the interspecies competition among electrotrophic communities (Molenaar et al. 2017).

Biohydrogen

Microbial biocathodic biofilm has been found to improve the yielding capacity of hydrogen using MFCs, although its growth is slower than the bioanodic biofilm, which is possibly an explanation for the question why biocathodic MFCs take more time to achieve maximum current density than bioanodic ones (Rozendal et al. 2006, 2008). Biological production of hydrogen is reportedly enhanced by biocathodes (Cheng and Logan 2007, Rozendal et al. 2008, Jeremiasse et al. 2010) using protons as electron acceptors. Besides protons, CO_2 can also be used as an electron acceptor in the biocathode of an MES, to produce commodity compounds, such as methane, volatile fatty acids (VFA), or alcohols (Nevin et al. 2010, Rabaey and Rozendal 2010). Expenses involved with the hydrogen production from water by direct electrolysis is far cheaper than other technologies (such as wave and solar technologies). In 2004, Logan coined a term 'hydrogen electricity' and envisioned the bioelectrochemical technologies as the potentially useful methods of recovering energy from wastes (Logan 2004). In addition, cathodic hydrogen production has also been reported, but using metal catalysts, such as platinum, doubles the fuel production cost, so the conquest to find the alternative cathode catalyst was started, which was efficient, sustainable, and inexpensive at the same time. Ultimately, microorganisms came out as the best sustainable alternatives to expensive metal catalysts with promising advantages, such as cost-effectiveness, self-reproducibility, and its metabolic derisiveness, which could be used to produce valuable compounds (He and Angenent 2006). Also, the ability of *Desulfovibrio paquesii* on microbial biocathodes have assessed the ability to accept electrons directly, and to catalyse H_2 production by reducing H^+ (Aulenta et al. 2012).

Bioethanol

Bioelectrochemical mode of biomass conversion into ethanol can be achieved by biocathodic microbial communities by catalysing the biological reduction of acetate in a bioelectrochemical system. Additionally, effect of mediator i.e., methyl viologen has found to affect the ethanol output positively, thereby validating its role in improving the efficiency of the reaction (Steinbusch et al. 2010). However, the microbiological analysis of the responsible microbial communities has not been conducted, and this gap is yet to be filled to know about the mechanism involved with the bioethanol production.

Future Perspective

The applications of electrotrophs, such as bioremediation, biosensing, and biosynthesis make them valuable biocommodity producers (Table 5.3). Initially, the only application of biocathodic electrotrophs was thought to be bioelectricity production in a bioelectrochemical system (air cathode MFCs) by efficient oxygen reduction reaction; however, later many other applications, such as bioremediation of nitrate, perchlorate, and sulphate have paved the path for a sustainable and healthy future. Use of the above described biocommodity compounds as fuel alternatives will surely help in reducing our dependence on fossil fuels, thereby making our environment pollution-free. More in-depth research into the cathodic core microbiome of bioelectrochemical system will help to discover many other novel compounds in the coming future.

Table 5.3. Informative researches worth referring to to gain more insight into biocathodic microbiome

Reviews/Research articles	Reference
Application of bacterial biocathodes in microbial fuel cells	He and Angenent 2006
Recent developments in microbial fuel cell technologies for sustainable bioenergy	Watanabe 2008
Electron transfer mechanisms, new applications, and performance of biocathode microbial fuel cells	Huang et al. 2011
Microbial fuel cells and microbial electrolysis cells for the production of bioelectricity and biomaterials	Zhou et al. 2013a
Recent advances in microbial fuel cells (MFCs) and microbial electrolysis cells (MECs) for wastewater treatment, bioenergy and bioproducts	Zhou et al. 2013b
Biocathode application in microbial fuel cells: Organic matter removal and denitrification	Kizilet et al. 2015
Delving through electrogenic biofilms: From anodes to cathodes to microbes	Semenec and Franks 2015
Role of biocathodes in bioelectrochemical systems	Prakasam et al. 2017

Acknowledgements

The authors would like to acknowledge the support provided by the Universiti Malaysia Pahang, Gambang, Malaysia and Oregon State University, USA.

REFERENCES

Aldrovandi, A., E. Marsili, L. Stante, P. Paganin, S. Tabacchioni and A. Giordano. 2009. Sustainable power production in a membrane-less and mediator-less synthetic wastewater microbial fuel cell. Bioresour. Technol. 100(13): 3252-3260.

Aulenta, F., L. Catapano, L. Snip, M. Villano and M. Majone. 2012. Linking bacterial metabolism to graphite cathodes: Electrochemical insights into the H₂ producing capability of *Desulfovibrio* sp. Chem. Sus. Chem. 5: 1080-1085.

Batlle-Vilanova, P., S. Puig, R. Gonzalez-Olmos, A. Vilajeliu-Pons, M.D. Balaguera and J. Colprim. 2015. Deciphering the electron transfer mechanisms for biogas upgrading to biomethane within a mixed culture biocathode. RSC Adv. 5: 52243.

Bergel, A., D. Feron and A. Mollica. 2005. Catalysis of oxygen reduction in PEM fuel cell by seawater biofilm. Electrochem. Commun. 7: 900-904.

Blanchet, E., S. Pecastaings, B. Erable, C. Roques and A. Bergel. 2014. Protons accumulation during anodic phase turned to advantage for oxygen reduction during cathodic phase in reversible bioelectrodes. Bioresour. Technol. 173: 224-230.

Blasco-Gomez, R., P. Batlle-Vilanova, M. Villano, M.D. Balaguer, J. Colprim and S. Puig. 2017. On the edge of research and technological application: a critical review of electromethanogenesis. Int. J. Mol. Sci. 18(4): 874.

Blazquez, E., D. Gabriel, J.A. Baeza and A. Guisasola. 2016. Treatment of high-strength sulfate wastewater using an autotrophic biocathode in view of elemental sulfur recovery. Water Res. 105: 395-405.

Butler, C.S., P. Clauwaert, S.J. Green, W. Verstraete and R. Nerenberg. 2010. Bioelectrochemical perchlorate reduction in a microbial fuel cell. Environ. Sci. Technol. 44(12): 4685-4691.

Cao, X., X. Huang, P. Liang, N. Boon, M. Fan, L. Zhang and X. Zhang. 2009. A completely anoxic microbial fuel cell using a photo-biocathode for cathodic carbon dioxide reduction. Energy Environ. Sci. 2: 441-548.

Chen, G.W., S.J. Choi, T.H. Lee, G.Y. Lee, J.H. Cha and C.W. Kim. 2008. Application of biocathode in microbial fuel cells: cell performance and microbial community. Appl. Microbiol. Biotechnol. 79(3): 379-388.

Chen, G.W., S.J. Choi, J.H. Cha, T.H. Lee and C.W. Kim. 2010. Microbial community dynamics and electron transfer of a biocathode in microbial fuel cells. Korean J. Chem. Eng. 27: 1513-1520.

Cheng, S. and B.E. Logan. 2007. Sustainable and efficient biohydrogen production via electrohydrogenesis. Proc. Natl. Acad. Sci. 104(47): 18871-18873.

Choi, O. and B.-I. Sang. 2016. Extracellular electron transfer from cathode to microbes: application for biofuel production. Biotechnol. Biofuel 9: 11.

Chung, K., I. Fujiki and S. Okabe. 2011. Effect of formation of biofilms and chemical scale on the cathode electrode on the performance of a continuous two-chamber microbial fuel cell. Bioresour. Technol. 102: 355-360.

Clauwaert, P., K. Rabaey, P. Aelterman, L. de Schamphelaire, T.H. Pham, P. Boeckx, N. Boon and W. Verstraete. 2007a. Biological denitrification in microbial fuel cells. Environ. Sci. Technol. 41(9): 3354-3360.

Clauwaert, P., D. Van-der Ha, N. Boon, K. Verbeken, M. Verhaege, K. Rabaey and W. Verstraete. 2007b. Open air biocathode enables effective electricity generation with microbial fuel cells. Environ. Sci. Technol. 41(21): 7564-7569.

Cournet, A., M.L. Delia, A. Bergel, C. Roques and M. Berge. 2010. Electrochemical reduction of oxygen catalyzed by a wide range of bacteria including Gram-positive. Electrochem. Commun. 12: 505-508.

Cucu, A., A. Tiliakos, I. Tanase, C.E. Serban, I. Stamatin, A. Ciocanea and C. Nichita. 2016. Microbial fuel cell for nitrate reduction. Energy Procedia. 85: 156-161.

Das, S. and H.R. Dash. 2017. Handbook of metal-microbe interactions and bioremediation. pp. 445-454. *In*: Handbook of Metal-Microbe Interactions and Bioremediation. CRC Press.

Debuy, S., S. Pecastaings, A. Bergel and B. Erable. 2015. Oxygen-reducing biocathodes designed with pure cultures of microbial strains isolated from seawater biofilms. Int. Biodeterior. Biodegradation 103: 16-22.

Dennis, P.G., F. Harnisch, Y.K. Yeoh, G.W. Tyson and K. Rabaey. 2013. Dynamics of cathode-associated microbial communities and metabolite profiles in a glycerol-fed bioelectrochemical system. Appl. Environ. Microbiol. 79(13): 4008-4014.

De-Schamphelaire, L., P. Boeckx and W. Verstraete. 2010. Evaluation of biocathodes in freshwater and brackish sediment microbial fuel cells. Appl. Microbiol. Biotechnol. 87: 1675-1687.

Dolch, K., J. Danzer, T. Kabbeck, B. Bierer, J. Erben, A.H. Forster, J. Maisch, P. Nick, S. Kerzenmacher and J. Gescher. 2014. Characterization of microbial current production as a function of microbe-electrode-interaction. Bioresour. Technol. 157: 284-292.

Du, Y., Y. Feng, Y. Dong, Y. Qu, J. Liu, X. Zhou and N. Rena. 2014. Coupling interaction of cathodic reduction and microbial metabolism in aerobic biocathode of microbial fuel cell. RSC Adv. 4: 34350-34355.

Eddie, B.J., Z. Wang, W.J. Hervey-IV, D.H. Leary, A.P. Malanoski, L.M. Tender, B. Lin and S.M. Strycharz-Glaven. 2017. Metatranscriptomics supports the mechanism for biocathode electroautotrophy by *"Candidatus Tenderia* electrophaga". mSystems 2(2): e00002-17.

Erable, B., N.M. Duţeanu, M.M. Ghangrekar, C. Dumas and K. Scott. 2010. Application of electro-active biofilms. Biofouling 26(1): 57-71.

Fredrickson, J.K., M.F. Romine, A.S. Beliaev, J.M. Auchtung, M.E. Driscoll, T.S. Gardner, K.H. Nealson, A.L. Osterman, G. Pinchuk, J.L. Reed, D.A. Rodionov, J.L. Rodrigues, D.A. Saffarini, M.H. Serres, A.M. Spormann, I.B. Zhulin and J.M. Tiedje. 2008. Towards environmental systems biology of *Shewanella*. Nat. Rev. Microbiol. 6: 592-603.

He, Z. and L.T. Angenent. 2006. Application of bacterial biocathodes in microbial fuel cells. Electroanalysis 18(19-20): 2009-2015.

Huang, L., J.M. Regan and X. Quan. 2011. Electron transfer mechanisms, new applications, and performance of biocathode microbial fuel cells. Bioresour. Technol. 102: 316-323.

Jang, J.K., J. Kan, O. Bretschger, Y.A. Gorby, L. Hsu, B.H. Kim and K.H. Nealson. 2013. Electricity generation by microbial fuel cell using microorganisms as catalyst in cathode. J. Microbiol. Biotechnol. 23(12): 1765-1773.

Jeremiasse, A.W., H.V.M. Hamelers and C.J.N. Buisman. 2010. Microbial electrolysis cell with a microbial biocathode. Bioelectrochemistry 78(1): 39-43.

Jiang, Q., D. Xing, R. Sun, L. Zhang, Y. Feng and N. Ren. 2016. Anode biofilm communities and the performance of microbial fuel cells with different reactor configurations. RSC Adv. 6: 85149-85155.

Juang, D.F., P.C. Yang, H.Y. Chou and L.J. Chiu. 2011. Effects of microbial species, organic loading and substrate degradation rate on the power generation capability of microbial fuel cells. Biotechnol. Lett. 33(11): 2147-2160.

Kanani, B. 2017. Microbial fuel cell, new technologies in the field of green energy and wastewater treatment. Anatomy Physiol. Biochem. Int. J. 2(5): 1-4.

Ki, D., J. Park, J. Lee and K. Yoo. 2008. Microbial diversity and population dynamics of activated sludge microbial communities participating in electricity generation in microbial fuel cells. Water Sci. Technol. 58(11): 2195-2201.

Kiely, P.D., G. Rader, J.M. Regan and B.E. Logan. 2011. Long-term cathode performance and the microbial communities that develop in microbial fuel cells fed different fermentation endproducts. Bioresour. Technol. 102: 361-366.

Kim, H.J., M.S. Hyun, I.S. Chang and Kim B.H. 1999. A microbial fuel cell type lactate biosensor using a metal-reducing bacterium, *Shewanella putrefaciens*. J. Microbiol. Biotechnol. 9: 365-367.

Kim, B.H., I.S. Chang, G.C. Gil, H.S. Park and H.J. Kim. 2003. Novel BOD (biological oxygen demand) sensor using mediator-less microbial fuel cell. Biotechnol. Lett. 25: 541-545.

Kizilet, A., D. Akman, V. Akgul, K. Cirik and O. Cinar. 2015. Biocathode application in microbial fuel cells: organic matter removal and denitrification. Environment and Electrical Engineering (EEEIC), 2015 IEEE 15th International Conference. Rome, Italy.

Kumar, R., L. Singh, Z.A. Wahid and M.F.M. Din. 2015. Exoelectrogens in microbial fuel cells toward bioelectricity generation: a review. Int. J. of Energy Research. 39: 1048-1067.

Kumar, R., L. Singh and A.W. Zularisam. 2016. Exoelectrogens: recent advances in molecular drivers involved in extracellular electron transfer and strategies used to improve it for microbial fuel cell applications. Renew. Sustain. Energy Rev. 56: 1322-1336.

Kumar, G., R.G. Saratale, A. Kadier, P. Sivagurunathan, G. Zhen, S.H. Kim and G.D. Saratale. 2017. A review on bioelectrochemical systems (BESs) for the syngas and value added biochemicals production. Chemosphere 177: 84-92.

Kumar, R., L. Singh, Z.A. Wahid, D.M. Mahapatra and H. Liu. 2018. Novel mesoporous $MnCo_2O_4$ nanorods as oxygen reduction catalyst at neutral pH in microbial fuel cells. Bioresour. Technol. 254: 1-6.

Logan, B.E. 2004. Extracting hydrogen and electricity from renewable resources. Environ. Sci. Technol. 38(9): 160A-167A.

Logan, B.E. and J.M. Regan. 2006. Electricity-producing bacterial communities in microbial fuel cells. Trends Microbiol. 14: 512–518.

Logan, B.E. and K. Rabaey. 2012. Conversion of wastes into bioelectricity and chemicals by using microbial electrochemical technologies. Science 337: 686-690.

Lovely, D.R. 2006. Microbial fuel cells: novel microbial physiologies and engineering approaches. Curr. Opini. Biotechnol. 17: 327-332.

Lovley, D.R. 2008. The microbe electric: conversion of organic matter to electricity. Curr. Opin. Biotechnol. 19: 564-571.

Lovley, D.R. 2011. Powering microbes with electricity: direct electron transfer from electrodes to microbes. Environ. Microbio. Reports 3(1): 27-35.

Lovley, D.R. 2012. Electromicrobiology. Annu. Rev. Microbiol. 66: 391-409.

Lovley, D.R. and K.P. Nevin. 2013. Electrobiocommodities: powering microbial production of fuels and commodity chemicals from carbon dioxide with electricity. Curr. Opin. Biotechnol. 24: 385-390.

Lu, M., S. Fong and Y. Li. 2012. Cathode reactions and applications in microbial fuel cells: a review. Critic. Rev. Environ. Sci. Technol. 42(23): 2504-2525.

Lyautey, E., A. Cournet, S. Morin, S. Boulêtreau, L. Etcheverry, J.Y. Charcosset, F. Delmas, A. Bergel and F. Garabetian. 2011. Electroactivity of phototrophic river biofilms and constitutive cultivable bacteria. Appl. Environ. Microbiol. 77(15): 5394-5401.

Marshall, C.W., D.E. Ross, E.B. Fichot, R.S. Norman and H.D. May. 2012. Electrosynthesis of commodity chemicals by an autotrophic microbial community. Appl. Environ. Microbiol. 78: 8412-8420.

Michaelidou, U., A. Ter-Heijne, G.J.W. Euverink, H.V.M. Hamelers, A.J.M. Stams and J.S. Geelhoed. 2011. Microbial communities and electrochemical performance of titanium-based anodic electrodes in a microbial fuel cell. Appl. Environ. Microbiol. 77(3): 1069-1075.

Milner, E.M., D. Popescu, T. Curtis, I.M. Head, K. Scott and E.H. Yu. 2016. Microbial fuel cells with highly active aerobic biocathodes. J. Power Sources 324(30): 8-16.

Molenaar, S.D., P. Saha, A.R. Mol, T.H.J.A. Sleutels, A. Ter-Heijne and J.N.C. Buisman. 2017. Competition between methanogens and acetogens in biocathodes: a comparison between potentiostatic and galvanostatic control. Int. J. Mol. Sci. 18(1): 204.

Morita, M., N.S. Malvankar, A.E. Franks, Z.M. Summers, L. Giloteaux, A.E. Rotaru, C. Rotaru and D.R. Lovley. 2011. Potential for direct interspecies electron transfer in methanogenic wastewater digester aggregates. mBio. 2: e00159-11.

Mostafazadeh, A.K., P. Drogui, S.K. Barar, R.D. Tyagi, Y.L. Bihan, G. Buelna and S.-D. Rasolomanana. 2016. Enhancement of biobutanol production by electromicrobial glucose conversion in a dual chamber fermentation cell using C. *pasteurianum*. Energ. Conver. Manag. 130: 165-175.

Nevin, K.P., H. Richter, S.F. Covalla, J.P. Johnson, T.L. Woodard, A.L. Orloff et al. 2008. Power output and coulombic efficiencies from biofilms of *Geobacter sulfurreducens* comparable to mixed community microbial fuel cells. Environ. Microbiol. 10: 2505-2514.

Nevin, K.P., T.L. Woodard, A.E. Franks, Z.M. Summers and D.R. Lovley. 2010. Microbial electrosynthesis: feeding microbes electricity to convert carbon dioxide and water to multicarbon extracellular organic compounds. MBio 1: e00103-00110.

Newman, D.K. and R. Kolter. 2000. A role for excreted quinones in extracellular electron transfer. Nature 405: 94-97.

Pisciotta, J.M., Z. Zaybak, D.F. Call, J.-Y. Nam and B.E. Logan. 2012. Enrichment of microbial electrolysis cell biocathodes from sediment microbial fuel cell bioanodes. Appl. Environ. Microbiol. 78(15): 5212.

Prakasam, V., S.G.F. Bagh, S. Ray, B. Fifield, L.A. Porter and J.A. Lalman. 2017. Role of biocathodes in bioelectrochemical systems. pp. 165-187. *In*: Microbial Fuel cell. Springer, Cham.

Rabaey, K., N. Boon, S.D. Siciliano, M. Verhaege and W. Verstraete. 2004. Biofuel cells select for microbial consortia that self-mediate electron transfer. Appl. Environ. Microbiol. 70: 5373-5382.

Rabaey, K., N. Boon, M. Hofte and W. Verstraete. 2005. Microbial phenazine production enhances electrontransfer in biofuel cells. Environ. Sci. Technol. 39: 3401-3408.

Rabaey, K., S.T. Read, P. Clauwaert, S. Freguia, P.L. Bond, L.L. Blackall and J. Keller. 2008. Cathodic oxygen reduction catalysed by bacteria in microbial fuel cells. ISME J. 2(5): 519-527.

Rabaey, K. and R.A. Rozendal. 2010. Microbial electrosynthesis-revisiting the electrical route for microbial production. Nat. Rev. Micro. 8(10): 706-716.

Rago, L., P. Cristiani, F. Villa, S. Zecchin, A. Colombo, L. Cavalca and A. Schievano. 2017. Influences of dissolved oxygen concentration on biocathodic microbial communities in microbial fuel cells. Bioelectrochem. 116: 39-51.

Reimers, C., P. Girguis, H. Stecher, L. Tender, N. Ryckelynck and P. Whaling. 2006. Microbial fuel cell energy from an ocean cold seep. Geobiology 4(2): 123-136.

Rhoads, A., H. Beyenal and Z. Lewandowshi. 2005. Microbial fuel cell using anaerobic respiration as an anodic reaction and biomineralized manganese as a cathodic reactant. Environ. Sci. Technol. 39: 4666-4671.

Rimboud, M., E.D.-L. Quemener, B. Erable, T. Bouchez and A. Bergel. 2015. The current provided by oxygen-reducing microbial cathodes is related to the composition of their bacterial community. Bioelectrochemistry 102: 42-49.

Rismani-Yazdi, H., S.M. Carver, A.D. Christy and O.H. Tuovinen. 2008. Cathodic limitations in microbial fuel cells: an overview. J. Power Sources 180: 683-694.

Rosenbaum, M., U. Schroder and F. Scholz. 2005. In situ electrooxidation of photobiological hydrogen in a photobioelectrochemical fuel cell based on *Rhodobacter sphaeroides*. Environ. Sci. Technol. 39: 6328-6333.

Rosenbaum, M., F. Aulenta, M. Villano and L.T. Angenent. 2011. Cathodes as electron donors for microbial metabolism: which extracellular electron transfer mechanisms are involved? Bioresour. Technol. 102(1): 324-333.

Rotaru, A.E., P.M. Shrestha, F. Liu, B. Markovaite, S. Chen, B. Markovaite, S. Chen, K.P. Nevin and D.R. Lovley. 2014. Direct interspecies electron transfer between *Geobacter metallireducens* and *Methanosarcina barkeri*. Appl. Environ. Microbiol. 80: 4599-4605.

Rozendal, R.A., H.V.M. Hamelers, G.J.W. Euverink, S.J. Metz and C.J.N. Buisman. 2006. Principle and perspectives of hydrogen production through biocatalyzed electrolysis. Int. J. Hydrogen Energy 31: 1632-1640.

Rozendal, R.A., A.W. Jeremiasse, H.V.M. Hamelers and C.J.N. Buisman. 2008. Hydrogen production with a microbial biocathode. Environ. Sci. Technol. 42: 629-634.

Schroder, U., F. Harnisch and L.T. Angenent. 2015. Microbial electrochemistry and technology: terminology and classification. Energy Environ. Sci. 8: 513-519.

Semenec, L. and A.E. Franks. 2015. Delving through electrogenic biofilms: from anodes to cathodes to microbes. AIMS Bioengineering 2(3): 222-248.

Sevda, S. and T.R. Sreekrishnan. 2013. Removal of organic matters and nitrogenous pollutants simultaneously from two different wastewaters using biocathode microbial fuel cell. J. Environ. Sci. Health 49(11): 1265-1275.

Singh, L., M.F. Siddiqui, A. Ahmad, M.H.A. Rahim, M. Sakinah and Z.A. Wahid. 2013. Application of polyethylene glycol immobilized *Clostridium* sp. LS2 for continuous hydrogen production from palm oil mill effluent in upflow anaerobic sludge blanket reactor. J. Biochem. Eng. 70: 158-165.

Singh, L. and Z.A. Wahid. 2015. Methods for enhancing bio-hydrogen production from biological process: a review. J. Indus. Eng. Chem. 21: 70-80.

Stams, A.J.M. and C.M. Plugge. 2009. Electron transfer in syntrophic communities of anaerobic bacteria and archaea. Nat. Rev. Microbiol. 7: 568-577.

Steinbusch, K.J.J., H.V.M. Hamelers, J.D. Schaap, C. Kampman and C.J.N. Buisman. 2010. Bioelectrochemical ethanol production through mediated acetate reduction by mixed cultures. Environ. Sci. Technol. 44: 513-517.

Sun, M., G. Sheng, L. Zhang, C. Xia, Z. Mu, X. Liu, H.-L. Wang, H.-Q. Yu, R. Qi, T. Yu and M. Yang. 2008. An MEC-MFC-coupled system for biohydrogen production from acetate. Environ. Sci. Technol. 42: 8095-8100.

Sun, Y., J. Wei, P. Liang and X. Huang. 2012. Microbial community analysis in biocathode microbial fuel cells packed with different materials. AMB Express 2: 21.

Tandukar, M., S.J. Huber, T. Onodera and S.G. Pavlostathis. 2009. Biological chromium(VI) reduction in the cathode of a microbial fuel cell. Environ. Sci. Technol. 43: 8159-8165.

Ter-Heijne, A., H.V.M. Hamelers and C.J.N. Buisman. 2007. Microbial fuel cell operation with continuous biological ferrous iron oxidation of the catholyte. Environ. Sci. Technol. 41(11): 4130-4134.

Tran, H.-T., D.-H. Kim, S.-J. Oh, K. Rasool, D.-H. Park, R.-H. Zhang and D.-H. Ahn. 2009. Nitrifying biocathode enables effective electricity generation and sustainable wastewater treatment with microbial fuel cell. Water Sci. Technol. 59(9): 1803-1808.

Ucar, D., Y. Zhang and I. Angelidaki. 2017. An overview of electron acceptors in microbial fuel cells. Front. Microbiol. 8: 643.

Van Eerten-Jansen, M.C.A.A., A.B. Veldhoen, C.M. Plugge, A.J.M. Stams, C.J.N. Buisman and A. Ter-Heijne. 2013. Microbial community analysis of a methane-producing biocathode in a bioelectrochemical system. Archaea 2013: 481784.

Van Eerten-Jansen, M.C.A.A., N.C. Jansen, C.M. Plugge, V. de Wilde, C.J.N. Buisman and A. Ter-Heijne. 2015. Analysis of the mechanisms of bioelectrochemical methane production by mixed cultures. J. Chem. Technol. Biotechnol. 90(5): 963-970.

Villano, M., F. Aulenta, C. Ciucci, T. Ferri, A. Giuliano and M. Majone. 2010. Bioelectrochemical reduction of CO_2 to CH_4 via direct and indirect extracellular electron transfer by a hydrogenophilic methanogenic culture. Bioresour. Technol. 101: 3085-3090.

Wang, Z., Y. Zheng, Y. Xiao, S. Wu, Y. Wu, Z. Yang and F. Zhao. 2013. Analysis of oxygen reduction and microbial community of air-diffusion biocathode in microbial fuel cells. Bioresour. Technol. 144: 74-79.

Watanabe, K. 2008. Recent developments in microbial fuel cell technologies for sustainable bioenergy. J. Biosci. Bioeng. 106(6): 528-536.

Wrighton, K.C. and J.D. Coates. 2009. Microbial fuel cells: plug-in and power-on microbiology. Microbe 4 (6): 2081-2287.

Xiao, B., Y. Han, X. Liu and J. Liu. 2014. Relationship of methane and electricity production in two-chamber microbial fuel cell using sewage sludge as substrate. Inte. J. Hydrogen Energy 39(29): 16419-16425.

Zaybak, Z., J.M. Pisciotta, J.C. Tokash and B.E. Logan. 2013. Enhanced start-up of anaerobic facultatively autotrophic biocathodes in bioelectrochemical systems. J. Biotechnol. 168(4): 478-485.

Zhang, G.D., D. Pant and B.E. Logan. 2011. Long-term performance of activated carbon air cathodes with different diffusion layer porosities in microbial fuel cells. Biosens. Bioelectron 30(1): 49-55.

Zhang, S.H., C.H. Qiu, C.F. Fang, Q.L. Ge, Y.X. Hui, B. Han and S. Pang. 2017. Characterization of bacterial communities in anode microbial fuel cells fed with glucose, propyl alcohol and methanol. Appl. Biochem. Microbio. 53(2): 250-257.

Zhao, J., C. Zhang, C. Sun, W. Li, S. Zhang, S. Li and D. Zhang. 2018. Electron transfer mechanism of biocathode in a bioelectrochemical system coupled with chemical absorption for NO removal. Bioresour. Technol. 254: 16-22.

Zhou, M.H., M.L. Chi, J.M. Luo, H.H. He and T. Jin. 2011. An overview of electrode materials in microbial fuel cells. J. Power Sources 196: 4427-4435.

Zhou, M., H. Wang, D.J. Hassettand and T. Gu. 2013a. Recent advances in microbial fuel cells (MFCs) and microbial electrolysis cells (MECs) for wastewater treatment, bioenergy and bioproducts. Chem. Technol. Biotechnol. 88(4): 508-518.

Zhou, M., J. Yang, H. Wang, T. Jin, D. Xu and T. Gu. 2013b. Microbial fuel cells and microbial electrolysis cells for the production of bioelectricity and biomaterials. Environ. Technol. 34(13-16): 1915-1928.

Role of Biocatalyst in Microbial Fuel Cell Performance

M. Amirul Islam[1], Ahasanul Karim[2], Puranjan Mishra[2], Che Ku Mohammad Faizal[2], Md. Maksudur Rahman Khan[1] and Abu Yousuf[3]*

[1] Universiti Malaysia Pahang Faculty of Chemical and Natural Resources Engineering Gambang, Pahang Malaysia, 26300
[2] Universiti Malaysia Pahang Faculty of Engineering Technology Gambang, Pahang Malaysia, 26300
[3] Department of Chemical Engineering & Polymer Sciences, Shahjalal University of Science & Technology, Sylhet - 3114, Bangladesh

Introduction

The contemporary world is suffering from several problems, such as excessive population, air pollution, water pollution, wastewater generation, and energy crisis (Hanjra and Qureshi 2010). Among those, the excessive wastewater generation and energy depletion are most critical cruxes, since those cannot be controlled due to rapid industrialization and population growth (Schneider et al. 2013). In the past decade, intensive research efforts have been dedicated to find alternative renewable energy sources, particularly energy generation through the sustainable treatment of wastewater (Nayak et al. 2016). Renewable energy sources are considered sustainable and carbon neutral substitutes to fossil fuels are highly needed to relieve the global environmental deterioration and energy crisis (Kothari et al. 2010), as shown in Fig. 6.1. According to the European Renewable Energy Council (EREC), around 50% of the global energy supply will be supported by renewable energy in 2040 (Sun et al. 2016). Generally, various wastes, particularly wastewater, contain abundant organic matter, thus they could be considered as potential renewable energy reservoirs (Baranitharan et al. 2015). The microbial fuel cells (MFCs) are the bio electrochemical system which provide conversion of chemical energy available in substrate into electricity using catalytic activity of electrogens (Mohan et al. 2014). MFCs have been one of the most studied systems for applications in electricity production and

*Corresponding author: ayousufcep@yahoo.com, ayousuf-cep@sust.edu

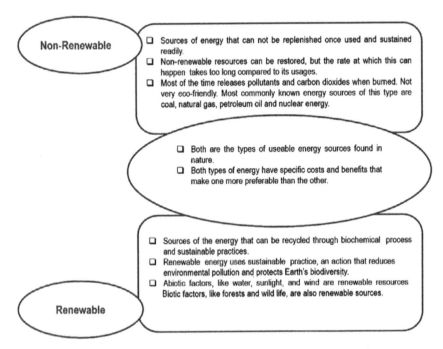

Figure 6.1. Comparison of advantages between renewable
and non-renewable energy sources.

simultaneous wastewater treatment as compared to other available bio-electrochemical systems because they can simultaneously accomplish both energy production and resolve environmental issues (McCarty et al. 2011).

MFCs are considered a promising sustainable technology for combating energy crisis using wastewater as substrates. This is due to their ability to simultaneously solve electricity generation and wastewater treatment issues (Ivanov et al. 2013, Logan and Rabaey 2012). Moreover, it was reported that the energy efficiency of common renewable energy sources, such as wind turbine (15-40%) (Habash et al. 2011) and biogas plant (5-20.7%) (Naegele et al. 2012) are low compared to that of MFC (77.6%) (Schröder 2007). Besides, MFCs can also utilize several substrates, ranging from simple to complex sources. Several simple substrates, such as glucose and acetate (Pant et al. 2010), as well as other complex organic substrates, such as industrial and domestic (Liu et al. 2014, Pandey et al. 2016) wastewater, have been tested in MFC. However, a major limitation of its application in power generators has been the low and unstable power output.

A number of new strains of microorganism have been reported in the past decade which can generate electrical current through MFCs (Wang and Ren 2013). However, among them, the majority of microbes are electrochemically less active due to the fact that the proteins involved with electron transport are contained within the cell membrane, and the absence of electron transport molecules in the cell membrane does not allow the electron transfer to electrodes (Debabov 2008). Few exo-electrogens, such as *Shewanella oneidensis* MR-1, *Shewanella putrefaciens* IR-1, *Geobacter metallireducens*, and *Geobacter sulfurreducens* have been reported as potential anode biocatalysts due to their direct electron transfer (DET) capability (Zhi et al. 2014). However, those species were not able to achieve high performance while operated with real wastewater which might be due to the inability of those microorganisms to utilize complex substrates. Therefore, the bacteria possessing electrogenic properties with ability to utilize wide variety of substrates present in complex wastewater could bring a breakthrough in enhancing the performance of MFCs (Islam et al. 2016).

Bacteria as a Biocatalyst

The use of bacteria is preferred as inoculum in MFCs due to the capability of direct electron transfer from microbial cell membrane to the surface of electrodes (Chaudhuri and Lovley 2003). Till date, some pure culture strains belonging to all classes of *Firmicutes, Proteobacteria*, and *Acidobacteria* (Table 6.1) have been reported as electrogen. A number of microbes have reported as potential electrogen to be used in MFCs; however, *Geobacter sulfurreducens* (*G. sulfurreducens*) is considered as a model organism to elucidate the molecular basis (genes and proteins associated with electron transfer) of electron transfer. Till date, the *G. sulfurreducens* produces maximum current densities among the reported pure culture (Yi et al. 2009). Although *G. sulfurreducens* contains a number of genes (Methe et al. 2003), it has been reported that c-type cytochrome associated with OmcZ is the only cytochrome which plays crucial role in producing maximum current production (Nevin et al. 2009).

For practical application of MFCs, mixed culture inoculums are more suitable to degrade complex substrate, such as wastewaters. Furthermore, it was reported that in a single-chamber air cathode, MFC showed lower internal resistance while it was inoculated with mixed cultures. Besides, the mixed cultures produced around 22% more power (576 mW/m^2) compared to pure culture of *G. sulfurreducens* (Ishii et al. 2008). The analysis of microbial community in anode biofilms has shown that the microorganisms are not confined in all anode biofilms and there is no defined microbial community in MFC. Several factors, such as inoculum (e.g., AS, aerobic sludge, activated sludge), the configuration (e.g., single-chamber, two-chamber, tubular), substrate (e.g., pure, complex, wastewater) and operational conditions (e.g., temperature, pH, time) severely influence the microbial community composition in MFCs. For example, in marine sediment MFC, *Geobacteraceae* was observed as a predominant microorganism (Holmes et al. 2004). However, in thermophilic MFC, *Firmicutes* belonging to *Thermincola* spp., and *Deferribacteres*, were found as the predominant bacterial groups in the anode of MFCs (Mathis et al. 2008). In another study, Chae et al. (2009) found the variation in the microbial community, while MFC operated with different substrates. Likewise, the substrate concentrations can also influence the microbial community composition in the anode bulk as well as biofilm

Table 6.1. Some commonly studied pure culture microorganisms in MFCs

Microbes	Higher taxonomic level	Substrate	References
Pseudomonas aeruginosa	γ-proteobacteria	Glucose	Rabaey et al. 2004
Proteus vulgaris	γ-proteobacteria	Glucose, Fructose	Kim et al. 2000
Clostridium butyricum	Firmicutes	Glucose	Park et al. 2001
Geobacter sulfurreducens	δ-proteobacteria	Acetate	Bond and Lovley 2003
Shewanella oneidensis MR-1	γ-proteobacteria	Lactate	Bretschger et al. 2007
Escherichia coli	γ-proteobacteria	Glucose	Zhang et al. 2006
Shewanella oneidensis DSP10	γ-proteobacteria	Glucose	Zhang et al. 2006
Bacillus subtilis	Firmicutes	Glucose	Nimje et al. 2009
Klebsiella pneumoniae L17	γ-proteobacteria	Glucose	Zhang et al. 2008
Enterobacter cloacae	γ-proteobacteria	Acetate	Rezaei et al. 2009
Citrobacter sp.	γ-proteobacteria	Acetate	Xu and Liu 2011
Arthrospira maxima	Cyanophyceae	Acetate	Inglesby et al. 2012

(Choo et al. 2006). Therefore, the influence of inoculum on the performance of MFC is hard to determine.

Anaerobic sludge (AS) is one of the most widely used inoculums in MFC for electricity generation and treatment of wastewater (Baranitharan et al. 2015). However, it is difficult to reproduce electricity generation from sludges because they are not confined to a specific location. In addition, when using mixed microbial cultures, it is hard to ascertain the mechanism and role of individual microorganisms to power generation. MFCs operated with complex substrates using mixed microbial cultures usually show high power densities compared to pure cultures (Logan et al. 2006) (Table 6.2), because most of the pure cultures are not able to utilize wide range of substrates. So far, only few pure cultures have been used in MFC operated with wastewater. Nevertheless, several new strains of pure culture have been stated in the past decade, which can generate electricity through their metabolic cycle in MFCs (Table 6.2). However, among them, only few strains are capable of producing significant current in complex substrates (e.g., cassava mill, POME, brewery, synthetic wastewater) fed MFCs (Li et al. 2014). Apart from that, AS contains some fermentative bacteria; however, those are not electrochemically active because of the unavailability of proteins associated with electron transfer within the cellular membrane (Baranitharan et al. 2015).

This happens because the final electron acceptors used by these cells, such as oxygen or soluble metals, can diffuse all the way inside the cells till the

Table 6.2. Performance MFC using mixed culture inoculums in wastewater fed MFCs

Substrate (Wastewater)	COD (mg/L)	Inoculum	Electrode material	Power density (W/m³)	COD removal efficiency (%)	CE (%)	Reference
Dairy	1,600	AS	Graphite	2.7	91	17	Elakkiya and Matheswaran 2013
Food waste	1,000	AS	Carbon felt	0.43	92	20	Li et al. 2013a
Municipal	-	AS	Graphite	1.43	94	-	Wang et al. 2013
Piggery	3,250	AS	Stainless steel	1.41	-	1.7	Ryu et al. 2013
Domestic	238	AS	Carbon fibre	0.48	78	14	Jiang et al. 2013
Carcass	1,118	AS	Graphite	2.19	50	0.25	Li et al. 2013b
Dairy	4,440	AS	Graphite	1.10	95	4.3	Mohan et al. 2010
Rice mill	2,250	AS	Stainless steel	2.30	96	21	M. Behera et al. 2010

The values were recalculated to show in W/m³ but originally it was reported in W/m².

electron transport chain (Ilbert and Bonnefoy 2013). The absence of electron transport molecules in the cell membrane does not allow the electron transfer to electrodes. In some cases, to overcome this problem, artificial mediators could facilitate the transfer from the microbial membrane to the electrodes, but the use of artificial mediators is not so advantageous because they are toxic to microbes at high concentrations, expensive, and environmentally unacceptable (Hubenova et al. 2010). Moreover, the performance of artificial mediated MFC is lower compared to the one obtained with electrogenic bacteria (Table 6.2). The special group of bacteria called exo-electrogenic or simply electrogenic, possess in their membrane special electron transporters that allow electron transfer to an external electron acceptor, such as insoluble metals, fumarate, or nitrate (Islam et al. 2016).

Microbial Electron Transfer Mechanism

The electron transfer system during MFCs performance can be affected by the presence or absence of biocatalyst (Liu and Logan 2004). The electron transfer process is performed either mediator-less (direct electron transfer) or mediator-based in MFCs. During direct electron transfer system, the electron is transferred using nano wires, transmembrane proteins, or using different microbial cyto-chromes (Thauer et al. 2007). During mediator-based electron transfer process, the electron shuttles play an important role and researchers suggested that the mediator-based electron transfers system has ability to increase the electron transfer significantly (Sharma and Kundu 2010).

Direct electron transfer system

The electron transfer process between anode and bacterial cell membrane is based on their physical contact (biofilm, pilli, etc.) (Schröder 2007). The c-type cytochromes in *G. sulfurreducens* and *S. oneidensis* have been suggested as terminal outer-membrane redox proteins required for electron transport to the anode (Kim et al. 2008) (Fig. 6.2). Metal-reducing microorganisms, e.g., Geobacter, Rhodoferax, and Shewanella, use solid terminal electron acceptors, such as Fe (III) oxides in their natural habitats, but can switch to using an anode electrode as their terminal electron acceptor in MFCs (Chaudhuri and Lovley 2003). *G. sulfurreducens* and *S. oneidensis* are also known to produce electrically conductive pilus-like structures, referred to as 'nanowires', that allow electron transport to more distant solid electron acceptors (Gorby et al. 2006, Reguera et al. 2006) (Fig. 6.2). *G. sulfurreducens* cells in the anodic biofilm even at relatively long distance from the anode surface are shown to contribute to electric current generation (Nevin et al. 2008). Long range electron transfer is also suggested to occur via cytochromes released into the biofilm matrix (Lovley 2008). The metal reducing bacteria, particularly *Cyanobacterium synechocystis* PCC6803) and the thermophilic fermentative (*Pelotomaculum thermopropionicum*) have been shown to evolve nanowires (Gorby et al. 2006).

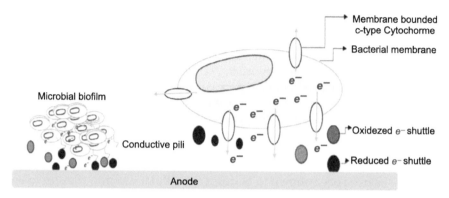

Figure 6.2. Different types of electron transfer mechanisms from microbes to electrode.

Mediated electron transfer

Mediated electron transfer has been proposed to occur via three different types of redox compounds: exogenous (artificial) redox mediators, microbially produced redox mediators, or the cells in the biofilm close to the electrode transfer electrons through the membrane bound cytochromes, whereas distant cells from the electrode could use a conductive pilli network and electron shuttling mediators to transfer electrons from the cells to the electrode primary metabolites (Fig. 6.2) (Schröder 2007). A good mediator should have a redox potential as negative as possible, but yet higher than the potential of substrate and sufficiently lower than that of the anode electrode, and is readily re-oxidized on the anode surface under the conditions of MFC (Schröder 2007). Artificial mediators, such as thionine, methylene blue, and neutral red have been used to assist anodic electron transfer in MFCs (Ieropoulos et al. 2005). These mediators penetrate the bacterial cell and scavenge electrons from the reducing agents, such as NADH. The reduced mediators then diffuse out of the cells and are oxidized on the electrode surface. Some microorganisms can also produce electron shuttling compounds. Pyocyanine and other phenazine compounds produced by *Pseudomonas* sp. have been shown to assist their own extracellular electron transfer or improve current generation by other species or communities (Pham et al. 2008). Riboflavins and quinones released by *S. oneidensis* and *E. coli* have also been proposed to act as electron shuttles (Qiao et al. 2008). The metabolic pathways of anaerobic respiration and fermentation produce metabolites that can also serve as MFC mediators. The sulphate/sulphide redox couple has been exploited e.g., in MFCs based on an anaerobic sulphate-reducing bacterium, *Desulfovibrio desulfuricans* (Cooney et al. 1996). Sulphate is reduced by bacteria to hydrogen sulphide, which in turn is abiotically oxidized on the anode electrode. Microbial fermentation products, such as hydrogen, lactate, and formate can be successfully oxidized on MFC anodes, but catalysts are required on anode electrodes to facilitate the direct oxidation of these compounds (Rosenbaum et al. 2007). Electron transfer to an anode by Gram-positive bacteria was first thought to be restricted by their rigid cell wall containing peptidoglycan layers (Pham et al. 2008). However, redox mediators produced by Gram-negative bacteria have been shown to enhance current generation by Gram-positive species (Pham et al. 2008). Three Gram-positive species, *Clostridium butyricum*, *Thermincola* sp., and *Lactococcus lactis*, capable of electricity production in pure culture have been reported (Freguia et al. 2009). Only the exo-cellular electron transfer mechanism of *Lactococcus* sp. has been reported so far (Freguia et al. 2009). *L. lactis* produces mainly lactic acid from the fermentation of carbohydrates, and is also capable of coupling carbohydrate metabolism to the reduction of Fe^{3+}, Cu^{2+}, and O_2, mediated by membrane-bound quinones (Rezaïki et al. 2008). Pure cultures of *L. lactisti*s have been shown to perform exocellular electron transfer to an MFC anode via excretion of soluble quinones with a simultaneous shifting of metabolism to the production of acetate and pyruvate (Freguia et al. 2009).

Cyanobacteria as biocatalyst

Cyanobacteria, in particular, have great potential as biocatalyst for electricity production. During the last few years the biotechnological development of cyanobacteria has experienced an unprecedented increase and the use of these photosynthetic organisms for electricity production is becoming a tangible reality (Behera and Varma 2016). Cyanobacteria have been used as biocatalysts as they used broad spectrum of high value compound and minimal nutritional requirement (Gudmundsson and Nogales 2015). The minimal nutritional requirement for growth and its efficient exoelectrogenic properties have led cyanobacteria to great interest in using MFCs as biocatalysts (Pandit et al. 2012). The incorporation of cyanobacteria provides the active electron acceptors for the cathode, as well as dissolved oxygen for ORR in MFCs. The use of cyanobacteria as biocatalyst in double chamber MFCs and single chamber MFCs have been studied extensively. Previously intensified research (1985-2010) has been repowered on photo MFCs, especially application of cyanobacterial species in including anabaena and synechocystis as biocatalysts. More recently, several researches have been reported successfully on bioelectricity generation using cyanobacteria as electron producer, when various wastewaters were subjected as sole carbon sources, as presented in Table 6.3. Moreover, successive efforts have been focused on demonstrating the photosynthetic exoelectrogenic activities in terms of electric current, the range of substrate utilization, maximum COD utilization of substrate, biofilm synthesis, electrode material, etc. There is a syntrophic relationship between phototrophic and provided energy rich compounds on electrogenic activities. The maximum utilization of energy-rich compounds (COD) accelerates the current generation efficiency. For instance, recently *Spirulina*, a representative biomass of cyanobacteria (blue green algae), was investigated for bioelectricity potential with maximum utilization of COD using cathodic compartments of photo-microbial fuel cells. Their reports suggested the sustainable cathodic reactions to ensure the electrochemical performance and COD removal in P-MFCs, extant to air-cathode MFCs. Maximum nutrients recovery into the cathodic chamber, with minimal COD leakage through the porous separator were observed with maximum current production ~5 A/m² for A-MFCs and ~4 A/m² for P-MFCs. They reported the higher COD removal and anodic chamber, with average COD removal of 89 ±1% for A-MFC and 87±1% for W-MFC. Furthermore, in MFCs, the development of biofilm with electroactive microbes influences the electrogenic pathways. Therefore, development of biofilm on electrode (anode) enhances the electron transfer between solid electrode surface and extracellular transfer chain of microbes. Recently, attempts have been made for bioelectricity generation with the development of cyanobacterial biofilms technologies. The novel silk-fibroin nanocomposite matrix endowed with biocompatible, optoelectronic, and electroactive properties suitable for bio-anode fabrication for a P-MFC has been developed. By using *Synechococcus* sp. as biocatalyst, the maximum current density of photo-MFC was reported with the nanocomposite of bianode (1.89 A/m²), which was estimated ~5.7 time more than the corresponding blank graphite anode.

Table 6.3. Comparisons of MFC performance using different algae

Microbes	Anodic volume (mL)	Maximum power	COD removal efficiency	References
Scenedesmus obliquus	250	30 mW/m²	-	Kakarla and Min 2014
C. vulgaris	150	187 mW/m²	-	Liu et al. 2015
Anabaena sp.	225	57.8 mW/m²	-	Pandit et al. 2012
Spirulina platensis	16	1.64 mW/m²	-	Menicucci Jr 2010
C. vulgaris	50	62.7 mW/m²	80%	Fu et al. 2009
C. vulgaris	800	14.40 mW/m²	-	del Campo et al. 2013
Chlamydomonas reinhardtii	500	12.95 mW/m²	-	Lan et al. 2013
C. vulgaris	220	5.6 W/m³	-	Wang et al. 2010
C. vulgaris	400	2.26 W/m³	81.4%	Zhou et al. 2012
Chlorella and *Phormidium*	797	2.01 mW/m²	54%	Juang et al. 2012
C. pyrenoidosa	250	6.36 W/m³	87%	Jadhav et al. 2017

Yeast as Biocatalyst

The application of yeast cells in various biotechnological processes, including food industry have been known for centuries (Ferreira et al. 2010). Among the different species of yeast, the *Saccharomyces cerevisiae* is a representative species as it comprises with non-pathogenic in nature and easier to cultivate (Pant et al. 2012). The development of MFCs concept for harvesting the microbial electricity prompts the researchers to investigate the yeast-mediated bio-catalytic activity (Lin et al. 2017). The existence of bio-catalytic activity would be related to the presence of various natural electron shuttles, such as cytochromes, ferredoxin, and azurin which can be used by the redox enzymes for electron transfer process in MFCs (Sayed and Abdelkareem 2017).

The yeast cytochromes of yeast cells are located in the mitochondria, which facilitates the electron transport system via trans-mem0p10brane proteins (Srivastava et al. 2017). The electron transfer during the metabolism of organic matter by yeast cells under anaerobic condition is shown in Fig. 6.3. The electron is produced from the oxidation of glucose into pyruvate in glycolysis process. Since the glycolysis takes place in the cytosol rather than in the mitochondria, reduced NADH is easily accessible to a mediator molecule that is attached to the cell membrane. After liberating the electron by NADH to anode, oxidized back into the NAD+, the MFC which operates using *S. cerevisiae* extracts the energy using NADH/NAD+ redox cycle (Behera and Varma 2016). *S. cerevisiae* has been extensively studied for bioelectricity generation and many approaches have been employed to increase the power of MFCs by facilitating the extracellular electron transfer mechanism (Christwardana and Kwon 2017). During the electron transfer process in MFCs, it is imperative to build a "electroactive-biofilm" to produce efficient electricity as the electricity generation also depends upon microbial biofilm and surface type of electrodes (Kumar et al. 2016). The electroactive biofilms include both electrochemically and inactive microorganism. Previously, *S. cerevisiae* was known to be electrochemically inactive or less active for electromagnetic biofilm formation in MFCs and it was restricted with methylene blue as mediator (Raghavulu et al. 2011). However, its electrogenic properties were considered too low to implement in a system when chemical mediator was not added. These limitations had been overcome, when Rahimnejad et al. (2012) successfully reported the increased electricity production in MFC by *S. cerevisiae* by using thionine as mediator. Addition of 500 mM thionine with *S. cerevisiae* in MFC can provide the maximum voltage and current generation of 420 mV and 700 mA/m^2, respectively.

Further, the addition of yeast extract to the medium is suggested as the influencer in yeast cell adhesion (yeast-based biofilm) on the surface of plain and gold sputtered carbon paper anodes. With the addition of yeast extract with *S. cerevisiae* when plain carbon paper was used as anode, the increased current and power densities from 94 to 190 and 12.9 to 32.6 mA/cm^2, respectively have been reported. With the addition of the yeast extract to the medium with gold-sputtered anode, increased current and power densities from 25 to 300 and 2 to 70 mA/cm^2 , respectively have been reported (Rahimnejad et al. 2012). Several strains of yeast have been studied as

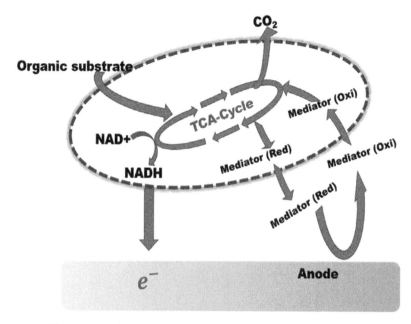

Figure 6.3. Schematic illustration of microbial mediator-based
electron transfer system.

biocatalysts in which some of species used mediator as "electron-shuttle" (to carry electrons from inside the cell to the electrode), while others performed "mediator-less" electrochemically electricity generation (Table 6.4) including *S. cerevisiae, Hansenula anomala* (Prasad et al. 2007), *Arxula adeninivorans* (Haslett et al. 2011), *Candida melibiosica* (Hubenova and Mitov 2010), etc. The electron transfer process in MFCs, when performed without using or addition of external mediator's source, is considered as mediator-less MFCs. The mediators-less electron transfer in the baker's yeast *S. cerevisiae* has been well-described by Sayed et al. (2015). They confined the yeast biofilm, which adhered to the anode and its surface had potential as active biocatalyst for oxidation of glucose in mediator-less MFCs. Further, it was suggested that the electron transfer takes place through the surface confined species and the species in anolyte solution including dispersed *S. cerevisiae* did not participate in power generation (Thauer et al. 2007). In another study, authors studied the electron transfer system from the external surface of yeast cell wall to the electrode using *S. cerevisiae* as biocatalyst. They reported that the electrode in MFCs, which exhibits a thinner stable mediator layer than the yeast cell wall, makes the direct contact of yeast cell redox centre to the electrode difficult. At the same time, applying modified-electrodes electron transfer was observed between yeast cell wall and electrode (Sayed and Abdelkareem 2017). On the other hand, when another MFC using a fresh anode was operated with the filtered anolyte solution, i.e., no yeast cells, neither cell voltage nor anode potential changed. The same conclusions for the direct electron transfer

Table 6.4. Mediator-based electricity generation from yeast

Biocatalyst	Anode materials	Substrate	Catholyte	Maximum power density	References
C. melibiosica	Graphite rods	Fructose	$K_3[Fe(CN)_6]$	60 mW/m^3	Hubenova and Mitov 2010
C. melibiosica	Carbon felt	Fructose	$K_3[Fe(CN)_6]$	20 mW/m^2	Babanova et al. 2011
A. adeninivorans	Carbon cloth	Glucose	$KMnO_4$	70 mW/m^2	Haslett et al. 2011
H. anomala	Graphite	Glucose	$K_3[Fe(CN)_6]$	0.7 W/m^3	Prasad et al. 2007
S. cerevisiae	Graphite	Dextrose	-	30 mW/m^2	Raghavulu et al. 2011
C. melibiosica	Carbon felt	Fructose	$K_3[Fe(CN)_6]$	5 mW/m^2	Hubenova et al. 2014

and no role of the mediator in the electron transfer of the *S. cerevisiae* were confirmed by Rawson et al. (2012), who studied the direct electron transfer from the *S. cerevisiae* cells attached to the anode surface. The authors modified the anode surface with a mediator, osmium bipyridine complex, layer that hindered the mediator from penetrating the cell wall and reacting with the internal redox species. Results showed that the electron transferred from the yeast cells to the electrode surface through the yeast cell wall and no involvement of the endogenous mediator in this electron transfer.

The mediator-based yeast MFCs has been investigated by various researchers to improve the electro-activity potential of MFCs using different type of external electron shuttles, as illustrated in Table 6.4. Being efficient electron shuttles, it should be electrochemically active and fast in electron transfers to the electrode surface. Li et al. (2018) investigated the comparable power-output efficiency of *S. cerevisiae* in both manners including mediator-less and mediator-based system in dual-chamber MFCs. Their results suggested the better power-out, high cell voltage, and slightly lower glucose consumption in comparison to mediator-less MFCs performance. In another study of mediator-based MFCs, rotating discs were used as electrodes and methylene blue as mediator, to determine the chronoamperometric effect. Their experimental results suggested that power density of MFCs can be enhanced by using high concentrations of mediators under the control of mediator adsorption.

Enzyme as Biocatalyst

Many enzymes have been used as biocatalysts in fuel due to the availability of an appropriate electron-acceptor. The active redox organic cofactor, such as $FAD/FADH_2$, wrapped within the protein matrix makes the active site of enzyme. The direct electron transfer processes to the electrode are slow processes due to the proportional interment of enzymes active site into the polypeptides.

The reduction and oxidation couple mediator is usually required to increase the anodic glucose oxidation rate by oxidases enzyme. The redox mediators can be defined as a small electroactive molecule which is able to make stability in two inter convertible redox states. Therefore, it is intelligible to diffuse in and out from the protein channels, hence easy to transfer electrons between the anodic electrode and enzymes active site. To facilitate the electron-transfer processes, significant driving forces are very important. Besides, to enhance the cell performance, the redox potential of the electron shuttle mediator should be close to the active site of redox-enzymes. The electron transfer mechanism from glucose to the anodic biofuel cells is illustrated in Fig. 6.4, along with demonstrations of the redox potentials of the several redox couples. It is important to note that finally, the redox potential of the mediator dominates the other potentials at which the anodic reactions occur.

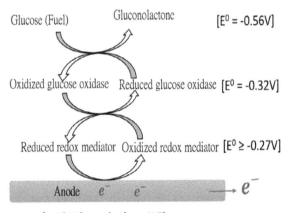

(Verses Ag/AgCl KCl standard at pH 7)

Figure 6.4. Flow chart of mediator-based electron transfer between glucose oxidase and anode electrode.

The anodic and cathodic reactions occur in two different compartments inside the fuel cell that are separated by an exchange membrane to filter the unwanted reactive compounds between them. This strategy relives the utilization of membrane because of its specificity of the enzyme to the given enzyme and its coating on electrode surface along with a redox mediator abstains the cross-response problems. Therefore, the enzyme-based MFCs comprise simply two modified electrodes, which are directly connected with external load. The different varieties of commonly used efficient mediators have been illustrated in Fig. 6.4. Nonetheless, the enzyme-based MFCs required a membrane as separator between both of the electrodes compartments since it has to be used in solution. Meanwhile, a large amount of recovered energy is dissipated due to the high redox potential as compared to the enzyme. During the process, first the refinement step includes co-ordination of redox active electron shuttling mediator to a polymer backbone. For instance, a Cis-Os (bipy) reacted with PVI to yield the redox polymer, where the metal is coordinated with an imidazole group of polymer backbone. Hence, it allows the crosslinking of polymer to enzyme and aggregation on the surface of electrode, which removes the requirement of membrane inside the fuel cells. The observed redox potential (E_0 = 0.20 V) is, higher than that of cis-Os (bipy)Cl$_2$ and therefore less interesting in this prospect. The chemistry of classical co-ordination allows an appropriate tuning of the redox potential through the ability of ligand's donating and accepting properties. Here, the electron donating para-amino substitute bipy, the observed redox polymer of (E_0 = −0.11 V), which is satisfying the two criteria as defined above: low redox potential but more than that of the enzyme. The further step of refinement involves the "advent of tether" between the redox-active moiety and polymer backbone. Firstly, it is very important to bring a larger flexibility to the immobilized redox-mediator which can move in and out to the enzymes and accelerate the "hopping mechanism", which is responsible for electron transfer to the electrode. Based on redox-polymer PVP ($E_0 =$ –0.11 V) is considered as one of the well-refined anodic mediators of these type of MFCs with the demand of a membrane-less enzyme-based biofuels.

Conclusion

MFC is considered as a potential method for bioelectricity generation as well as wastewater treatment though significant technical obstacles exist. The chemical energy existing in organic wastes can be directly converted to electric energy. Among the different parameters that are involved for the rate limiting steps, efficient electron transferring biocatalysts are one of them. Hence, the improvement in the biocatalyst's electron transfer rate allows controlling the electron transfer cascades in the assemblies. Hence, the development of new methodologies in order to facilitate electron transfer is emergent to obtain practical output from MFC.

REFERENCES

Babanova, S., Y. Hubenova and M. Mitov. 2011. Influence of artificial mediators on yeast-based fuel cell performance. J. Biosci. Bioeng. 112: 379-387.

Baranitharan, E., M.R. Khan, A. Yousuf, W.F.A. Teo, G.Y.A. Tan and C.K. Cheng. 2015. Enhanced power generation using controlled inoculum from palm oil mill effluent fed microbial fuel cell. Fuel 143: 72-79.

Behera, B.K. and A. Varma. 2016. Microbial fuel cell (MFC). pp. 181-221. *In*: Helmut Koenig (ed.). Microbial Resources for Sustainable Energy. Springer International Publishing, Switzerland.

Behera, M., P.S. Jana, T.T. More and M. Ghangrekar. 2010. Rice mill wastewater treatment in microbial fuel cells fabricated using proton exchange membrane and earthen pot at different pH. Bioelectrochemistry. 79: 228-233.

Bond, D.R. and D.R. Lovley. 2003. Electricity production by *Geobacter sulfurreducens* attached to electrodes. Appl. Environ. Microb. 69: 1548-1555.

Bretschger, O., A. Obraztsova, C.A. Sturm, I.S. Chang, Y.A. Gorby, S.B. Reed, . . . M.F. Romine. 2007. Current production and metal oxide reduction by *Shewanella oneidensis* MR-1 wild type and mutants. Appl. Environ. Microb. 73: 7003-7012.

Chae, K.-J., M.-J. Choi, J.-W. Lee, K.-Y. Kim and I.S. Kim. 2009. Effect of different substrates on the performance, bacterial diversity, and bacterial viability in microbial fuel cells. Bioresource Technol. 100: 3518-3525.

Chaudhuri, S.K. and D.R. Lovley. 2003. Electricity generation by direct oxidation of glucose in mediatorless microbial fuel cells. Nat. Biotechnol. 21: 1229-1232.

Choo, Y.F., J. Lee, I.S. Chang and B.H. Kim. 2006. Bacteria Communities in microbial fuel cells enriched with high concentrations of glucose and glutamate. J. Microbiol. Biotechn. 16: 1481-1484.

Christwardana, M. and Y. Kwon. 2017. Yeast and carbon nanotube based biocatalyst developed by synergetic effects of covalent bonding and hydrophobic interaction for performance enhancement of membraneless microbial fuel cell. Bioresource Technol. 225: 175-182.

Cooney, M.J., E. Roschi, I.W. Marison, C. Comminellis and U. von Stockar. 1996. Physiologic studies with the sulfate-reducing bacterium Desulfovibrio desulfuricans: evaluation for use in a biofuel cell. Enzyme Microb. Tech. 18: 358-365.

Debabov, V. 2008. Electricity from microorganisms. Microbiology. 77: 123-131.

del Campo, A.G., P. Cañizares, M.A. Rodrigo, F.J. Fernández and J. Lobato. 2013. Microbial fuel cell with an algae-assisted cathode: a preliminary assessment. J. Power Sources. 242: 638-645.

Elakkiya, E. and M. Matheswaran. 2013. Comparison of anodic metabolisms in bioelectricity production during treatment of dairy wastewater in Microbial Fuel Cell. Bioresource Technol. 136: 407-412.

Ferreira, I., O. Pinho, E. Vieira and J. Tavarela. 2010. Brewer's *Saccharomyces* yeast biomass: characteristics and potential applications. Trends Food Sci. Tech. 21: 77-84.

Freguia, S., M. Masuda, S. Tsujimura and K. Kano. 2009. *Lactococcus lactis* catalyses electricity generation at microbial fuel cell anodes via excretion of a soluble quinone. Bioelectrochemistry. 76: 14-18.

Fu, C.-C., C.-H. Su, T.-C. Hung, C.-H. Hsieh, D. Suryani and W.-T. Wu. 2009. Effects of biomass weight and light intensity on the performance of photosynthetic microbial fuel cells with *Spirulina platensis*. Bioresource Technol. 100: 4183-4186.

Gorby, Y.A., S. Yanina, J.S. McLean, K.M. Rosso, D. Moyles, A. Dohnalkova, . . . K.S. Kim. 2006. Electrically conductive bacterial nanowires produced by *Shewanella oneidensis* strain MR-1 and other microorganisms. P. Natl A. Sci. 103: 11358-11363.

Gudmundsson, S. and J. Nogales. 2015. Cyanobacteria as photosynthetic biocatalysts: a systems biology perspective. Mol. BioSyst. 11: 60-70.

Habash, R.W., V. Groza, Y. Yang, C. Blouin and P. Guillemette. 2011. Performance of a contrarotating small wind energy converter. ISRN Mechanical Engineering 2011: 1-10.

Hanjra, M.A. and M.E. Qureshi. 2010. Global water crisis and future food security in an era of climate change. Food Policy. 35: 365-377.

Haslett, N.D., F.J. Rawson, F. Barriëre, G. Kunze, N. Pasco, R. Gooneratne and K.H. Baronian. 2011. Characterisation of yeast microbial fuel cell with the yeast *Arxula adeninivorans* as the biocatalyst. Biosens. Bioelectron. 26: 3742-3747.

Holmes, D., D. Bond, R. O'neil, C. Reimers, L. Tender and D. Lovley. 2004. Microbial communities associated with electrodes harvesting electricity from a variety of aquatic sediments. Microb. Ecol. 48: 178-190.

Hubenova, Y., D. Georgiev and M. Mitov. 2014. Enhanced phytate dephosphorylation by using *Candida melibiosica* yeast-based biofuel cell. Biotechnol. Lett. 36: 1993-1997.

Hubenova, Y. and M. Mitov. 2010. Potential application of Candida melibiosica in biofuel cells. Bioelectrochemistry 78: 57-61.

Hubenova, Y.V., R.S. Rashkov, V.D. Buchvarov, M.H. Arnaudova, S.M. Babanova and M.Y. Mitov. 2010. Improvement of yeast – biofuel cell output by electrode modifications. Ind. Eng. Chem. Res. 50: 557-564.

Ieropoulos, I.A., J. Greenman, C. Melhuish and J. Hart. 2005. Comparative study of three types of microbial fuel cell. Enzyme Microb. Tech. 37: 238-245.

Ilbert, M. and V. Bonnefoy. 2013. Insight into the evolution of the iron oxidation pathways. BBA-Bioenergetics 1827: 161-175.

Inglesby, A.E., D.A. Beatty and A.C. Fisher. 2012. *Rhodopseudomonas palustris* purple bacteria fed *Arthrospira maxima* cyanobacteria: demonstration of application in microbial fuel cells. RSC Adv. 2: 4829-4838.

Ishii, S.I., K. Watanabe, S. Yabuki, B.E. Logan and Y. Sekiguchi. 2008. Comparison of electrode reduction activities of *Geobacter sulfurreducens* and an enriched consortium in an air-cathode microbial fuel cell. Appl. Environ. Microb. 74: 7348-7355.

Islam, M.A., C.W. Woon, B. Ethiraj, C.K. Cheng, A. Yousuf and M.M.R. Khan. 2016. Ultrasound driven biofilm removal for stable power generation in microbial fuel cell. Energ. Fuel. 31: 968-976.

Ivanov, I., L. Ren, M. Siegert and B.E. Logan. 2013. A quantitative method to evaluate microbial electrolysis cell effectiveness for energy recovery and wastewater treatment. Int. J. Hydrogen Energ. 38: 13135-13142.

Jadhav, D.A., S.C. Jain and M.M. Ghangrekar. 2017. Simultaneous wastewater treatment, algal biomass production and electricity generation in clayware microbial carbon capture cells. Appl Biochem Biotech. 183: 1076-1092.

Jiang, H.-m., S.-j. Luo, X.-s. Shi, M. Dai and R.-b. Guo. 2013. A system combining microbial fuel cell with photobioreactor for continuous domestic wastewater treatment and bioelectricity generation. J. Cent. South Univ. 20: 488-494.

Juang, D.-F., C.-H. Lee, S.-C. Hsueh and H.-Y. Chou. 2012. Power generation capabilities of microbial fuel cells with different oxygen supplies in the cathodic chamber. Appl. Biochem. Biotech. 167: 714-731.

Kakarla, R. and B. Min. 2014. Photoautotrophic microalgae *Scenedesmus obliquus* attached on a cathode as oxygen producers for microbial fuel cell (MFC) operation. Int. J. Hydrogen Energ. 39: 10275-10283.

Kim, B.-C., B.L. Postier, R.J. DiDonato, S.K. Chaudhuri, K.P. Nevin and D.R. Lovley. 2008. Insights into genes involved in electricity generation in *Geobacter sulfurreducens* via whole genome microarray analysis of the OmcF-deficient mutant. Bioelectrochemistry 73: 70-75.

Kim, N.J., Y.J. Choe, S.H. Jeong and S.H. Kim. 2000. Development of microbial fuel cells using *Proteus vulgaris*. B. Kor. Chem. Soc. 21: 44-48.

Kothari, R., V. Tyagi and A. Pathak. 2010. Waste-to-energy: a way from renewable energy sources to sustainable development. Renew. Sust. Energ. Rev. 14: 3164-3170.

Kumar, R., L. Singh and A. Zularisam. 2016. Exoelectrogens: recent advances in molecular drivers involved in extracellular electron transfer and strategies used to improve it for microbial fuel cell applications. Renew. Sust. Energ. Rev. 56: 1322-1336.

Lan, J.C.-W., K. Raman, C.-M. Huang and C.-M. Chang. 2013. The impact of monochromatic blue and red LED light upon performance of photo microbial fuel cells (PMFCs) using *Chlamydomonas reinhardtii* transformation F5 as biocatalyst. Biochem Eng. J. 78: 39-43.

Li, M., M. Zhou, X. Tian, C. Tan, C.T. McDaniel, D.J. Hassett and T. Gu. 2018. Microbial fuel cell (MFC) power performance improvement through enhanced microbial electrogenicity. Biotechnol. Adv. (in press).

Li, N., L. Liu and F. Yang. 2014. Power generation enhanced by a polyaniline-phytic acid modified filter electrode integrating microbial fuel cell with membrane bioreactor. Sep. Purif. Technol. 132: 213-217.

Li, X., N. Zhu, Y. Wang, P. Li, P. Wu and J. Wu. 2013a. Animal carcass wastewater treatment and bioelectricity generation in up-flow tubular microbial fuel cells:

effects of HRT and non-precious metallic catalyst. Bioresource Technol. 128: 454-460.

Li, X.M., K.Y. Cheng, A. Selvam and J.W. Wong. 2013b. Bioelectricity production from acidic food waste leachate using microbial fuel cells: effect of microbial inocula. Process Biochem. 48: 283-288.

Lin, T., X. Bai, Y. Hu, B. Li, Y.J. Yuan, H. Song, . . . J. Wang. 2017. Synthetic *Saccharomyces cerevisiae-Shewanella oneidensis* consortium enables glucose-fed high-performance microbial fuel cell. AIChE J. 63: 1830-1838.

Liu, H. and B.E. Logan. 2004. Electricity generation using an air-cathode single chamber microbial fuel cell in the presence and absence of a proton exchange membrane. Environ. Sci. Technol. 38: 4040-4046.

Liu, J., L. Liu, B. Gao, F. Yang, J. Crittenden and N. Ren. 2014. Integration of microbial fuel cell with independent membrane cathode bioreactor for power generation, membrane fouling mitigation and wastewater treatment. Int. J. Hydrogen Energ. 39: 17865-17872.

Liu, T., L. Rao, Y. Yuan and L. Zhuang. 2015. Bioelectricity generation in a microbial fuel cell with a self-sustainable photocathode. The Sci. World J. 2015: 1-8.

Logan, B.E., B. Hamelers, R. Rozendal, U. Schröder, J. Keller, S. Freguia, . . . K. Rabaey. 2006. Microbial fuel cells: methodology and technology. Environ. Sci. Technol. 40: 5181-5192.

Logan, B.E. and K. Rabaey. 2012. Conversion of wastes into bioelectricity and chemicals by using microbial electrochemical technologies. Science 337: 686-690.

Lovley, D.R. 2008. The microbe electric: conversion of organic matter to electricity. Curr Opin. Biotech. 19: 564-571.

Mathis, B., C. Marshall, C. Milliken, R. Makkar, S. Creager and H. May. 2008. Electricity generation by thermophilic microorganisms from marine sediment. Appl. Microbiol. Biot. 78: 147-155.

McCarty, P.L., J. Bae and J. Kim. 2011. Domestic wastewater treatment as a net energy producer – can this be achieved? Environ. Sci. Techol. 45: 7100-7106.

Menicucci Jr, J.A. 2010. Algal biofilms, microbial fuel cells, and implementation of state-of-the-art research into chemical and biological engineering laboratories. ProQuest Dissertations Publishing.

Methe, B., K.E. Nelson, J. Eisen, I.T. Paulsen, W. Nelson, J. Heidelberg, . . . M. Beanan. 2003. Genome of *Geobacter sulfurreducens*: metal reduction in subsurface environments. Science 302: 1967-1969.

Mohan, S.V., G. Mohanakrishna, G. Velvizhi, V.L. Babu and P. Sarma. 2010. Bio-catalyzed electrochemical treatment of real field dairy wastewater with simultaneous power generation. Biochem. Eng. J. 51: 32-39.

Mohan, S.V., G. Velvizhi, J.A. Modestra and S. Srikanth. 2014. Microbial fuel cell: critical factors regulating bio-catalyzed electrochemical process and recent advancements. Renew. Sust. Energ. Rev. 40: 779-797.

Naegele, H.-J., A. Lemmer, H. Oechsner and T. Jungbluth. 2012. Electric energy consumption of the full scale research biogas plant "Unterer Lindenhof": Results of longterm and full detail measurements. Energies 5: 5198-5214.

Nayak, M., A. Karemore and R. Sen. 2016. Performance evaluation of microalgae for concomitant wastewater bioremediation, CO_2 biofixation and lipid biosynthesis for biodiesel application. Algal Res. 16: 216-223.

Nevin, K.P., B.-C. Kim, R.H. Glaven, J.P. Johnson, T.L. Woodard, B.A. Methé, . . . A. Liu. 2009. Anode biofilm transcriptomics reveals outer surface components essential for high density current production in *Geobacter sulfurreducens* fuel cells. PloS one. 4: e5628.

Nevin, K.P., H. Richter, S. Covalla, J. Johnson, T. Woodard, A. Orloff, . . . D. Lovley. 2008. Power output and columbic efficiencies from biofilms of *Geobacter sulfurreducens* comparable to mixed community microbial fuel cells. Environ. Microbiol. 10: 2505-2514.

Nimje, V.R., C.-Y. Chen, C.-C. Chen, J.-S. Jean, A.S. Reddy, C.-W. Fan, . . . J.-L. Chen. 2009. Stable and high energy generation by a strain of *Bacillus subtilis* in a microbial fuel cell. J. Power Sources 190: 258-263.

Pandey, P., V.N. Shinde, R.L. Deopurkar, S.P. Kale, S.A. Patil and D. Pant. 2016. Recent advances in the use of different substrates in microbial fuel cells toward wastewater treatment and simultaneous energy recovery. Appl Energ. 168: 706-723.

Pandit, S., B.K. Nayak and D. Das. 2012. Microbial carbon capture cell using cyanobacteria for simultaneous power generation, carbon dioxide sequestration and wastewater treatment. Bioresource Technol. 107: 97-102.

Pant, D., A. Singh, G. Van Bogaert, S.I. Olsen, P.S. Nigam, L. Diels and K. Vanbroekhoven. 2012. Bioelectrochemical systems (BES) for sustainable energy production and product recovery from organic wastes and industrial wastewaters. Rsc Adv. 2: 1248-1263.

Pant, D., G. Van Bogaert, L. Diels and K. Vanbroekhoven. 2010. A review of the substrates used in microbial fuel cells (MFCs) for sustainable energy production. Bioresource Technol. 101: 1533-1543.

Park, H.S., B.H. Kim, H.S. Kim, H.J. Kim, G.T. Kim, M. Kim, . . . H.I. Chang. 2001. A novel electrochemically active and Fe (III)-reducing bacterium phylogenetically related to *Clostridium butyricum* isolated from a microbial fuel cell. Anaerobe. 7: 297-306.

Pham, H.T., N. Boon, P. Aelterman, P. Clauwaert, L. De Schamphelaire, P. Van Oostveldt, . . . W. Verstraete. 2008. High shear enrichment improves the performance of the anodophilic microbial consortium in a microbial fuel cell. Microb Biotechnol. 1: 487-496.

Prasad, D., S. Arun, M. Murugesan, S. Padmanaban, R. Satyanarayanan, S. Berchmans and V. Yegnaraman. 2007. Direct electron transfer with yeast cells and construction of a mediatorless microbial fuel cell. Biosens. Bioelectron. 22: 2604-2610.

Qiao, Y., C.M. Li, S.-J. Bao, Z. Lu and Y. Hong. 2008. Direct electrochemistry and electrocatalytic mechanism of evolved *Escherichia coli* cells in microbial fuel cells. Chem. Commun. 11: 1290-1292.

Rabaey, K., N. Boon, S.D. Siciliano, M. Verhaege and W. Verstraete. 2004. Biofuel cells select for microbial consortia that self-mediate electron transfer. Appl. Environ. Microbiol. 70: 5373-5382.

Raghavulu, S.V., R.K. Goud, P. Sarma and S.V. Mohan. 2011. *Saccharomyces cerevisiae* as anodic biocatalyst for power generation in biofuel cell: influence of redox condition and substrate load. Bioresource Technol. 102: 2751-2757.

Rahimnejad, M., G.D. Najafpour, A.A. Ghoreyshi, F. Talebnia, G.C. Premier, G. Bakeri, . . . S.-E. Oh. 2012. Thionine increases electricity generation from microbial fuel cell using *Saccharomyces cerevisiae* and exoelectrogenic mixed culture. J. Microbiol. 50: 575-580.

Rawson, F.J., A.J. Gross, D.J. Garrett, A.J. Downard and K.H. Baronian. 2012. Mediated electrochemical detection of electron transfer from the outer surface of the cell wall of *Saccharomyces cerevisiae*. Electrochem. Commun. 15: 85-87.

Reguera, G., K.P. Nevin, J.S. Nicoll, S.F. Covalla, T.L. Woodard and D.R. Lovley. 2006. Biofilm and nanowire production leads to increased current in *Geobacter sulfurreducens* fuel cells. Appl. Environ. Microbiol. 72: 7345-7348.

Rezaei, F., D. Xing, R. Wagner, J.M. Regan, T.L. Richard and B.E. Logan. 2009. Simultaneous cellulose degradation and electricity production by *Enterobacter cloacae* in a microbial fuel cell. Appl. Environ. Microbiol. 75: 3673-3678.

Rezaïki, L., G. Lamberet, A. Derré, A. Gruss and P. Gaudu. 2008. *Lactococcus lactis* produces short-chain quinones that cross-feed Group B *Streptococcus* to activate respiration growth. Mol. Microbiol. 67: 947-957.

Rosenbaum, M., F. Zhao, M. Quaas, H. Wulff, U. Schröder and F. Scholz. 2007. Evaluation of catalytic properties of tungsten carbide for the anode of microbial fuel cells. Appl. Catal. B-Environ. 74: 261-269.

Ryu, J., H. Lee, Y. Lee, T. Kim, M. Kim, D. Anh, . . . D. Ahn. 2013. Simultaneous carbon and nitrogen removal from piggery wastewater using loop configuration microbial fuel cell. Process Biochem. 48: 1080-1085.

Sayed, E.T. and M.A. Abdelkareem. 2017. Yeast as a Biocatalyst in Microbial Fuel Cell. pp. 41-65. *In*: Célia Pais (ed.). Old Yeasts-New Questions. InTech.

Sayed, E.T., N.A. Barakat, M.A. Abdelkareem, H. Fouad and N. Nakagawa. 2015. Yeast extract as an effective and safe mediator for the baker's-yeast-based microbial fuel cell. Ind. Eng. Chem. Res. 54: 3116-3122.

Schneider, T., S. Graeff-Hönninger, W. French, R. Hernandez, N. Merkt, W. Claupein, . . . P. Pham. 2013. Lipid and carotenoid production by oleaginous red yeast *Rhodotorula glutinis* cultivated on brewery effluents. Energy 61: 34-43.

Schröder, U. 2007. Anodic electron transfer mechanisms in microbial fuel cells and their energy efficiency. Phys. Chem. Chem. Phys. 9: 2619-2629.

Sharma, V. and P. Kundu. 2010. Biocatalysts in microbial fuel cells. Enzyme Microb. Tech. 47: 179-188.

Srivastava, A., S. Sircaik, F. Husain, E. Thomas, S. Ror, S. Rastogi, . . . C.J. Nobile. 2017. Distinct roles of the 7-transmembrane receptor protein Rta3 in regulating the asymmetric distribution of phosphatidylcholine across the plasma membrane and biofilm formation in *Candida albicans*. Cell. Microbiol. 19: e12767.

Sun, M., L.-F. Zhai, W.-W. Li and H.-Q. Yu. 2016. Harvest and utilization of chemical energy in wastes by microbial fuel cells. Chem. Soc. Rev. 45: 2847-2870.

Thauer, R.K., E. Stackebrandt and W.A. Hamilton. 2007. Energy metabolism and phylogenetic diversity of sulphate-reducing bacteria. pp. 1-37. *In*: L.L. Barton and W.A. Hamilton (eds.). Sulphate-reducing Bacteria: Environmental and Engineered Systems. Cambridge University Press, New York, NY, USA.

Wang, H. and Z.J. Ren. 2013. A comprehensive review of microbial electrochemical systems as a platform technology. Biotechnol. Adv. 31: 1796-1807.

Wang, X., Y. Feng, J. Liu, H. Lee, C. Li, N. Li and N. Ren. 2010. Sequestration of CO_2 discharged from anode by algal cathode in microbial carbon capture cells (MCCs). Biosens. Bioelectron. 25: 2639-2643.

Wang, Y.-K., W.-W. Li, G.-P. Sheng, B.-J. Shi and H.-Q. Yu. 2013. In-situ utilization of generated electricity in an electrochemical membrane bioreactor to mitigate membrane fouling. Water Res. 47: 5794-5800.

Xu, S. and H. Liu. 2011. New exoelectrogen *Citrobacter* sp. SX-1 isolated from a microbial fuel cell. J. Appl. Microbiol. 111: 1108-1115.

Yi, H., K.P. Nevin, B.-C. Kim, A.E. Franks, A. Klimes, L.M. Tender and D.R. Lovley. 2009. Selection of a variant of *Geobacter sulfurreducens* with enhanced capacity for current production in microbial fuel cells. Biosens. Bioelectron. 24: 3498-3503.

Zhang, L., S. Zhou, L. Zhuang, W. Li, J. Zhang, N. Lu and L. Deng. 2008. Microbial fuel based on *Klebsiella pneumoniae* biofilm. Electrochem. Commun. 10: 1641-1643.

Zhang, T., C. Cui, S. Chen, X. Ai, H. Yang, P. Shen and Z. Peng. 2006. A novel mediatorless microbial fuel cell based on direct biocatalysis of *Escherichia coli*. Chem. Commun. 2257-2259.

Zhi, W., Z. Ge, Z. He and H. Zhang. 2014. Methods for understanding microbial community structures and functions in microbial fuel cells: a review. Bioresource Technol. 171: 461-468.

Zhou, M., H. He, T. Jin and H. Wang. 2012. Power generation enhancement in novel microbial carbon capture cells with immobilized *Chlorella vulgaris*. J. Power Sources 214: 216-219.

Exoelectrogenic Bacteria: A Candidate for Sustainable Bio-electricity Generation in Microbial Fuel Cells

Shraddha Shahane[1], Payel Choudhury[2], O.N. Tiwari[3], Umesh Mishra[1] and Biswanath Bhunia[4]*

[1] Department of Civil Engineering, National Institute of Technology Agartala, Agartala - 799046, Tripura, India
[2] Department of Electrical Engineering, National Institute of Technology Agartala, Agartala - 799046, Tripura, India
[3] Department of Bio Engineering, National Institute of Technology Agartala, Agartala - 799046, Tripura, India
[4] Centre for Conservation and Utilisation of Blue Green Algae, Division of Microbiology, ICAR-Indian Agricultural Research Institute, New Delhi - 110012, India

Introduction

Energy crisis owing to development and industrialization makes it obligatory to produce power from available resources to fulfill the world's increasing energy demand. The fossil fuels, such as natural gas and coal are natural resources of energy production. The natural sources are not sustainable and their level is diminishing owing to a high consumption rate. Hence, development of an alternate long term energy source is necessary. The fossil fuels are efficient sources of power production. The other energy sources are sunlight, wind, and water which are renewable sources. The increase in urbanization, industrialization and rapid socio-economic development has increased the utilization of anthropogenic sources producing environmental pollution. The prevention of environmental water pollution has become a problem in various areas. The effluents discharged from various industries and domestic areas contain toxic compounds, heavy metals, organic substrate, nitrogen, phosphorus, and phenolic compounds. Thus, by considering high energy requirement and wastewater treatment method for different

*Corresponding author: bbhunia@gmail.com

wastewater, it is essential to find an efficient alternative to fulfill the huge energy demand continuously.

Microbial fuel cells have been used as a promising technology of power generation through the wastewater process by a few researchers (HaoYu et al. 2007, Salgado 2009). MFCs behave like bioreactors, where the conversion of chemical energy takes place in the presence of bacteria as catalysts. The microbial fuel cell reactor oxidizes biodegradable wastes, thus producing energy. Therefore, researchers are trying to utilize MFCs not only for wastewater treatment, but also for energy generation (He et al. 2017, Qin et al. 2017). There are different conventional wastewater methods, such as chemical precipitation, coagulation, iron exchange, and adsorption. Besides these available wastewater treatment methods, no technology was established to be sustainable. Many physico-chemical techniques have been used extensively for treating wastewater, but they have several drawbacks, such as high operational cost and large production of undesirable by products. Biological wastewater treatment processes are considered environmentally friendly in comparison to other treatment processes. The energy generation from MFCs by utilization of wastes is an alternative sustainable approach to address the problem of high energy demand and environmental pollution.

Introduction to Microbial Fuel Cells

MFC is a recent technology which produces direct electrical energy from organic wastes by transferring electrons produced from respiratory chain of microbes. MFCs are devices which utilize microbes as a catalyst which converts stored chemical energy into electrical energy. MFCs not only give energy benefit, but also have a good operational stability with resistance to environmental stresses (Ledezma et al. 2013). MFC creates energy from utilization of organic substrates. MFC plays an important role as it is a good wastewater treatment process and energy generation device. It is an eco friendly viable system of generation of electricity by utilization of biomass through the chain of respiratory enzyme in microbes. In MFC, electro-chemically active bacteria are proficient to produce electrical energy from chemical energy through microbial respiration of inhabitants. The anaerobic bacteria species are competent in transforming organic compounds from wastewater to electricity. The layer of microorganisms on anode surface acts as a biocatalyst, which helps convert the organic matter in wastewater into electricity through the processes of bio-oxidation. The microorganisms which are in the anode chamber of MFCs are termed as anode respiring bacteria, and oxidize the organic matter present in wastewater. The oxidation process of substrate is carried out by ARB generating electrons and proton with production of electricity from the obtained biomass. The several potent microbial species have been studied to produce power in MFC mentioned in Table 7.1. The power production by utilization of diversity microbial system, for example iron reducing bacteria Geobacter sulfurredicen, Geobacter toluenoxydans, and Geobacter metallireducens have been studied earlier

(Liu et al. 2005), and metal reducing species *Shewanella japonica* (Ivanova et al. 2001). Geobacter sulfurredicen is identified for producing electricity from acetate. MFC may include pure bacterial or mixed bacterial culture. Rabaey has observed that microbial fuel cells using pure microbial culture produce less power as compared to fuel cells with mixed culture (Rabaey et al. 2005, Rabaey et al. 2004).

Table 7.1. Energy recovery with MFCs from utilization of different wastewater sources

Wastewater (Substrate)	Microbia fuel cell	Power production	References
Acetate	Dual chamber	$0.08 \, mA/cm^2$	Min et al. 2008
Domestic wastewater	Single chamber	$1.7, 3.7 \, W/m^3$	Liu & Logan 2004
Domestic wastewater	Dual chamber	$0.06 \, mA/cm^2$	Wang et al. 2011
Paper wastewater	Single chamber	$8 \, mA/m^2$	Nimje et al. 2012
Swine wastewater	Single chamber	$0.015 \, mA/cm^2$	Min et al. 2005
Synthetic wastewater	Single chamber	$0.017 \, mA/cm^2$	Aldrovandi et al. 2009
Dairy wastewater	Single chamber	$25 \, mA/m^2$	Velasquez-Orta et al. 2011
Chocolate-industry wastewater	Dual chamber	$0.302 \, mA/cm^2$	Patil et al. 2009
Alcohol distillery	Dual chamber	$1000 \, mA/m^2$	Ha et al. 2012
Landfill leachate	Single chamber	$114 \, mA/m^2$	Ganesh & Jambeck 2013

Electrogenic Bacteria as Biocatalyst

Electrogens are the microorganisms which are capable of exocellular electron transfer. Electrochemically active bacteria or anode respiring bacteria have the potential to convert the biochemical energy into ATP through respiration. The exoelectrogens are capable of producing biofilm, which is electrochemically active through oxidation of organics in wastewater. The biofilm produced by electrogenic bacteria is of great importance since it is a good source of bioelectricity production. The most common forms of respiration involve consumption of soluble compounds, such as oxygen, nitrate, and sulfates, as an electron acceptor, whereas there are a few microorganisms which use solid electron acceptors, such as metal oxides, metal electrodes, and carbon for respiration to gain energy (Marcus et al. 2010). Even though there are variety of microorganisms which produce power in MFCs, a few microbial strain exhibit low power generation when developed in pure culture (Rabaey et al. 2004, Zuo et al. 2008).

Components of MFC

Electrode Material

Anode

Anode used in MFC performs a significant role in terms of energy production and stable output of MFC. The anodic material must be conductive, chemically stable, and non-corrosive in the reactor solution. The anode surface must be large and porous with good stability. Carbon in the variety of graphite rod or plates is the simple material for anode owing to sufficient surface area and being cost-effective. A larger surface area can be acquired by using graphite felt electrodes (Logan 2004, Park and Zeikus 1999). The use of electro catalytic material, such as Polyaniline/Pt shows improved anode performance, thus current generation. Various carbon-based electrodes, such as carbon cloth, graphite felt, and carbon felt are generally used due to more delivery of power output per unit surface area when they are used to operate MFC (Feng et al. 2010). For efficient power output, large porous surface area of electrode and electrical conductivity for charge transfer are required for attachment. Several modified anode materials have been recently introduced to improve microbial adhesion and superior electron flow by surface modification. The spiral-shaped anode is more helpful for effective utilization of MFC because it offers large surface area for attachment of microbes, superior electrical conductivity of electrons, and good current collection potential. It has been observed that improvement of bacterial adhesion on spiral-shaped anode is a favorable condition for gaining highest energy output using MFCs. Instead of using plain nickel foam, graphene oxide–nickel (rGO–Ni) foam was operated as an anode in MFC to get the high volumetric power generation. The obtained power densities are higher than conventionally used carbon electrode. The maximum volumetric power density was achieved with MFC with pure culture of *Shewanella oneidensis* MR-1 (Wang et al. 2013). Graphene sponge has been stated as a high performance, inexpensive anode material in order to maximize MFC performance. The use of graphene sponge composite and stainless steel (G-S-SS) electrode was found to be effective for extensive applications (Xie et al. 2012). The dual-chambered MFC fabricated with active anode (ruthenium oxide-coated carbon electrode) was found to be successful to increase electron transfer and power output (Lv et al. 2012). Carbon brush is a material containing twisted carbon fibers in another anode material for operation of MFC, and it gives quite a high surface area with the most favorable area to volume ratio (Liao et al. 2015). Carbon veil, a carbonaceous material with high porosity and electrical conductivity, gives the great pores permit to have bacteria through the porous structure and enhance the biofilm attachment (Gajda et al. 2016). Granular activated carbon (GAC) is porous, low cost, and a biocompatible material which can be used as anode material (Yasri and Nakhla 2017). Sulfate reducing bacteria enriched system MFC was developed for investigating removal of heavy metals. This study revealed the impact of Cu^{2+} on MFC performance, by using SRB enriched anode (Miran et al. 2017).

Cathode

The selection of cathode material strongly affects the working principle of MFC. The cathode material may differ corresponding to the application for which it is required to be used. Oxygen has the highest redox potential and it is abundant in nature, and is thus a commonly-used electron acceptor (Zhao et al. 2006). There is no requirement of a separate cathode compartment when oxygen is used as a cathode in MFCs. Much higher power densities may be achieved with the use of air cathode instead of aqueous cathode, when MFC operated with oxygen is used as cathode (Logan and Regan 2006). Fe-based catalysts are cheap substitutes instead of using Pt-based cathodes in MFC. The oxygen reduction reaction (ORR) at the cathode is frequently the restrictive aspect observed in microbial fuel cells, which has to focus on boosting the performances and reducing the cost (Kodali et al. 2017). Other electron acceptors are bicarbonate, oxygen, nitrate, nitrites, permanganate, iron, copper and manganese dioxide (Choudhury et al. 2017). Experimentally, it has been estimated that the porosity of the cathode governs the overall performance of MFCs (Ahn et al. 2014). Carbonaceous materials based on graphene are generally used as cathode catalysts (Yuan and He 2015). Various carbonaceous materials, such as activated carbon nanofiber and nanotube, have been widely used in MFCs as cathode electrode (Ghasemi et al. 2011, Santoro et al. 2013, Yang et al. 2014a). Ferricynide is another popular electron acceptor used in MFCs, due to its high-quality performance. It can be regenerated or replaced. It is found to be a promising cathode material owing to low overpotential using plain carbon cathode resulting in cathode working potential near to its open circuit potential (Park and Zeikus 2003). The use of multiple electron acceptors enhanced the kinetics of electron transport across the biofilm of anode significantly. Addition of carbon black in AC greatly advances the electron transfer mechanism in MFC, generating considerable output (Zhang et al. 2014). The nitrogen-doped carbon nanosheet on graphene porous was found to be an excellent alternative electron collector instead of using Pt cathode in fuel cell reactor (Wen et al. 2014). This modified catalyst has been proven to enhance the catalytic activities for oxygen reduction reaction in the cell. The study revealed that when the temperature treatment applied to the activated carbon (AC) was used as an electron acceptor catalyst for microbial fuel cell, it significantly changed the surface properties of activated carbon (Santoro et al. 2014). The comparative study of low cost (FePc)-MnOx composite and Pt catalysts on the cathode were performed by Burkitt (Burkitt et al. 2016). Graphite and MnO_2 and MoS_2 as ORR as cathode catalyst has shown stable power output, which is higher than MFC without catalyst (Jiang et al. 2017). Several researches with CoNi-based cathode catalyst have been executed successfully to enhance the electrolytic activity in MFC to get effective energy recovery with the use of inexpensive catalyst (Hou et al. 2016). In recent years, M-N-C catalyst is getting huge attention to get efficient energy recovery from fuel cell in terms of durability of cathode (Yuan et al. 2016).

Membrane

Microbial fuel cells require the separator for anode and cathode chamber, which may be provided by use of cation exchange membrane. In single compartment MFCs / Sediment MFCs, the separation takes place naturally. The CEM used in MFCs must be capable of passing the ferricynide, oxygen, ions, and organic substrate. Absence of CEM causes dispersion of oxygen from cathode to anode, leading to lowering the coulombic efficiency. By keeping practical applications in mind, the membrane material plays a vital role in making developed and sustainable MFC. The selection of suitable membrane material, design of MFC governs the overall operating cost and performance of MFC. The most popular CEM used is Nafion, because of its mechanical properties, good chemical stability, and conductivity (Mook et al. 2012, Smitha et al. 2005). Since this is an expensive material, majority of research work has then focused in order to find a substitute with extensive materials, such as biodegradable bags and ceramic (Winfield et al. 2013a), nylon fiber, and glass fiber (Zhang et al. 2010). The power density and coulombic efficiency were found to be substantially increased in air-cathode MFC when j-cloth layer was applied on the air cathode (Fan et al. 2007). Conventional membrane material causes membrane fouling, which is a common problem in MFC. The uses of natural rubber membrane have been studied, which enhanced the power generation compared to CEM, along with the exclusion of membrane fouling problem (Winfield et al. 2013b). Canvas (Zhuang et al. 2009), nylon infused membrane (Hernández-Fernández et al. 2015), photocopy paper (Winfield et al. 2015)—these membranes have been reported and studied comparatively by Kondaveeti et al. (2014). The membranes have some drawbacks, such as membrane fouling. To overcome this drawback, introduction of hydrophilic functional groups, such as amine, hydroxyl, carboxyl, and sulfone was carried out in membranes (Blanco et al. 2001). This membrane prepared with polymer was found to act as a hydrophilic membrane for nanofiltration. Recently studied polyethersulfone (PES) blend proton exchange membranes prepared with various composites showed production of $58.726 \, mWm^{-2}$ power output experimentally, which is higher than Nafion membrane (Zinadini et al. 2017).

Working Principle of MFC

The electrons and protons are generated during the process of oxidation at the anode of MFCs. The generated electrons will travel towards cathode through the external circuit, whereas protons will transmit through the ion exchange membrane. Thus, depending on the nature of the association and the substrates, potential difference is created, which will further lead to the removal of pollutants in MFC (Barsky et al. 2014). The fundamental design of MFC includes anode, cathode, and electrolyte medium connected by cationic exchange membrane to electrodes. The basic design of microbial fuel cell is shown in Fig. 7.1. The bacteria present in wastewater converts the organic compound into carbon dioxide, protons, and electrons. Figure 7.2 illustrates the potential and power densities in microbial fuel cells. The various commonly used electron donors and acceptors used in MFC are mentioned in Table 7.2. The exoelectrogenic microbe utilizes organic and inorganic substrate available in wastewater as fuel and produces current. The electrodes are linked by an external resister, which allows electron to flow from anode to cathode creating a current. Anaerobic degradation of organic matter causes the production of protons, which diffuse through the CEM to the cathode. Further it reacts with oxygen and electrons, thus forming water as an end product (Bond et al. 2002). Exoelectrogenic microbes produce electricity through transfer of electrons generated from metabolism activities to the anode of microbial fuel cell system. However besides these improved parameters, high residence time is required for power generation, as the rate of biodegradation of organic matter in wastewater is slower.

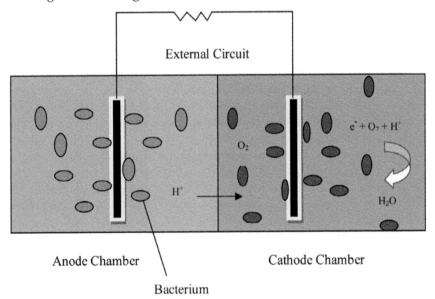

Figure 7.1. Microbial fuel cell (Saba et al. 2017, Zinadini et al. 2017).

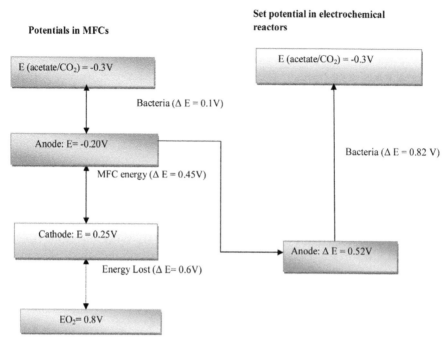

Figure 7.2. Potentials and power densities in microbial fuel cells (Logan 2009).

Table 7.2. The commonly used electron donor and acceptor in microbial fuel cells

Chamber	Electron donor/acceptor
Anode	Glucose
	Malate
	Glycerol
	Sulfur
	Citrate
	Acetate
Cathode	Oxygen
	Nitrate
	Nitrite
	Bicarbonate
	Iron
	Copper
	Ferricynide
	Manganese dioxide
	Potassium persulfate
	Acetate

Electron Transfer Mechanisms

The mechanisms for electron transfer from the substrates to electrodes may be via direct or indirect electron transfers (Sun et al. 2016). The direct electron transfer by contact of outer membrane cytochromes are also known as shuttles (Myers and Myers 1992, Rabaey et al. 2005). The electron transfer mechanisms in MFCs are shown in Fig. 7.3.

Direct electron transfer mechanism

The direct electron transfer mechanism indicates direct transfer of electron between electron carrier in the microbes and the cathode electrode in microbial fuel cells. In this transfer mechanism, the biofilm is formed on the anode, through which electron transfer takes place. As the transfer of electron is direct, generation of additional energy takes place in direct electron transfer process (Chaudhuri and Lovley 2003, Kim et al. 2004). When electrons reach the electrode surface, electrochemical reaction must occur, which liberates these electrons into the anode (electron donor). Direct electron transfer mechanisms are taking place in the presence of outer membrane (OM) cytochromes, which can act together directly with the solid surface to carry out respiration (Beliaev et al. 2001, Myers and Myers 1992).

The direct electron transfer doesn't include diffusional redox species in direct mechanism of electron transfer. Such type of electron transfer has been found in Geobacter species or mixed culture (Pant et al. 2010). There are various microorganisms which transfer electrons from inside the cell to extracellular acceptors through cytochromes, biofilms and well conductive nanowires (Lovley 2008). *Shewanella putrefaciens, Geobacter sulferreducens,* and *Rhodoferax ferrireducens* etc. are some examples of microbes reported for having direct contact mechanism for electron transfer. *Shewanella putrefaciens* have first reported a microbial strain to generate electricity in absence of mediator (Park and Zeikus1999). *Shewanella putrefaciens* has outer membrane cytochromes to transfer electron directly as well as extract electrically conductive nanowires (Gorby et al. 2006, Myers and Myers 1992).

Indirect electron transfer mechanisms

Indirect transfer of electrons takes place by mediators, which may be produced by microorganisms or externally added. This mediator-based mechanism takes place in the presence of soluble electron shuttles (Logan and Regan 2006, Marcus et al. 2010). Electron shuttles are the compounds which act as electron carriers from microbes by transferring to the surface of electrode. Further, it reacts with it to give its electrons. Simply, we can say that the soluble electron shuttle acts as a mediator to transfer the electrons from microorganisms to the anode. The shuttles disperse back to the cell when it is in an oxidized state, which should be used as the same compound repeatedly, hence known to be shuttles. The microorganisms known to generate mediator which serves as electron shuttle includes melanin, quinines, phenazines, and flavins (Hernandez et al. 2004, Newman and Kolter 2000, Turick et al. 2002, Von Canstein et al. 2008). The third mechanism is supported by conductive or semi-conductive nanowires, in which ARB are able to carry electrons by using these nanowires. The nanowires have recently been discovered to be electrically conductive material by bacteria. Pili formed by some bacteria have been revealed to be electrically conductive (Gorby et al. 2006). The other components characterized for their conductive properties help in electron transfer, such as bound electron mediators or extracellular cytochromes (Lovley 2008, Rittmann et al. 2008).

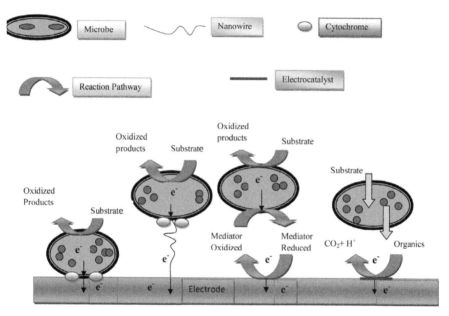

Figure 7.3. Types of electron transfer mechanism from microbes to electrode (modified figure from Patil et al. 2012).

Chemical Reactions in MFC

The following equations (1-4) represent the oxidation-reduction reactions in MFCs. It shows possible bioelectric-chemical reactions in MFC while generating electricity. The microbe utilizes wastewater as substrate so as to perform their metabolic activities, which would further lead to generate electrons.

Oxidation Reaction (Anode)

Glucose

$$C_6H_{12}O_6 + 12H_2O \rightarrow 6HCO_3^- + 3OH^+ + 24e^- \tag{1}$$

Reduction Reactions (Cathode)

$$O_2 + 4H^+ + 4e^- \rightarrow 2H_2O \tag{2}$$

$$O_2 + 2H^+ + 2e^- \rightarrow H_2O_2 \tag{3}$$

$$NO_3^- + 2e^- + 2H^+ \rightarrow NO_2^- + H_2O \tag{4}$$

Progress in MFC Research with Bacteria for Sustainable Electricity Generation

Since MFC is considered a bioreactor, optimum balance of nutrients is required to achieve the maximum yield. Therefore, medium design is a prerequisite, as it helps in proper utilization of nutrients during the process of biochemical reaction. There are various statistical packages available for bioprocess design, as they have documented advantages (Mandal et al. 2013). In the last few years, researchers are putting in efforts on MFC configuration as well as surface modifications of electrodes material in fuel cells to generate electricity inexpensively. Moreover, less attention has been given on different process parameters, such as pH and temperature, which must be considered to make cost-effective operation of fuel cell. It is obvious that utilization of organic substrate depends on microorganisms, as it is their intrinsic property. Therefore, the structure of the microbial community is greatly affected, as they can use various types of substrate in MFCs (Wang et al. 2012). Although there are various types of substrate used in MFC, fermentable substrate is easily taken by the microbial community, and hence electricity is produced easily (Yi et al. 2009). Since domestic wastewater contains various types of substrate, bacteria requires additional enzymes to hydrolyse and be subsequently used for their growth. Therefore, electricity is produced less after inoculation at MFC. Park et al. (2017) reported pre-acclimation strategies on microbial community, which is beneficial as the lag phase becomes shortened.

Since there are no specific nutrients or medium which exist to obtain high power densities with variety of microorganisms in MFCs, medium design is of utmost importance for all biochemical engineers. It is indeed accepted that to perform the biochemical process, every microbe requires varying nutrients according to its need. Usually microorganisms receive energy from available carbon and nitrogen sources in MFC (Santoro et al. 2017, Uday et al. 2016). The available carbon and nitrogen sources influence the bioelectricity production, and this is controlled by different parameters since the biochemical processes are complex in nature. Metabolic pathway for utilization of nutrients for electricity generation is dependent on types of organisms along with type of substrate used. In this regard, several researchers experimented with optimization of substrate with microorganisms and reported in various publications. Chaudhary et al. (2016) studied the available carbon and nitrogen sources in single-chambered fuel cell to get high voltage generation. Although simulated media was used for this experiment, the changing one-variable-at-time approach was used to find out best carbon and nitrogen sources to obtain voltage in MFC. It was evident from their report that the level of effect of various sources significantly influences the electricity generation (Choudhury et al. 2016). It is obvious that not only medium size, but also operating conditions, such as temperature and size of inoculums have significant influence on electricity production of MFC because of drastic change in metabolic pathway (Çalık et al. 2001). It has been found that the percentage of inoculums has a key role in the process of electricity generation. Like temperature, pH of the medium affects bioprocesses by interfering transfer mechanism through the cell membrane. Since pH alters balances in acid-base, it inhibits the microbial growth (Moon & Parulekar 1991). The rate of circulation of feed also facilitates effective mass transfer during operation at MFC, as it influences extent of mixing in MFCs and effective distribution of nutrients. Fed batch operation is more suitable for sustainable power generation (Choudhury et al. 2017). Few researchers used a different buffer in order to maintain the pH of medium in fuel cells, as pH of anolyte controls the process of power generation in MFCs (Ye et al. 2016). On the contrary, few researchers demonstrated that it is unrealistic because of their operation cost.

Similar to the anode chamber, cathode chamber also influences the overall performance of microbial fuel cells (Kumar et al. 2017). They reported that the buffering of catholyte offered stability to the fuel cell system with less internal resistance. The porous structure of electrode, electrode material, surface area, and flow of electron from bacteria to electrodes are the significant parameters on which the overall performance of MFC is dependent. Platinum-based bicathodes were examined by different researchers to develop high power input through improved catalytic activities. On the other hand, the Pt-based cathodes are expensive (Huang et al. 2011). Several attempts were made to bacteria with metal nanoparticles to enhance the mechanism of flow of electrons from microorganism to anode. Many experiments were performed on the nickel (Ni) NP-dispersed carbon nanofiber (CNF), which was grown over an activated carbon fiber substrate. The nanofiber was fabricated as electrodes of MFC directly with no post treatment for bio energy production (Singh and Verma 2015). Recently, the nitrogen (N)-doped CNF/CNT-based electrode material was implemented to improve the output of MFCs (Yang et al. 2014b). The spacing between electrodes in fuel cell is an important factor to obtain high performance MFCs through continued electron transfer mechanism. It has been observed that power output can be enlarged by minimizing the spacing in the electrodes, because there is reduction in ohmic losses. Less spacing facilitates improved transfer of electrons, but electrodes with very close space have shown reduced output of MFCs, owing to cross from the cathode to the anode (Ren et al. 2014). The ion exchange membrane is used to separate anode and cathode chambers. The membrane permits flow of protons through it to the cathode electrode. Use of IEM delays transfer of protons from anode to cathode chamber (Choudhury et al. 2017), as it offers significant internal resistance. Ion Exchange membrane decreases diffusion of oxygen and substrate, which causes increase in catalytic activity of microbes. Various anode materials, such as carbon felt, carbon cloth, graphite felt, and carbon mesh are commonly used due to high power output achieved. In recent years, highest power density of 4200 mW/m^3 in MFC could be achieved (Sharma and Li 2010). Although high energy production is possible with MFC, it is not feasible to use this system on a large scale. The maintenance of MFC, which is an energy recovery device, and durability of system are necessary to implement this technology on a large scale application. If MFC is included in wastewater treatment plant, it will act as an energy producer and thus reduce extra cost for aeration and less sludge production. Attention must be given to suitable electrode material and design of MFC to minimize the expenditure to make the system cost-effective. It is difficult to obtain high quality effluents as well as energy production through a single unit simultaneously. MFC possesses some drawbacks; it is accessible for remote power generation. Thus, there is a need to make use of this largest growing technology on a large scale to generate power along with treatment of wastewater.

Conclusion

MFC is a bioremediation-technique that emphasizes two vital problems the world is facing nowadays—non availability of clean water and the growing energy demand. Microbial fuel cell has become popular due to having dual advantages of good wastewater technology and energy recovery, rather than energy consumption for its operation. MFCs are promising candidates as they utilize exoelectrogen to remove pollutants from the wastes along with production of electricity. In the upcoming years, development of MFCs to generate electricity will be restricted by the pH, design of MFCs, conductivity of solution, efficiency, and cost of electrode material. To make MFCs an efficient source, it is important to study the kinetics of electron transfer and develop the electrode surfaces to achieve better colonization, which leads to producing maximum power output. The modifications in the surface properties of electrodes in fuel cell system are challenging to make fuel cells more efficient and sustainable. Furthermore, increasing anode surface area in fuel cell system influences the overall performance of microbial fuel cells. The MFCs operated with air cathode automatically reduce the energy demand required for aeration. MFC is a good technology of wastewater treatment process as well as power generation; moreover it needs lot of modifications to use this technology practically. The cathode material is the main parameter which limits this approach to be feasible in large scale applications. Hence, the adaptation of effective electrode material has demonstrated an effective way of improving the performance of MFC by producing high power output. It is possible one day that MFC will be the only source of power production when cathodic limitations in MFC are overcome. It may be possible to make MFCs sustainable through energy recovery from wastewater and biomass. The IEM are expensive and increase the total cost of the process. So there is a need to find out another inexpensive separator for maximum power generation by using cost-effective MFC technology in the future. Therefore, power generation can be accelerated by utilizing microbes (exoelectrogens), MFC design, surface and electrode material as well as different process parameters which would improve the usability of this technology in the coming years. With the proper use of electrode material, modified MFC configuration and suitable microbial strain would definitely produce high power densities. Thus, MFC can be cost-effective and a sustainable energy source.

REFERENCES

Ahn, Y., I. Ivanov, T.C. Nagaiah, A. Bordoloi and B.E. Logan. 2014. Mesoporous nitrogen-rich carbon materials as cathode catalysts in microbial fuel cells. Journal of Power Sources 269: 212-215.

Aldrovandi, A., E. Marsili, L. Stante, P. Paganin, S. Tabacchioni and A. Giordano. 2009. Sustainable power production in a membrane-less and mediator-less synthetic wastewater microbial fuel cell. Bioresource Technology 100: 3252-3260.

Barsky, E., G. Dolnikova, Y.V. Savanina and E. Lobakova. 2014. Generation of electric potential difference across the electrodes of the microbial fuel cell in the anaerobic oxidation of substrates by microbial associations. Moscow University Biological Sciences Bulletin 69: 113-117.

Beliaev, A.S., D.A. Saffarini, J.L. McLaughlin and D. Hunnicutt. 2001. MtrC, an outer membrane decahaem c cytochrome required for metal reduction in Shewanella putrefaciens MR-1. Molecular Microbiology 39: 722-730.

Blanco, J., Q. Nguyen and P. Schaetzel. 2001. Novel hydrophilic membrane materials: sulfonated polyethersulfone Cardo. Journal of Membrane Science 186: 267-279.

Bond, D.R., D.E. Holmes, L.M. Tender and D.R. Lovley. 2002. Electrode-reducing microorganisms that harvest energy from marine sediments. Science 295: 483-485.

Burkitt, R., T. Whiffen and E.H. Yu. 2016. Iron phthalocyanine and MnOx composite catalysts for microbial fuel cell applications. Applied Catalysis B: Environmental 181: 279-288.

Çalık, P., G. Çalık and T.H. Özdamar. 2001. Bioprocess development for serine alkaline protease production: a review. Reviews in Chemical Engineering 17: 1-62.

Chaudhuri, S.K. and D.R. Lovley. 2003. Electricity generation by direct oxidation of glucose in mediatorless microbial fuel cells. Nature Biotechnology 21: 1229.

Choudhury, P., U.S. Prasad Uday, T.K. Bandyopadhyay, R.N. Ray and B. Bhunia. 2017. Performance improvement of microbial fuel cell (MFC) using suitable electrode and bioengineered organisms: a review. Bioengineered 8: 471-487.

Choudhury, P., R.N. Roy and T.K. Bandyopadhyaya. 2016. To study the effect of process parameters on power generation from waste waters containing single chamber microbial fuel cell. Proceeding of International Conference of Biotechnology and Environment, Alexandria, Egypt. pp. 62-73.

Fan, Y., H. Hu and H. Liu. 2007. Enhanced Coulombic efficiency and power density of air-cathode microbial fuel cells with an improved cell configuration. Journal of Power Sources 171: 348-354.

Feng, Y., Q. Yang, X. Wang and B.E. Logan. 2010. Treatment of carbon fiber brush anodes for improving power generation in air–cathode microbial fuel cells. Journal of Power Sources 195: 1841-1844.

Gajda, I., J. Greenman, C. Melhuish, C. Santoro and I. Ieropoulos. 2016. Microbial Fuel Cell-driven caustic potash production from wastewater for carbon sequestration. Bioresource Technology 215: 285-289.

Ganesh, K. and J.R. Jambeck. 2013. Treatment of landfill leachate using microbial fuel cells: alternative anodes and semi-continuous operation. Bioresource Technology 139: 383-387.

Ghasemi, M., S. Shahgaldi, M. Ismail, B.H. Kim, Z. Yaakob and W.R.W. Daud. 2011. Activated carbon nanofibers as an alternative cathode catalyst to platinum in a two-chamber microbial fuel cell. International Journal of Hydrogen Energy 36: 13746-13752.

Gorby, Y.A., S. Yanina, J.S. McLean, K.M. Rosso, D. Moyles, A. Dohnalkova, T.J. Beveridge, I.S. Chang, B.H. Kim and K.S. Kim. 2006. Electrically conductive bacterial nanowires produced by Shewanella oneidensis strain MR-1 and other microorganisms. Proceedings of the National Academy of Sciences 103: 11358-11363.

Ha, P.T., T.K. Lee, B.E. Rittmann, J. Park and I.S. Chang. 2012. Treatment of alcohol distillery wastewater using a Bacteroidetes-dominant thermophilic microbial fuel cell. Environmental Science & Technology 46: 3022-3030.

HaoYu, E., S. Cheng, K. Scott and B. Logan. 2007. Microbial fuel cell performance with non-Pt cathode catalysts. Journal of Power Sources 171: 275-281.

He, L., P. Du, Y. Chen, H. Lu, X. Cheng, B. Chang and Z. Wang. 2017. Advances in microbial fuel cells for wastewater treatment. Renewable and Sustainable Energy Reviews 71: 388-403.

Hernández-Fernández, F., A.P. de los Ríos, F. Mateo-Ramírez, C. Godínez, L. Lozano-Blanco, J. Moreno and F. Tomás-Alonso. 2015. New application of supported ionic liquids membranes as proton exchange membranes in microbial fuel cell for waste water treatment. Chemical Engineering Journal 279: 115-119.

Hernandez, M.E., A. Kappler and D.K. Newman. 2004. Phenazines and other redox-active antibiotics promote microbial mineral reduction. Applied and Environmental Microbiology 70: 921-928.

Hou, Y., H. Yuan, Z. Wen, S. Cui, X. Guo, Z. He and J. Chen. 2016. Nitrogen-doped graphene/CoNi alloy encased within bamboo-like carbon nanotube hybrids as cathode catalysts in microbial fuel cells. Journal of Power Sources 307: 561-568.

Huang, L., J.M. Regan and X. Quan. 2011. Electron transfer mechanisms, new applications, and performance of biocathode microbial fuel cells. Bioresource Technology 102: 316-323.

Ivanova, E.P., T. Sawabe, N.M. Gorshkova, V.I. Svetashev, V.V. Mikhailov, D.V. Nicolau and R. Christen. 2001. Shewanella japonica sp. nov. International Journal of Systematic and Evolutionary Microbiology 51: 1027-1033.

Jiang, B., T. Muddemann, U. Kunz, H. Bormann, M. Niedermeiser, D. Haupt, O. Schläfer and M. Sievers. 2017. Evaluation of microbial fuel cells with graphite plus MnO_2 and MoS_2 paints as oxygen reduction cathode catalyst. Journal of the Electrochemical Society 164: H3083-H3090.

Kim, B.H., H. Park, H. Kim, G. Kim, I. Chang, J. Lee and N. Phung. 2004. Enrichment of microbial community generating electricity using a fuel-cell-type electrochemical cell. Applied Microbiology and Biotechnology 63: 672-681.

Kodali, M., R. Gokhale, C. Santoro, A. Serov, K. Artyushkova and P. Atanassov. 2017. High performance platinum group metal-free cathode catalysts for microbial fuel cell (MFC). Journal of the Electrochemical Society 164: H3041-H3046.

Kondaveeti, S., J. Lee, R. Kakarla, H.S. Kim and B. Min. 2014. Low-cost separators for enhanced power production and field application of microbial fuel cells (MFCs). Electrochimica Acta 132: 434-440.

Kumar, S.S., S. Basu and N.R. Bishnoi. 2017. Effect of cathode environment on bioelectricity generation using a novel consortium in anode side of a microbial fuel cell. Biochemical Engineering Journal 121: 17-24.

Ledezma, P., J. Greenman and I. Ieropoulos. 2013. MFC-cascade stacks maximise COD reduction and avoid voltage reversal under adverse conditions. Bioresource Technology 134: 158-165.

Liao, Q., J. Zhang, J. Li, D. Ye, X. Zhu and B. Zhang. 2015. Increased performance of a tubular microbial fuel cell with a rotating carbon-brush anode. Biosensors and Bioelectronics 63: 558-561.

Liu, H., S. Cheng and B.E. Logan. 2005. Production of electricity from acetate or butyrate using a single-chamber microbial fuel cell. Environmental Science & Technology 39: 658-662.

Liu, H. and B.E. Logan. 2004. Electricity generation using an air-cathode single chamber microbial fuel cell in the presence and absence of a proton exchange membrane. Environmental Science & Technology 38: 4040-4046.

Logan, B.E. 2004. Peer reviewed: extracting hydrogen and electricity from renewable resources. ACS Publications. Environ Sci Technol. 38: 160A-167A.

Logan, B.E. 2009. Exoelectrogenic bacteria that power microbial fuel cells. Nature Reviews Microbiology 7: 375.

Logan, B.E. and J.M. Regan. 2006. Electricity-producing bacterial communities in microbial fuel cells. Trends in Microbiology 14: 512-518.

Lovley, D.R. 2008. The microbe electric: conversion of organic matter to electricity. Current opinion in Biotechnology 19: 564-571.

Lv, Z., D. Xie, X. Yue, C. Feng and C. Wei. 2012. Ruthenium oxide-coated carbon felt electrode: a highly active anode for microbial fuel cell applications. Journal of Power Sources 210: 26-31.

Mandal, S., B. Bhunia, A. Kumar, D. Dasgupta, T. Mandal, S. Datta and P. Bhattacharya. 2013. A statistical approach for optimization of media components for phenol degradation by Alcaligenes faecalis using Plackett–Burman and response surface methodology. Desalination and Water Treatment 51: 6058-6069.

Marcus, A.K., C.I. Torres and B.E. Rittmann. 2010. Evaluating the impacts of migration in the biofilm anode using the model PCBIOFILM. Electrochimica Acta 55: 6964-6972.

Min, B., J. Kim, S. Oh, J.M. Regan and B.E. Logan. 2005. Electricity generation from swine wastewater using microbial fuel cells. Water Research 39: 4961-4968.

Min, B., Ó.B. Román and I. Angelidaki. 2008. Importance of temperature and anodic medium composition on microbial fuel cell (MFC) performance. Biotechnology Letters 30: 1213-1218.

Miran, W., J. Jang, M. Nawaz, A. Shahzad, S.E. Jeong, C.O. Jeon and D.S. Lee. 2017. Mixed sulfate-reducing bacteria-enriched microbial fuel cells for the treatment of wastewater containing copper. Chemosphere 189: 134-142.

Mook, W., M. Chakrabarti, M. Aroua, G. Khan, B. Ali, M. Islam and M.A. Hassan. 2012. Removal of total ammonia nitrogen (TAN), nitrate and total organic carbon

(TOC) from aquaculture wastewater using electrochemical technology: a review. Desalination 285: 1-13.

Moon, S.H. and S.J. Parulekar. 1991. A parametric study ot protease production in batch and fed-batch cultures of Bacillus firmus. Biotechnology and Bioengineering 37: 467-483.

Myers, C.R. and J.M. Myers. 1992. Localization of cytochromes to the outer membrane of anaerobically grown Shewanella putrefaciens MR-1. Journal of Bacteriology 174: 3429-3438.

Newman, D.K. and R. Kolter. 2000. A role for excreted quinones in extracellular electron transfer. Nature 405: 94.

Nimje, V.R., C.-Y. Chen, H.-R. Chen, C.-C. Chen, Y.M. Huang, M.-J. Tseng, K.-C. Cheng and Y.-F. Chang. 2012. Comparative bioelectricity production from various wastewaters in microbial fuel cells using mixed cultures and a pure strain of Shewanella oneidensis. Bioresource Technology 104: 315-323.

Pant, D., G. Van Bogaert, L. Diels and K. Vanbroekhoven. 2010. A review of the substrates used in microbial fuel cells (MFCs) for sustainable energy production. Bioresource Technology 101: 1533-1543.

Park, D. and J. Zeikus. 1999. Utilization of electrically reduced neutral Red by Actinobacillus succinogenes: physiological function of neutral Red in membrane-driven fumarate reduction and energy conservation. Journal of Bacteriology 181: 2403-2410.

Park, D.H. and J.G. Zeikus. 2003. Improved fuel cell and electrode designs for producing electricity from microbial degradation. Biotechnology and Bioengineering 81: 348-355.

Park, Y., H. Cho, J. Yu, B. Min, H.S. Kim, B.G. Kim and T. Lee. 2017. Response of microbial community structure to pre-acclimation strategies in microbial fuel cells for domestic wastewater treatment. Bioresource Technology 233: 176-183.

Patil, S.A., C. Hägerhäll and L. Gorton. 2012. Electron transfer mechanisms between microorganisms and electrodes in bioelectrochemical systems. pp. 71-129. *In:* Frank-Michael Matysik (ed.). Advances in Chemical Bioanalysis. Springer International Publishing, New York City, NY.

Patil, S.A., V.P. Surakasi, S. Koul, S. Ijmulwar, A. Vivek, Y. Shouche and B. Kapadnis. 2009. Electricity generation using chocolate industry wastewater and its treatment in activated sludge based microbial fuel cell and analysis of developed microbial community in the anode chamber. Bioresource Technology 100: 5132-5139.

Qin, M., E.A. Hynes, I.M. Abu-Reesh and Z. He. 2017. Ammonium removal from synthetic wastewater promoted by current generation and water flux in an osmotic microbial fuel cell. Journal of Cleaner Production 149: 856-862.

Rabaey, K., N. Boon, M. Höfte and W. Verstraete. 2005. Microbial phenazine production enhances electron transfer in biofuel cells. Environmental Science & Technology 39: 3401-3408.

Rabaey, K., N. Boon, S.D. Siciliano, M. Verhaege and W. Verstraete. 2004. Biofuel cells select for microbial consortia that self-mediate electron transfer. Applied and Environmental Microbiology 70: 5373-5382.

Ren, L., Y. Ahn and B.E. Logan. 2014. A two-stage microbial fuel cell and anaerobic fluidized bed membrane bioreactor (MFC-AFMBR) system for effective domestic wastewater treatment. Environmental Science & Technology 48: 4199-4206.

Rittmann, B.E., R. Krajmalnik-Brown and R.U. Halden. 2008. Pre-genomic, genomic and post-genomic study of microbial communities involved in bioenergy. Nature Reviews Microbiology 6: 604.

Saba, B., A.D. Christy and Z. Yu. 2017. Sustainable power generation from bacterio-algal microbial fuel cells (MFCs): an overview. Renewable and Sustainable Energy Reviews 73: 75-84.

Salgado, C.A. 2009. Microbial fuel cells powered by Geobacter sulfurreducens. Warning: get_class expects parameter 1 to be object, array given in/home/vhosts/ejournal/user-dir/htdocs/classes/cache/GenericCache. inc. php on line 63 MMG 445 Basic Biotechnology eJournal 5: 96-101.

Santoro, C., C. Arbizzani, B. Erable and I. Ieropoulos. 2017. Microbial fuel cells: From fundamentals to applications: a review. Journal of Power Sources 356: 225-244.

Santoro, C., K. Artyushkova, S. Babanova, P. Atanassov, I. Ieropoulos, M. Grattieri, P. Cristiani, S. Trasatti, B. Li and A.J. Schuler. 2014. Parameters characterization and optimization of activated carbon (AC) cathodes for microbial fuel cell application. Bioresource Technology 163: 54-63.

Santoro, C., A. Stadlhofer, V. Hacker, G. Squadrito, U. Schröder and B. Li. 2013. Activated carbon nanofibers (ACNF) as cathode for single chamber microbial fuel cells (SCMFCs). Journal of Power Sources 243: 499-507.

Sharma, Y. and B. Li. 2010. Optimizing energy harvest in wastewater treatment by combining anaerobic hydrogen producing biofermentor (HPB) and microbial fuel cell (MFC). International Journal of Hydrogen Energy 35: 3789-3797.

Singh, S. and N. Verma. 2015. Fabrication of Ni nanoparticles-dispersed carbon micro-nanofibers as the electrodes of a microbial fuel cell for bio-energy production. International Journal of Hydrogen Energy 40: 1145-1153.

Smitha, B., S. Sridhar and A. Khan. 2005. Solid polymer electrolyte membranes for fuel cell applications—a review. Journal of Membrane Science 259: 10-26.

Sun, H., S. Xu, G. Zhuang and X. Zhuang. 2016. Performance and recent improvement in microbial fuel cells for simultaneous carbon and nitrogen removal: a review. Journal of Environmental Sciences 39: 242-248.

Turick, C.E., L.S. Tisa and F. Caccavo Jr. 2002. Melanin production and use as a soluble electron shuttle for Fe (III) oxide reduction and as a terminal electron acceptor by Shewanella algae BrY. Applied and Environmental Microbiology 68: 2436-2444.

Uday, U.S.P., T.K. Bandyopadhyay and B. Bhunia. 2016. Bioremediation and detoxification technology for treatment of dye(s) from textile effluent. *In*: Textile Wastewater Treatment. IntechOpen Limited, London, UK.

Velasquez-Orta, S., I. Head, T. Curtis and K. Scott. 2011. Factors affecting current production in microbial fuel cells using different industrial wastewaters. Bioresource Technology 102: 5105-5112.

Von Canstein, H., J. Ogawa, S. Shimizu and J.R. Lloyd. 2008. Secretion of flavins by Shewanella species and their role in extracellular electron transfer. Applied and Environmental Microbiology 74: 615-623.

Wang, H., G. Wang, Y. Ling, F. Qian, Y. Song, X. Lu, S. Chen, Y. Tong and Y. Li. 2013. High power density microbial fuel cell with flexible 3D graphene-nickel foam as anode. Nanoscale 5: 10283-10290.

Wang, S., L. Huang, L. Gan, X. Quan, N. Li, G. Chen, L. Lu, D. Xing and F. Yang. 2012. Combined effects of enrichment procedure and non-fermentable or fermentable co-substrate on performance and bacterial community for pentachlorophenol degradation in microbial fuel cells. Bioresource Technology 120: 120-126.

Wang, Y.-K., G.-P. Sheng, W.-W. Li, Y.-X. Huang, Y.-Y. Yu, R.J. Zeng and H.-Q. Yu. 2011. Development of a novel bioelectrochemical membrane reactor for wastewater treatment. Environmental Science & Technology 45: 9256-9261.

Wen, Q., S. Wang, J. Yan, L. Cong, Y. Chen and H. Xi. 2014. Porous nitrogen-doped carbon nanosheet on graphene as metal-free catalyst for oxygen reduction reaction in air-cathode microbial fuel cells. Bioelectrochemistry 95: 23-28.

Winfield, J., L.D. Chambers, J. Rossiter, J. Greenman and I. Ieropoulos. 2015. Urine-activated origami microbial fuel cells to signal proof of life. Journal of Materials Chemistry A 3: 7058-7065.

Winfield, J., L.D. Chambers, J. Rossiter and I. Ieropoulos. 2013a. Comparing the short and long term stability of biodegradable, ceramic and cation exchange membranes in microbial fuel cells. Bioresource Technology 148: 480-486.

Winfield, J., I. Ieropoulos, J. Rossiter, J. Greenman and D. Patton. 2013b. Biodegradation and proton exchange using natural rubber in microbial fuel cells. Biodegradation 24: 733-739.

Xie, X., G. Yu, N. Liu, Z. Bao, C.S. Criddle and Y. Cui. 2012. Graphene-sponges as high-performance low-cost anodes for microbial fuel cells. Energy & Environmental Science 5: 6862-6866.

Yang, X., S. Zhu, Y. Dou, Y. Zhuo, Y. Luo and Y. Feng. 2014a. Novel and remarkable enhanced-fluorescence system based on gold nanoclusters for detection of tetracycline. Talanta 122: 36-42.

Yang, X., W. Zou, Y. Su, Y. Zhu, H. Jiang, J. Shen and C. Li. 2014b. Activated nitrogen-doped carbon nanofibers with hierarchical pore as efficient oxygen reduction reaction catalyst for microbial fuel cells. Journal of Power Sources 266: 36-42.

Yasri, N.G. and G. Nakhla. 2017. The performance of 3-D graphite doped anodes in microbial electrolysis cells. Journal of Power Sources 342: 579-588.

Ye, Y., X. Zhu and B.E. Logan. 2016. Effect of buffer charge on performance of air-cathodes used in microbial fuel cells. Electrochimica Acta 194: 441-447.

Yi, H., K.P. Nevin, B.-C. Kim, A.E. Franks, A. Klimes, L.M. Tender and D.R. Lovley. 2009. Selection of a variant of Geobacter sulfurreducens with enhanced capacity for current production in microbial fuel cells. Biosensors and Bioelectronics 24: 3498-3503.

Yuan, H. and Z. He. 2015. Graphene-modified electrodes for enhancing the performance of microbial fuel cells. Nanoscale 7: 7022-7029.

Yuan, H., Y. Hou, I.M. Abu-Reesh, J. Chen and Z. He. 2016. Oxygen reduction reaction catalysts used in microbial fuel cells for energy-efficient wastewater treatment: a review. Materials Horizons 3: 382-401.

Zhang, X., S. Cheng, X. Huang and B.E. Logan. 2010. The use of nylon and glass fiber filter separators with different pore sizes in air-cathode single-chamber microbial fuel cells. Energy & Environmental Science 3: 659-664.

Zhang, X., X. Xia, I. Ivanov, X. Huang and B.E. Logan. 2014. Enhanced activated carbon cathode performance for microbial fuel cell by blending carbon black. Environmental Science & Technology 48: 2075-2081.

Zhao, F., F. Harnisch, U. Schröder, F. Scholz, P. Bogdanoff and I. Herrmann. 2006. Challenges and constraints of using oxygen cathodes in microbial fuel cells. Environmental Science & Technology 40: 5193-5199.

Zhuang, L., S. Zhou, Y. Wang, C. Liu and S. Geng. 2009. Membrane-less cloth cathode assembly (CCA) for scalable microbial fuel cells. Biosensors and Bioelectronics 24: 3652-3656.

Zinadini, S., A. Zinatizadeh, M. Rahimi, V. Vatanpour and Z. Rahimi. 2017. High power generation and COD removal in a microbial fuel cell operated by a novel sulfonated PES/PES blend proton exchange membrane. Energy 125: 427-438.

Zuo, Y., D. Xing, J.M. Regan and B.E. Logan. 2008. Isolation of the exoelectrogenic bacterium Ochrobactrum anthropi YZ-1 by using a U-tube microbial fuel cell. Applied and Environmental Microbiology 74: 3130-3137.

Cyanobacteria: A Biocatalyst in Microbial Fuel Cell for Sustainable Electricity Generation

Thingujam Indrama[1], O.N. Tiwari[2], Tarun Kanti Bandyopadhyay[3], Abhijit Mondal[3] and Biswanath Bhunia[4]*

[1] DBT-Institute of Bioresources and Sustainable Development, Imphal - 795001, Manipur, India

[2] Centre for Conservation and Utilisation of Blue Green Algae, Division of Microbiology, ICAR-Indian Agricultural Research Institute, New Delhi - 110012, India

[3] Department of Chemical Engineering, National Institute of Technology Agartala, Agartala - 799046, Tripura, India

[4] Department of Bio Engineering, National Institute of Technology Agartala, Agartala - 799046, Tripura, India

Introduction

Climate change has been found to be the greatest environmental threat, and is therefore considered to be the biggest challenge. It is caused by the build-up of anthropogenic CO_2 emissions in combination with water vapor and other gases (Doherty et al. 2009). Additionally, energy security is becoming the most challenging area due to unstable oil prices, exhaustion of resources for fossil fuels, as well as geopolitical tensions. It is urged for all researchers to find out the concept of sustainable energy and, therefore, it is considered to be the most important issue in the 21st century (Balzani et al. 2008). One solution to overcome the biggest challenge might be utilization of natural photosynthetic organisms, which could convert solar energy to chemical energy. The above hypothesis can create a huge scope to capture energy from renewable sources in an alternative way (Cho et al. 2008). Accordingly, bioenergy derived from photosynthetic organisms will contribute a considerable share of sustainable renewable energy in the near future (Ragauskas et al. 2006). Although lots of research has been carried out with

*Corresponding author: bbhunia@gmail.com

photosynthetic organisms, it is limited to production of biofuels (Gouveia and Oliveira 2009), as microalgae (including algae and cyanobacteria) is considered a more promising biomass for production of biofuels.

Apart from biofuel production, microalgae is also utilised for production of electrical energy using photosynthetic-microbial fuel cell (p-MFC). Although, non-photosynthetic microbes (i.e., bacteria) are traditionally used in microbial fuel cells (MFCs) for production of electrical energy through exploration of metabolic activity through conversion of chemically bonded energy inside the organic and inorganic matter in electrical energy. However, photosynthetic-microbial fuel cell (p-MFC) is becoming more popular in recent years due to carbon neutral behavior of microalgae (Du et al. 2007). It is also reported that microalgae even produce electricity from waste biomass and, therefore, the technology related to bioremediation along with value addition using microalgae has led to a dramatic increase of interest in recent years (Bond et al. 2002). Hence, the process will be more economical if MFC technology were to be used together with that for waste remediation and renewable energy generation by using microalgae. MFC usually contains two chambers dedicated for anode and cathode chambers. They are separated by a membrane which will facilitate the exchange of the proton (PEM). Anaerobic environment is maintained at the anodic chamber, which consists of high surface area required for high growth of microbial biofilm.

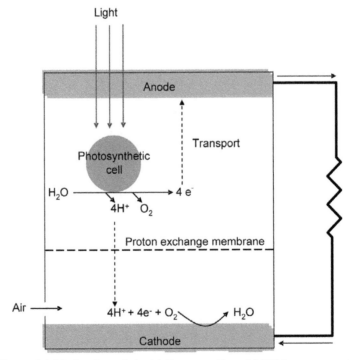

Figure 8.1. Mechanism of electricity generation at MFC in presence of photosynthetic microorganism.

The catalyst used in MFC catabolises organic complex molecules which are present in the medium and, subsequently, electrons are generated. These electrons can be transferred to the anode from the bacteria, where anode acts as the terminal electron acceptor in the redox chain. Additionally, protons generated due to metabolic activity diffuse through PEM and are reduced to water at the cathode in the presence of oxygen (Lovley 2006). There are several reports indicating the higher efficiency of MFCs in the presence of bacteria as catalyst; however, net production of carbon dioxide due to bacterial respiration is one drawback with regard to environmental concern (Lovley 2006). Therefore, researchers are looking for new catalysts which will be able to generate electricity without emission of carbon dioxide during their utilization at MFC. Although, a few researches have been carried out using photosynthetic microbes, and a satisfactory output was noticed. However, the same principles were used or existing principles with slight modifications were used for MFC containing bacteria. A schematic representation of photosynthetic microbial fuel cell (p-MFC) is illustrated in Fig. 8.1.

In p-MFC, microorganisms are able to generate electrons through both catabolism and photosynthesis process (Rosenbaum and Schröder 2010, Strik et al. 2008). It is obvious that p-MFC does not require additional feed of organic substrates as photosynthetic microorganisms are able to synthesize carbohydrates in the presence of light. The above scientific advantage makes the p-MFC device more attractive and might be a self-sustainable tool to create energy. Additionally, photosynthetic microorganisms can be explored for the production of carbon neutral energy due to their carbon sequestration properties.

Various photosynthetic materials used in the anodic chamber of p-MFC are classified into four groups. They are: sub-cellular, prokaryotic (cyanobacteria), and eukaryotic (algae), as well as mixed systems. In the sub-cellular organelles used so far are thylakoid membranes (Lam et al. 2006), photosystems (PS1 and PS2) (Badura et al. 2006), bacterial reaction centers (Janzen and Seibert 1980), and isolated chlorophyll and its derivatives (Amao and Komori 2004). Since organelles cannot self-repair, therefore, the lifetime of organelles present in p-MFCs is short. Owing to their disadvantage with regard to lifetimes of organelles, the whole cell is highly recommended in p-MFC for energy generation (Malik et al. 2009).

Source of Electrons in a p-MFC

Photosynthetic microorganisms have the capability to produce reductive electrons via the catabolism of complex carbon. Additionally, they can also contribute extra electrons for power generation during photosynthesis (Park and Zeikus 2000). Although two mechanisms are cleared for contributing electron for electricity generation in MFC, the relative contributions of each process are still not clear. It is well understood that two types of catabolism and photosynthesis, such as oxygenic and anoxygenic exist in eukaryotic and prokaryotic organisms. However, cyanobacteria usually follow oxygenic route; therefore, only oxygenic systems are discussed.

Photosynthesis

Cyanobacteria contain mainly two pigments. They are carotenoids and chlorophylls, which are considered antennae for light harvesting. These two antennae can absorb light and are considered centers for photochemical reaction (Colyer et al. 2005). Chloroplast is an organelle where photosynthesis reaction takes place (Fig. 8.2). Photosynthesis happens through light dependent reaction and light independent reaction. In light dependent reaction, adenosine diphosphate (ADP) and inorganic phosphate (P) are used for production of adenosine triphosphate (ATP). However, carbon dioxide is reduced to carbohydrate in light independent reaction (Kruse et al. 2005). In light dependent reactions, the light energy is adsorbed by the phycobilisome and transferred to PSII reaction centers unidirectionally (Arteni et al. 2009). Both photosystems, namely PSI and PSII, exist in cyanobacteria. The phycobilisome (PBS) is a protein complex which is acting as antenna for PSII for cyanobacteria. The PBS comprises rod and core cylinders, which are connected with various protein. They are called phycobilin-binding proteins, which are also linked with various colorless proteins. Generally, phycocyanin (PhC) and allophycocyanin (APC) are the major phycobiliprotein found in the rod and core sub-complex, respectively. The linker protein available between rod–core cylinder is cyanobacterial phycocyanin protein G (CpcG), which plays a crucial role during assembly of the PBS (Watanabe et al. 2014). The reaction center of PSI is composed of a dimer of chlorophylls, which has an absorption peak at 700 nm. However, chlorophyll-b, the only primary pigment, remains at the center of PSII.

Chlorophyll-b shows the absorption maximum at 680 nm. It is already established that both centers are boosted to higher energy levels during adsorption of light energy. However, it is further emitted and passed to electron carriers. It is obviously true that passed energy is reduced due to splitting of water, and electron is passed towards potential gradient through redox active species (Kruse et al. 2005). Therefore, the energy is released in the form of ATP, and release of protons takes place from the stroma into the thylakoid lumen due to pH gradient (Kruse et al. 2005). As the protons are diffused against the concentration gradient, they flow with ATP synthetase (Kruse et al. 2005). Cyclic phosphorylation can take place if large amounts of NADPH exist and subsequently, electrons drive from the electron transport chain to PSI. The above electron from PSI is further passed to PSII. However, during formation of ATP, the electrons return to PSI and make PSII redundant. Depending on conditions, electrons can reduce protons to molecular hydrogen or reduce oxygen to water (Maly et al. 2005). The stroma of chloroplasts is the place where light independent reactions occur; however, previously generated ATP for energy is required. In the Calvin cycle, rubisco catalysed the combination process between carbon dioxide and sugar ribulose-1,5- bisphosphate RuBP (Kruse et al. 2005).

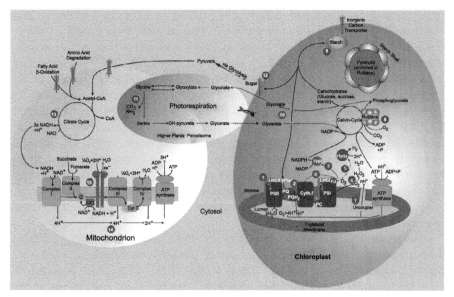

Figure 8.2. Biochemical pathways related to photon conversion efficiency in cyanobacteria (Kruse et al. 2005).

Respiration

In respiration, carbohydrate is broken down and formation of water and carbon dioxide takes place. In this process, energy is released in the form of ATP, which is measured as the general currency of energy in biological system (Gonzalez-Meler et al. 2004). Glycolysis, citrate cycle, and oxidative phosphorylation are considered aerobic respiration process. Glycolysis takes place at cytoplasm where sugar is converted to pyruvic acid and reduces NAD (NADH). In anaerobic respiration, pyruvic acid enters the mitochondrion and converts to acetyl CoA, along with further reduced NAD (NADH). Subsequently, reduced NAD (NADH) and reduced flavin adenine dinucleotide FAD (FADH2), along with ATP, are produced during the citrate pathway. In electron transport chain, ATP is finally produced from the reduced species NADH and FADH2. Therefore, electrons flow down from low to high redox potential during oxidative phosphorylation and ATP is produced from proton gradients (Mitchell 1961). This proton gradient creates the potential difference between the electron donor and the electron acceptor, which is turned into a process for the generation of energy. It is to be noted that although there are several electron acceptors used by different microbes; however, electron acceptor stimuluses energy for growth of bacteria if it carries the highest redox potential (Madigan 2000).

Cell to Anode Electron Transfer

The electrons are required to be drawn off and need to be deposited at the anode for generation of power. It is obvious that large amounts of reduced material will be produced in an MFC. Chiao et al. (2006) reported that electrons are collected either from NADPH or the electron transport chain. Since lots of possibilities are associated with electron transfer to anode, therefore, the efficiency of MFC varies a lot (Du et al. 2007, Lovley 2006). It is already reported that electrons can directly transfer to the anode via a conductive extracellular matrix involving pili and cytochrome c (cyt C) proteins (Reguera et al. 2005). In this process, microorganisms are required to grow on the anodic surface as a biofilm. Since huge complexity of biofilm structures exists, three-dimensional structures play a critical role in power generation. It is already established that electron transfer directly happens through cytochromes in the extracellular matrix (ECM) or through nanowires, which are bacterially conductive (Ryu et al. 2010, Tsujimura et al. 2001). It has been found that cyanobacteria transfer electron through direct transfer (Pisciotta et al. 2010).

Interestingly, the outer surface of most microorganisms contains non-conductive lipid membrane, which consists of peptidoglycans and lipopolysaccharides; therefore, they would not participate in electron transfer through direct transfer. However, during their growth on anode surface, they are able to generate power. In this case, microorganism secretes redox mediators which shuttle electrons to the anode (Davis and Higson 2007). It is not a compulsion that microorganisms need to be grown on anode surface; however, microorganisms can float on anode media if the redox mediator is dissolved in solution (Lovley 2006). In these systems, the source of electrons is supplied by cellular constituents, such as reduced form of NAD (NADH), which can act as a source of electrons (Yagishita et al. 1993). Although redox mediators may be natural or synthetic, however, redox mediators in synthetic form are normally used in p-MFCs to transfer electrons between the electrode and organism (Chiao et al. 2006, Rosenbaum and Schröder 2010). Since most of MFC operates in batch system, therefore, no redox mediators are used additionally (Rosenbaum and Schröder 2010). Alternatively immobilised redox mediator can be employed in the device for lifespan use. However, in that case, natural redox mediators are preferred (Bond and Lovley 2005). Since lots of complexity is associated with electron transfer, therefore, the mechanism for this case is not clear. It is hypothesised that mediators can cross the cell membrane. In cells, electron carriers will accept electrons. However, the reduced form usually exists in the cell and transfers to the electrode surface (Lovley 2006). Yagishita et al. (1997) reported that ferredoxin-NADP+ reductase and NAD(P)H dehydrogenase can couple with HNQ mediator in the dark condition.

Progress in MFC Research with Cyanobacteria for Sustainable Electricity Generation

Solar power contributed maximum capacity to renewable energy in 2013 (Sawin and Sverrisson 2014). Solar thermal power is dominating, as it contributes almost 70% to the solar power generation. However, the remaining 30% is almost completely provided by solar photovoltaic technologies. Electricity or biofuel derived from microbial photosynthetic organisms has been recognised as an emerging technology to contribute a substantial amount of renewable energy. Much more development is required before microbial fuel cells can significantly contribute (Hamelers et al. 2010). However, the research field on MFC is growing rapidly with an increasing diversity in its applications (Franks and Nevin 2010). Initially, several operational methods, such as exploiting soluble redox mediators, were abandoned for electrons transport from organisms to electrode. Presently, research is focused on electrode design as well as more suitable separator materials. Electrodes used in MFC are tailored towards a macroporous structure that supports cell adhesion and maximises the conductive interface between biofilm and electrode. Recently, carbon nanotube functionalised surfaces have been the predominant approach to enhance the electrode interface and also resulted in materials (Hussein et al. 2011, Yu et al. 2011). An alternative and less energy consuming strategy is osmium-based redox hydrogels containing co-immobilised biocatalysts (Boland et al. 2009). Such hydrogel coatings can be used to immobilize enzymes or whole cells, and act as an electron-mediating matrix, which effectively increases the conductive surface area. A further benefit from hydrogel matrices is the prolonged lifetime of exposed enzymes, which improves from several hours to days. The trend in electrode design is toward hierarchical structures, where a macroporous scaffold is coated in a mesoporous substrate for the functionalisation with nano-materials. Macroporous scaffolds utilise the compartment volume with as much surface area as possible, and simultaneously ensure a minimum spacing between the functionalised materials for sufficiently high flow rates through the electrode. Mesoporous coatings on these scaffolds expand the options for functionalisation of the surface, allow the embedding of biocatalysts, and further enlarge the surface area on the nanometre scale. Work on micrometer sized whole cells also benefits from surface modification on the nanometre scale, since it enables fine-tuning of the surface hydrophilicity and of the conductive interface. Direct electron transfer of the bacterium *Shewanella oneidensis* MR-1 was investigated on such a hierarchical electrode made from a fibrous carbon felt support, which was coated with carbon nanotube doped chitosan (Higgins et al. 2011). Carbon-based supports are generally the most applied macroporous structures, owing to the large variety of accessible geometries and comparatively simple modification routes, although metal foams have also been examined (Guan et al. 2012). Foam of reticulated vitreous carbon (RVC) is amongst the most popular macroporous supports, and exhibited promising results with a functional coating of carbon nanotubes in a chitosan matrix (Rincón et al. 2011a, Rincón et al. 2011b). Polymer electrolyte membranes are often used to increase the efficiency of MFCs, although their implementation generally involves a trade-off between coulombic efficiency and mass transport resistance in the MFC (Harnisch and Schröder 2009, Li et al., 2011). Insufficient proton selectivity additionally causes the acidification of the anode and simultaneous pH increase in the compartment of cathode also, which can result in non-physiological conditions in the whole MFC. Strik et al. (2009) addressed this problem through the use of a reversible bio-electrode that acts as a cathode under illumination and as an anode in the darkness. However, the low ionic mobility in membranes remains a problem. The most frequently employed separator is the proton exchange membrane Nafion®, since its high selectivity was shown to result in the largest power outputs. MFCs are operated at physiological pH to minimise cell damage and prevent concentration polarisation from diminishing the power output. The ionic conductivity of Nafion® is minimal in this pH range, which effectively increases the internal resistance inside the MFC and causes power losses (Jeremiasse et al. 2009). Identification of an efficient separator method between anodic and cathodic compartments is therefore a high priority in MFC research.

The confirmation of direct transfer of electron between microbial organisms and the electrode without the need for artificial electron mediators was a seminal discovery that fundamentally improved the sustainability of all MFC technologies (Malvankar et al. 2011). Several microbial electron transfer mechanisms have been suggested, from conductive pili to electron hopping between transmembrane proteins, which are elucidated in the theory section. There is an analysis of a single *Shewanella oneidensis* MR-1 pilus with nanofabricated electrodes, which conducts electron with a resistance. Applying a voltage of 100 mV resulted in an electron transport rate of 10^6-10^9. El-Naggar et al. (2010) compared these electron transfer rates with the specific respiration rate under cultivation conditions, and found that the entire supply of respiratory electrons can be discharged to a terminal acceptor by a single bacterial pilus. The length of bacterial pili is in the range of nanometres, although the electrical conduction range can be extended into the micrometer scale through networks of pili between cells. These length scales illustrate the importance of biofilm over the electrode, which is responsible for microbial exoelectrogenic activity. Furthermore, the electrical resistance of an external electricity consumer can no longer be adjusted according to the largest power generation alone, since the external resistance has also been shown to affect the structure of a biofilm (McLean et al. 2010, Ren et al. 2011). Currents obtained through direct electron transfer are comparatively low (Peng et al. 2010). Logan (2009) therefore concluded that, for MFC technologies developed in the near future, exoelectrogenic bacteria are more suitable for energy recovery during wastewater treatment than for electricity generation. However, El-Naggar et al. (2010) also found electrically conductive pili in the photosynthetic cyanobacterium *Synechocystis* PCC6803. Microorganisms that are both exoelectrogenic and photosynthetic drastically expand the field of applications, since carbon capture and solar energy utilisation improve the carbon neutrality and recovery of energy in conventional MFC systems. Strik et al. (2011) demonstrated a comprehensive overview of performance and efficiency of microbial solar cells (MSCs). It also illustrates the undefined parameters of the respective MSCs, showing the importance of a more systematic characterisation procedure. Strik et al. (2011) conclude from this that the accurate analysis of carbon and electron fluxes, which determine the coulombic efficiency, represents one of the greatest experimental challenges in modern MFC research.

Most of the bacteria can degrade the various organic compounds and therefore, electrons can be transferred to anode during their cultivation in wastewater. There are several researchers who experimented with either mixed culture or pure culture to evaluate the voltage generation efficiency in MFC (Cheng et al. 2006). From their observation, more resistance was noticed when mixed culture was used. Process disturbance along with substrate consumption and higher power output were significantly low when pure culture was used (Logroño et al. 2017, Rabaey and Verstraete 2005). It is obvious that each process has its own limitation; therefore, direct comparison between pure and mixed cultures is difficult with regard to process development for power generation (Kim et al. 2004, Parameswaran et al. 2009). Additionally, the exact mechanism for generation of electricity in pure and mixed culture is a complex process. Moreover, the strategy for generation of maximum energy through optimization of process parameters using pure culture is more significant in comparison to mixed culture. On the contrary, utilization of mixed culture is more significant as they can utilize the various substrates in different conditions (Chaudhuri and Lovley 2003). The process temperature plays a significant role as rate constant of any conversion depends on temperature. Therefore, it is obvious that bacteria are able to withstand a longer time in the presence of process temperature, and they should be active electrochemically (Logan, 2008, Rabaey and Verstraete 2005). It was reported that *K. pneumonia, D. Propionicus* and *R. palustris* are potential candidates for voltage generation when used in MFCs (Sharma and Kundu 2010). In this regard, it is of utmost importance to identify a potential microbial strain owing to their potentiality in power generation, which could be the future directional research. The optimization of process parameters are usually carried out to provide suitable environmental conditions to the power generation in the reactor. Additionally, reactor design along with the sizes of various compartmental allocations for anode and cathode has significant impact on electricity generation in MFCs. Furthermore, redox mediators producing microbe are usually preferred in batch reactors for electricity generation (Angenent et al. 2004, Rabaey and Verstraete, 2005). The continuous process is used for microbes, which have the capability to produce biofilm, where bacteria can directly use electrodes. Additionally, biofilms producing microorganisms can also grow on electrodes and through the biofilms, electrons can be transferred using shuttle molecule (Rabaey et al. 2005). In a continuous process, feed rate plays a pivotal role for uniform mass transfer inside MFCs. It is obvious that feeding rate influences the extent of mixing and, therefore, affects the nutrient availability in MFCs. Choudhuty et al. (2017) evaluated the efficiency of power generation through their observation in fed batch or semi batch study. Interestingly, electricity generation was found to be almost constant after 72 hours of operation (Fig. 8.3).

Since, during fed batch study, little time has been dedicated for batch observation, therefore, constant rate of power generation is not possible. However, as fed batch process operates at quasi steady state, therefore, sustainable electricity production is possible when fed batch operation was adopted (Choudhuty et al. 2017). Anolyte is the key factor, which creates the affinity of bacterial community; therefore, they act as electron donors. It is obvious that types of reactors along with types of substrate and their rate in fed batch process alter the efficiency of power generation. It is evident that anolyte has significant influence in MFC performance with regard to power generation (Chae et al. 2009, Kim et al. 2006). Furthermore, anode has an important role in MFCs with regard to power output. It is evident that large surface area of anode will help more attachment of bacteria in anode; therefore, it increases the electron collection efficiency, which leads to high power generation. Anode should be chemically inert, inexpensive, non-hazardous for bacteria, and able to withstand long hours during operation (Deng et al. 2010, Zhang et al. 2009). Since catholic reaction process is the slowest, it is considered a rate limiting step (Lu and Li 2012). Owing to their several advantages, the adjustment of materials in electrode can be an effective way for enhancement of performance in MFCs. The membrane placed between anode and cathode in MFCs is called an ion exchange membrane (IEM). It will help to physically separate and transport hydrogen ion only to cathode. The membrane fouling or resistance is one drawback for IEM, as electricity generation completely depends on proton movement from anode to cathode

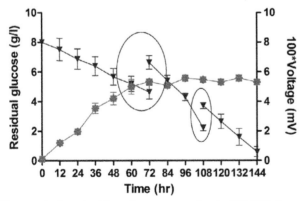

Figure 8.3. Sustainable voltage generation in SCMFC through fed batch operation (Choudhuty et al., 2017).

chamber (Gil et al. 2003, Rozendal et al. 2006). Additionally, microbial growth as well as power generation is significantly increased as the membrane resists the diffusion of oxygen and substrate (Zhang et al. 2010). Since the membrane is costly, therefore the total process cost is increased when IEM is used in MFCs. Therefore, researchers are looking for an alternative pathway which could be a substitute of membrane which is required to address the future development of MFCs. Temperature is considered an environmental parameter which influences bacterial growth as well as the power generation in MFC (Min et al. 2008). pH and ionic strength in anolyte have significant impact on electricity generation (Min et al. 2008). It is already reported that the mobility of proton to cathode is enhanced at optimum pH, and pH helps to minimize gradient of proton concentration. Furthermore, pH inhibits the growth of methanogens, which leads to an indirect increase in the performance of MFCs (Borole et al. 2011). Therefore, power generation in MFC can be increased by using potential microorganisms. However, suitable design of MFC with electrode materials and optimization of media and environmental parameters can increase the applicability of this technology in the near future.

Conclusion

The concept of microbial fuel cell is more than a century old. There are a few progresses, but it still has not reached a practical large-scale application. However, the current energy crisis makes them more promising today than ever, but prior to their practical implementation, they need to be improved further on their efficiency and design. However, in spite of progress in MFCs in the last few decades, most of them are still a few milliliters in volume in lab scale. Still, large-scale application of MFCs is missing. A few new research approaches have been tested recently for considering the development of practical large-scale application of MFCs, which include the integration of basic concept of MFCs with other established technologies. Such integration of MFCs not only improves the performance, but also generates higher electricity than normal MFCs. Since application of the integrated technology in the large scale is limited, various modifications and improvements are necessary for sustainable power generation. Over the past decade, researchers developed new processes and technologies with regard to deficiencies of previous technologies. However, various uncertainties associated with MFC operation deliberate the large-scale field applications for power generation. For this reason, the chapter demonstrates the current research towards sustainable power generation using cyanobacteria as a catalyst. Moreover, this chapter represents concrete information on the future perspective of research on the topic within an ever-growing diversity of uniqueness in MFCs research.

REFERENCES

Amao, Y. and T. Komori. 2004. Bio-photovoltaic conversion device using chlorine-e6 derived from chlorophyll from Spirulina adsorbed on a nanocrystalline TiO_2 film electrode. Biosensors and Bioelectronics 19: 843-847.

Angenent, L.T., K. Karim, M.H. Al-Dahhan, B.A. Wrenn and R. Domíguez-Espinosa. 2004. Production of bioenergy and biochemicals from industrial and agricultural wastewater. Trends in Biotechnology 22: 477-485.

Arteni, A.A., G. Ajlani and E.J. Boekema. 2009. Structural organisation of phycobilisomes from Synechocystis sp. strain PCC6803 and their interaction with the membrane. Biochimica et Biophysica Acta (BBA)-Bioenergetics 1787: 272-279.

Badura, A., B. Esper, K. Ataka, C. Grunwald, C. Wöll, J. Kuhlmann, J. Heberle and M. Rögner. 2006. Light-driven water splitting for (Bio-) hydrogen production: photosystem 2 as the central part of a bioelectrochemical device. Photochemistry and Photobiology 82: 1385-1390.

Balzani, V., A. Credi and M. Venturi. 2008. Photochemical conversion of solar energy. ChemSusChem 1: 26-58.

Boland, S., P. Jenkins, P. Kavanagh and D. Leech. 2009. Biocatalytic fuel cells: a comparison of surface pre-treatments for anchoring biocatalytic redox films on electrode surfaces. Journal of Electroanalytical Chemistry 626: 111-115.

Bond, D.R., D.E. Holmes, L.M. Tender and D.R. Lovley. 2002. Electrode-reducing microorganisms that harvest energy from marine sediments. Science 295: 483-485.

Bond, D.R. and D.R. Lovley. 2005. Evidence for involvement of an electron shuttle in electricity generation by Geothrix fermentans. Applied and Environmental Microbiology 71: 2186-2189.

Borole, A.P., G. Reguera, B. Ringeisen, Z.-W. Wang, Y. Feng and B.H. Kim. 2011. Electroactive biofilms: current status and future research needs. Energy & Environmental Science 4: 4813-4834.

Chae, K.-J., M.-J. Choi, J.-W. Lee, K.-Y. Kim and I.S. Kim. 2009. Effect of different substrates on the performance, bacterial diversity, and bacterial viability in microbial fuel cells. Bioresource Technology 100: 3518-3525.

Chaudhuri, S.K. and D.R. Lovley. 2003. Electricity generation by direct oxidation of glucose in mediatorless microbial fuel cells. Nature Biotechnology 21: 1229-1232.

Cheng, S., H. Liu and B.E. Logan. 2006. Power densities using different cathode catalysts (Pt and CoTMPP) and polymer binders (Nafion and PTFE) in single chamber microbial fuel cells. Environmental Science & Technology 40: 364-369.

Chiao, M., K.B. Lam and L. Lin. 2006. Micromachined microbial and photosynthetic fuel cells. Journal of Micromechanics and Microengineering 16: 2547.

Cho, Y., T. Donohue, I. Tejedor, M. Anderson, K. McMahon and D. Noguera. 2008. Development of a solar-powered microbial fuel cell. Journal of Applied Microbiology 104: 640-650.

Choudhuty, P., R.N. Roy and T.K. Bandyopadhyaya. 2017. To study the effect of glucose feeding rate for sustainable power generation from waste water containing single chamber microbial fuel cell. pp. 204. In: International Conference on New and Renewable Energy Resources for Sustainable Future. Jaipur, India.

Colyer, C.L., C.S. Kinkade, P.J. Viskari and J.P. Landers. 2005. Analysis of cyanobacterial pigments and proteins by electrophoretic and chromatographic methods. Analytical and Bioanalytical Chemistry 382: 559-569.

Davis, F. and S.P. Higson. 2007. Biofuel cells—recent advances and applications. Biosensors and Bioelectronics 22: 1224-1235.

Deng, Q., X. Li, J. Zuo, A. Ling and B.E. Logan. 2010. Power generation using an activated carbon fiber felt cathode in an upflow microbial fuel cell. Journal of Power Sources 195: 1130-1135.

Doherty, S.J., S. Bojinski, D. Goodrich, A. Henderson-Sellers, K. Noone, N.L. Bindoff, J.A. Church, K.A. Hibbard, T.R. Karl and L. Kajfez-Bogataj. 2009. Lessons learned from IPCC AR4: Scientific developments needed to understand, predict, and respond to climate change. Bulletin of the American Meteorological Society 90: 497-513.

Du, Z., H. Li and T. Gu. 2007. A state of the art review on microbial fuel cells: a promising technology for wastewater treatment and bioenergy. Biotechnology Advances 25: 464-482.

El-Naggar, M.Y., G. Wanger, K.M. Leung, T.D. Yuzvinsky, G. Southam, J. Yang, W.M. Lau, K.H. Nealson and Y.A. Gorby. 2010. Electrical transport along bacterial nanowires from Shewanella oneidensis MR-1. Proceedings of the National Academy of Sciences 107: 18127-18131.

Franks, A.E. and K.P. Nevin. 2010. Microbial fuel cells: a current review. Energies 3: 899-919.

Gil, G.-C., I.-S. Chang, B.H. Kim, M. Kim, J.-K. Jang, H.S. Park and H.J. Kim. 2003. Operational parameters affecting the performannce of a mediator-less microbial fuel cell. Biosensors and Bioelectronics 18: 327-334.

Gonzalez-Meler, M.A., L. Taneva and R.J. Trueman. 2004. Plant respiration and elevated atmospheric CO_2 concentration: cellular responses and global significance. Annals of Botany 94: 647-656.

Gouveia, L. and A.C. Oliveira. 2009. Microalgae as a raw material for biofuels production. Journal of Industrial Microbiology & Biotechnology 36: 269-274.

Guan, C., X. Li, Z. Wang, X. Cao, C. Soci, H. Zhang and H.J. Fan. 2012. Nanoporous walls on macroporous foam: rational design of electrodes to push areal pseudocapacitance. Advanced Materials 24: 4186-4190.

Hamelers, H.V., A. Ter Heijne, T.H. Sleutels, A.W. Jeremiasse, D.P. Strik and C.J. Buisman. 2010. New applications and performance of bioelectrochemical systems. Applied Microbiology and Biotechnology 85: 1673-1685.

Harnisch, F. and U. Schröder. 2009. Selectivity versus mobility: separation of anode and cathode in microbial bioelectrochemical systems. ChemSusChem 2: 921-926.

Higgins, S.R., D. Foerster, A. Cheung, C. Lau, O. Bretschger, S.D. Minteer, K. Nealson, P. Atanassov and M.J. Cooney. 2011. Fabrication of macroporous chitosan scaffolds doped with carbon nanotubes and their characterization in microbial fuel cell operation. Enzyme and Microbial Technology 48: 458-465.

Hussein, L., Y. Feng, N. Alonso-Vante, G. Urban and M. Krüger. 2011. Functionalized-carbon nanotube supported electrocatalysts and buckypaper-based biocathodes for glucose fuel cell applications. Electrochimica Acta 56: 7659-7665.

Janzen, A.F. and M. Seibert. 1980. Photoelectrochemical conversion using reaction-centre electrodes. Nature 286: 584.

Jeremiasse, A.W., H.V. Hamelers, J.M. Kleijn and C.J. Buisman. 2009. Use of biocompatible buffers to reduce the concentration overpotential for hydrogen evolution. Environmental Science & Technology 43: 6882-6887.

Kim, B.H., H. Park, H. Kim, G. Kim, I. Chang, J. Lee and N. Phung. 2004. Enrichment of microbial community generating electricity using a fuel-cell-type electrochemical cell. Applied Microbiology and Biotechnology 63: 672-681.

Kim, G., G. Webster, J. Wimpenny, B. Kim, H. Kim and A. Weightman. 2006. Bacterial community structure, compartmentalization and activity in a microbial fuel cell. Journal of Applied Microbiology 101: 698-710.

Kruse, O., J. Rupprecht, J.H. Mussgnug, G.C. Dismukes and B. Hankamer. 2005. Photosynthesis: a blueprint for solar energy capture and biohydrogen production technologies. Photochemical & Photobiological Sciences 4: 957-970.

Lam, K.B., E.A. Johnson, M. Chiao and L. Lin. 2006. A MEMS photosynthetic electrochemical cell powered by subcellular plant photosystems. Journal of Microelectromechanical Systems 15: 1243-1250.

Li, W.-W., G.-P. Sheng, X.-W. Liu and H.-Q. Yu. 2011. Recent advances in the separators for microbial fuel cells. Bioresource Technology 102: 244-252.

Logan, B.E. 2009. Exoelectrogenic bacteria that power microbial fuel cells. Nature Reviews Microbiology 7: 375.

Logan, B.E. 2008. Microbial Fuel Cells. John Wiley & Sons. New Jersey, United States.

Logroño, W., M. Pérez, G. Urquizo, A. Kadier, M. Echeverría, C. Recalde and G. Rákhely. 2017. Single chamber microbial fuel cell (SCMFC) with a cathodic microalgal biofilm: a preliminary assessment of the generation of bioelectricity and biodegradation of real dye textile wastewater. Chemosphere 176: 378-388.

Lovley, D.R. 2006. Bug juice: harvesting electricity with microorganisms. Nature Reviews Microbiology 4: 497.

Lu, M. and S.F.Y. Li. 2012. Cathode reactions and applications in microbial fuel cells: a review. Critical Reviews in Environmental Science and Technology 42: 2504-2525.

Madigan, M.T. 2000. Extremophilic bacteria and microbial diversity. Annals of the Missouri Botanical Garden 3-12.

Malik, S., E. Drott, P. Grisdela, J. Lee, C. Lee, D.A. Lowy, S. Gray and L.M. Tender. 2009. A self-assembling self-repairing microbial photoelectrochemical solar cell. Energy & Environmental Science 2: 292-298.

Malvankar, N.S., M. Vargas, K.P. Nevin, A.E. Franks, C. Leang, B.-C. Kim, K. Inoue, T. Mester, S.F. Covalla and J.P. Johnson. 2011. Tunable metallic-like conductivity in microbial nanowire networks. Nature Nanotechnology 6: 573.

Maly, J., J. Masojidek, A. Masci, M. Ilie, E. Cianci, V. Foglietti, W. Vastarella and R. Pilloton. 2005. Direct mediatorless electron transport between the monolayer of photosystem II and poly (mercapto-p-benzoquinone) modified gold electrode—new design of biosensor for herbicide detection. Biosensors and Bioelectronics 21: 923-932.

McLean, J.S., G. Wanger, Y.A. Gorby, M. Wainstein, J. McQuaid, S.I. Ishii, O. Bretschger, H. Beyenal and K.H. Nealson. 2010. Quantification of electron transfer rates to a solid phase electron acceptor through the stages of biofilm formation from single cells to multicellular communities. Environmental Science & Technology 44: 2721-2727.

Min, B., Ó.B. Román and I. Angelidaki. 2008. Importance of temperature and anodic medium composition on microbial fuel cell (MFC) performance. Biotechnology Letters 30: 1213-1218.

Mitchell, P. 1961. Coupling of phosphorylation to electron and hydrogen transfer by a chemi-osmotic type of mechanism. Nature 191: 144-148.

Parameswaran, P., C.I. Torres, H.S. Lee, R. Krajmalnik-Brown and B.E. Rittmann. 2009. Syntrophic interactions among anode respiring bacteria (ARB) and Non-ARB in a biofilm anode: electron balances. Biotechnology and Bioengineering 103: 513-523.

Park, D.H. and J.G. Zeikus. 2000. Electricity generation in microbial fuel cells using neutral red as an electronophore. Applied and Environmental Microbiology 66: 1292-1297.

Peng, L., S.-J. You and J.-Y. Wang. 2010. Carbon nanotubes as electrode modifier promoting direct electron transfer from Shewanella oneidensis. Biosensors and Bioelectronics 25: 1248-1251.

Pisciotta, J.M., Y. Zou and I.V. Baskakov. 2010. Light-dependent electrogenic activity of cyanobacteria. PLoS One 5: e10821.

Rabaey, K., N. Boon, M. Höfte and W. Verstraete. 2005. Microbial phenazine production enhances electron transfer in biofuel cells. Environmental Science & Technology 39: 3401-3408.

Rabaey, K. and W. Verstraete. 2005. Microbial fuel cells: novel biotechnology for energy generation. Trends in Biotechnology 23: 291-298.

Ragauskas, A.J., C.K. Williams, B.H. Davison, G. Britovsek, J. Cairney, C.A. Eckert, W.J. Frederick, J.P. Hallett, D.J. Leak and C.L. Liotta. 2006. The path forward for biofuels and biomaterials. Science 311: 484-489.

Reguera, G., K.D. McCarthy, T. Mehta, J.S. Nicoll, M.T. Tuominen and D.R. Lovley. 2005. Extracellular electron transfer via microbial nanowires. Nature 435: 1098.

Ren, Z., R.P. Ramasamy, S.R. Cloud-Owen, H. Yan, M.M. Mench and J.M. Regan. 2011. Time-course correlation of biofilm properties and electrochemical performance in single-chamber microbial fuel cells. Bioresource Technology 102: 416-421.

Rincón, R.A., C. Lau, K.E. Garcia and P. Atanassov. 2011a. Flow-through 3D biofuel cell anode for NAD+-dependent enzymes. Electrochimica Acta 56: 2503-2509.

Rincón, R.A., C. Lau, H.R. Luckarift, K.E. Garcia, E. Adkins, G.R. Johnson and P. Atanassov. 2011b. Enzymatic fuel cells: integrating flow-through anode and air-breathing cathode into a membrane-less biofuel cell design. Biosensors and Bioelectronics 27: 132-136.

Rosenbaum, M. and U. Schröder. 2010. Photomicrobial solar and fuel cells. Electroanalysis 22: 844-855.

Rozendal, R.A., H.V. Hamelers and C.J. Buisman. 2006. Effects of membrane cation transport on pH and microbial fuel cell performance. Environmental Science & Technology 40: 5206-5211.

Ryu, W., S.-J. Bai, J.S. Park, Z. Huang, J. Moseley, T. Fabian, R.J. Fasching, A.R. Grossman and F.B. Prinz. 2010. Direct extraction of photosynthetic electrons from single algal cells by nanoprobing system. Nano Letters 10: 1137-1143.

Sawin, J.L. and F. Sverrisson. 2014. Renewables 2014: Global status report. Paris, France.

Sharma, V. and P. Kundu. 2010. Biocatalysts in microbial fuel cells. Enzyme and Microbial Technology 47: 179-188.

Strik, D.P., H.V. Hamelers and C.J. Buisman. 2009. Solar energy powered microbial fuel cell with a reversible bioelectrode. Environmental Science & Technology 44: 532-537.

Strik, D.P., H. Terlouw, H.V. Hamelers and C.J. Buisman. 2008. Renewable sustainable biocatalyzed electricity production in a photosynthetic algal microbial fuel cell (PAMFC). Applied Microbiology and Biotechnology 81: 659-668.

Strik, D.P., R.A. Timmers, M. Helder, K.J. Steinbusch, H.V. Hamelers and C.J. Buisman. 2011. Microbial solar cells: applying photosynthetic and electrochemically active organisms. Trends in Biotechnology 29: 41-49.

Tsujimura, S., A. Wadano, K. Kano and T. Ikeda. 2001. Photosynthetic bioelectrochemical cell utilizing cyanobacteria and water-generating oxidase. Enzyme and Microbial Technology 29: 225-231.

Watanabe, M., D.A. Semchonok, M.T. Webber-Birungi, S. Ehira, K. Kondo, R. Narikawa, M. Ohmori, E.J. Boekema and M. Ikeuchi. 2014. Attachment of phycobilisomes in an antenna–photosystem I supercomplex of cyanobacteria. Proceedings of the National Academy of Sciences 111: 2512-2517.

Yagishita, T., T. Horigome and K. Tanaka. 1993. Effects of light, CO_2 and inhibitors on the current output of biofuel cells containing the photosynthetic organism Synechococcus sp. Journal of Chemical Technology and Biotechnology 56: 393-399.

Yagishita, T., S. Sawayama, K.-i. Tsukahara and T. Ogi. 1997. Effects of intensity of incident light and concentrations of Synechococcus sp. and 2-hydroxy-1, 4-naphthoquinone on the current output of photosynthetic electrochemical cell. Solar Energy 61: 347-353.

Yu, P., H. Zhou, H. Cheng, Q. Qian and L. Mao. 2011. Rational design and one-step formation of multifunctional gel transducer for simple fabrication of integrated electrochemical biosensors. Analytical Chemistry 83: 5715-5720.

Zhang, F., S. Cheng, D. Pant, G. Van Bogaert and B.E. Logan. 2009. Power generation using an activated carbon and metal mesh cathode in a microbial fuel cell. Electrochemistry Communications 11: 2177-2179.

Zhang, X., S. Cheng, X. Huang and B.E. Logan. 2010. The use of nylon and glass fiber filter separators with different pore sizes in air-cathode single-chamber microbial fuel cells. Energy & Environmental Science 3: 659-664.

9

Yeast and Algae as Biocatalysts in Microbial Fuel Cell

Neethu B.[1], G.D. Bhowmick[2] and M.M. Ghangrekar[1]*

[1] Department of Civil Engineering, Indian Institute of Technology Kharagpur - 721302, India

[2] Department of Agricultural and Food Engineering, Indian Institute of Technology Kharagpur - 721302, India

Introduction

Recent explosion of world's population leads to greater energy demands and deterioration of environment in an exponential way. Therefore, there is a trend in recent scientific discussion on the usage of alternative green energy sources, such as biomass to partially replace the fossil fuels, as it is on the verge of ending within this century. Despite huge availability of biomass in waste with reasonable energy densities, it is often a challenge to use this biomass directly as an energy source. Microbial fuel cells (MFCs), a cutting edge bio-electrochemical device, take advantage of anaerobic metabolic pathways of microorganisms to oxidise organic matter, allowing energy to be extracted from it. The transfer of electron in between the microorganism to electrode surface, to harvest energy in the form of electricity, can be classified into three broad categories: (1) redox-active proteins, such as c-type cytochromes on the outer cell membrane, let electroactive microorganism directly transfer electrons to the electrode surface; (2) electron shuttles or endogenous mediators to promote electron transfer to an electrode; and (3) direct electron transfer to the electrode surface through conductive locomotive organs of microorganisms, called nanowires (e.g., pili, fimbriae, etc.). Sometimes even microorganisms interconnect themselves through these nanowires to form electrical networks/channels, which efficiently transfer electrons directly to the anodic surface. Another possible type of electron transfer is said to be cell-cell communication or synonymously called quorum sensing, which is a positive hypothesis yet to be proven experimentally (Logan 2009). It says that microorganisms can communicate through quorum sensing chemicals, such

*Corresponding author: ghangrekar@civil.iitkgp.ernet.in

as fatty acyl-homoserine lactones (acyl-HSls) and p-counaroyl-HSl produced by *Pseudomonas aeruginosa* and *Rhodopseudomonas palustri*, respectively (Schaefer et al. 2008).

Microorganisms capable of exocellular electron transfer (EET) are generally called exoelectrogens. Sometimes, they are referred to as anode respiring bacteria or electrochemically active bacteria. Rich diversity of these types of bacteria determines the higher power outputs from the MFCs, though it often depends on architecture of MFC, electrode spacing, solution conductivity, etc. Hence, power production from one MFC cannot be directly compared to another due to the different exoelectrogen strain, unless the architecture of MFC, catalyst, and electrolytes used are the same (Logan et al. 2006).

Apart from the identified strains of electrogenic bacteria, such as *Rhodoferax ferrireducens* (Chaudhuri and Lovley 2003), *Shewanella putrefaciens* (Kim et al. 2002), *Shewanella oneidensis* (Hasan et al. 2017), etc., yeast species, such as *Saccharomyces cerevisiae* (Bennetto 1990, Rawson et al. 2012, Gunawardena et al. 2008, Raghavulu et al. 2011, Walker and Walker 2006), *Hansenula anomala* (Prasad et al. 2007), *Hansenula polymorpha* (Shkil et al. 2011), *Arxula adeninivorans* (Haslett et al. 2011), and *Candida melibiosica* (Hubenova and Mitov 2010), have also been reported as biocatalysts in MFC. Easy cultivation procedure, wide substrate spectrum, fast development, and tolerance to widespread environmental conditions are beneficial for the yeast-based biofuel cells.

On the other hand, cathodic reduction reaction takes place by reducing terminal electron acceptor (TEA). Optimum concentration of TEA helps in increasing the cathodic efficiency to accelerate the electron transfer from anode, thus reducing the electron losses (i.e., concentration losses) and yielding higher power output (Srikanth and Venkata Mohan 2012). The major limitation of traditional MFCs is the use of aeration to provide O_2, the most promising TEA available till date, which is energy consuming. A microbial carbon-capture cell (MCC) uses the photosynthetic oxygen producing capacity of microalgae as a potential alternative to the mechanical aeration, with an added advantage of CO_2 sequestration along with valuable by-product recovery, such as harvested algal biomass and electricity. This chapter deals with different biocatalysts, focusing mainly on yeast and algae to improve the overall performance of MFC/MCC to make it a green technology for its futuristic applications.

Yeast Microbial Fuel Cell

Bioelectrochemical devices, such as MFC can effectively utilize the microorganism's catalytic activity to harvest electrical energy from the organic matter present in the wastewater. Biocatalyst application in the anode has vital role in the performance of the MFC. Prokaryotic microorganisms utilize aerobic environment to grow through respiratory pathways and anaerobic environment for fermentation pathways. Yeast, a eukaryotic microorganism classified under fungus kingdom, can grow in both the environments. It can be easily actuated, by hydrating or heating, and does not require sterile environments, because of better tolerances in diverse environmental conditions for long-term applications; hence, it can be stored and placed in the dry state. Several strains of yeast have been tested in MFC as biocatalysts, *viz. C. melibiosica, S. cerevisiae, H. anomala, H. polymorpha, A. adeninvorans*, etc., so far with or without mediators. Different anode surface modifications further enhance the performance of yeast-based MFCs, leading to better wastewater treatment efficiency with simultaneous production of bioenergy. The electron transferring phenomenon of different yeast strains in the anodic chamber of MFC as biocatalysts is discussed in detail to improve the efficacy of yeast-based MFCs.

Electron Transferring Mechanism

Yeast's bio-catalytic activity is mainly attributed to the presence of various characteristic natural electron mediators or shuttles, for example, cytochromes, azurin, ferredoxin, etc., which can be utilized by the redox enzymes to enhance the flow of electrons from the cell to anode surface. Yeast's cytochromes and transmembrane proteins (tPMETs) are located inside its mitochondria and cell membrane, respectively. Hence, a mediator is necessary to travel through the cell wall or the cell membrane and the redox sites, such as NAD^+/NADH (nicotinamide-adenine dinucleotide phosphate) inside the cell. The electron transfer mechanism of yeast cells, while consuming organic substances in MFC, have been well described by Raghavulu et al. (2011) and Gunawardena et al. (2008). Inside the cytosol of the cell, electrons liberated through the oxidation of the substrate turn into the pyruvate called glycolysis procedure. In this fermentative respiration path, two pyruvate molecules are produced from one glucose molecule, which will then form two acetaldehyde molecules by the enzyme named pyruvate decarboxylase. NADH dependent enzyme, such as alcohol dehydrogenase transform acetaldehyde into alcohol. The electrons liberated by this process are received by NAD^+ to form NADH, which is again recycled by releasing electrons through NADH oxidation at the electrode surface (mostly anode surface), whether by tPMETs or by the internal/external mediators to again form NAD^+. This completes the NADH to NAD^+ cycle for anaerobic fermentation, which is necessary to continue the glycolysis process. As this process is taking place in cell's cytosol instead of mitochondria, mediator molecules can easily attach to the cell membrane to access NADH.

Yeast-based MFC – With Exogenous Mediators

A few investigations have been done to improve the exchange of electrons through an external mediator addition. External mediator must fulfil a few prerequisites, for example, should be electrochemically dynamic, quick release of electrons to the electrode, soluble in the anolyte, chemically stable at anolyte media, biocompatible to the microorganisms, effectively infiltrate the cell membrane, possess positive redox potential that is adequate to deliver fast electron exchange from microorganism's cell wall to the anode, while not very solid to dodge a major potential loss. Mediators, such as methylene blue (MB), thionine, neutral red (NR), and yeast extract (YE) improved the electron transfer in MFC with the *S. cerevisiae* yeast and its energy yield, as depicted in different studies (Table 9.1). *S. cerevisiae* is thought to be a decent MFC biocatalyst, because of its wide substrate range, simple and quick mass development, non-pathogenic, cheap cost, and longer storage time at dried state.

Early cases of yeast-based biofuel cells utilized different mediators as a way to increase electron transport kinetics over the quasi-permeable cell membrane. Potter (1911) reported first yeast-based bio-fuel cell, after that Bennetto et al. (1983) reported baker's yeast as a biocatalyst; wherein mediator resorufin was used for transferring electrons from yeast cells to the anode. Radioactive isotopes ^{14}C-labelled glucose was used in that study to demonstrate that the generated voltage was due to glucose degradation of $^{14}CO_2$.

Utilizing proton exchange sulfonated polyether ether ketone (SPEEK) membrane and copper electrodes, Permana et al. (2015) evaluated the performance of double chamber *S. cerevisiae* yeast-based MFCs with glucose as substrate with and without the presence of MB. With the presence of MB, MFC showed higher voltage and net energy yields, but slightly lowered consumption of the glucose without influencing the production of bioethanol compared to the mediator-less MFC. Ganguli and Dunn (2009) used rotating disc electrodes (RDEs) to determine the current generation due to the catalytic activity under dormant conditions and at low concentration of mediator. In view of the outcomes from anodic kinetic study, yeast-controlled MFC could effectively produce a power density of around 1500 mW/m².

The conditions for generation of electricity in a dual-chambered yeast-based biofuel cell were optimized by Walker and Walker (2006). The study showed that when a reasonable amount of oxygen is there, MFC could yield higher current output compared to the deoxygenated or oxygen-saturated solutions. Power production has been increased with more accessible net electrons produced by means of favorable aerobic respiration, and then again with a base rivalry for the electrons delivered between the oxygen and reduced exogenous mediator (MB). Hubenova and Mitov (2010) investigated *C. melibiosica 2491* yeast strain for biofuel cell capable of producing power even without an artificial mediator. The current, which was delivered, associated with the production of yeast species and the substrate assimilation rate, exhibited its association with the *in vivo* electrons produced. Critical increment of current values and power yields were observed by the MB addition up to a concentration of 0.9 mM. Additional increment of MB resulted in an abatement of power because of the cytotoxic impact of the mediator.

Gunawardena et al. (2008) explored the impact of the addition of MB on the electrical outputs, the internal resistances, and the efficacy of *S. cerevisiae*-based biofuel cell by utilizing reticulated vitreous carbon (RVC) anode and

Table 9.1. Comparative performance of yeast-based biofuel cells operated with artificial electron transfer mediators

Biocatalysts (Bioagent)	Carbon source	Anode material	Exogenous mediator	Max. power density (mW/m²)	Reference
S. cerevisiae	Glucose	Carbon rod and carbon fiber bundles	2-hydroxy-1,4-naphthoquinone	22	Kaneshiro et al. 2013
S. cerevisiae	Dextrose	RVC	MB	400	
			NR	100	Wilkinson et al. 2006
S. cerevisiae	Dextrose	Carbon felt	MB + NR	500	Ganguli and Dunn 2009
			MB	300	
S. cerevisiae	Glucose	RVC	MB + K$_3$[FeCN)$_6$]	147 (mW/m³)	Gunawardena et al. 2008
S. cerevisiae	Glucose	Glassy carbon	Resorufin	155	Bennetto et al. 1983
S. cerevisiae	Glucose	Platinum mesh	MB	65	Walker and Walker 2006
		Carbon paper		80	
S. cerevisiae	Glucose	Co-sputtered carbon paper	MB	148	Kasem et al. 2013
		Au-sputtered carbon paper		120	
C. melibiosa	Fructose	Graphite rods	MB	185 (mW/m³)	Hubenova and Mitov 2010
Arxulaadenini vorans	Dextrose + Fructose	Carbon fiber cloth	TMPD	1000	Haslett et al. 2011

(*Contd.*)

Table 9.1. (*Contd.*)

Biocatalysts (Bioagent)	Carbon source	Anode material	Exogenous mediator	Max. power density (mW/m²)	Reference
	D-xylose			39	
	D-glucose			31	Gal et al. 2016
S. cerevisiae	L-arabinose	Graphite plate	MB (0.1 M)	32	
	D-cellobiose			22	
	D-galactose			14	
			MB	400	
S. cerevisiae	Dextrose	Reticulated vitreous carbon	NR	80	Wilkinson et al. 2006
			MB + NR	500	
		Carbon paper		36	
S. cerevisiae	Glucose	Au-plated carbon paper	YE	70	Sayed et al. 2015
			BcG	46	
			NR	89	
Candida melibiosa	Fructose	Carbon felt	MR	113	Babanova et al. 2011
			MO	137	
			MB	640	

MB - Methylene blue; YE - Yeast extract; MR - Methyl red; MO - Methyl orange; BcG - Bromocresol green; TMPD - Tetramethyl-phenylenediamine; NR - Neutral red; RVC - Reticulated vitreous carbon.

cathode and potassium ferricyanide as TEA in the cathodic chamber. A maximum volumetric power density of 146.7 ± 7.7 mW/m^3 was recovered. The lower power output was because of the overpotential loss associated with the O$_2$ reduction at the RVC and the electron exchange inadequacies in between the cell walls of the microorganisms and the mediator. Presumably, mediator's cytotoxicity at concerned high concentration is another explanation behind this lower power yield.

Kasem et al. (2013) investigated the effect of modification of anode on the performance of *S. cerevisiae*-based MFC with mediator MB. Thin layer (30 nm) of Co (Co30) and/or Au (Au30) were used to sputter in anodic carbon paper, which enhanced the power production from 80 to 148 mW/m^2 and to 120 mW/m^2 with Co30 and Au30, respectively. In spite of the fact that the resistance of MFC in case of Co is higher than Au, the execution of the first one was better, and that was identified with the toxic effect of Au on any biotic cell. Subsequently, in the anolyte of Au30, only the yeast took part; whereas in case of Co30, analytic yeast cells and biofilm present on the surface of anode participated in the electron exchange. It was witnessed that highly conductive Co or Au made highly conductive ion transport between the mediator and the anode.

Rahimnejad et al. (2012) reported thionine as another effective mediator for *S. cerevisiae* yeast-based MFC. Adding thionine altogether expressively increased power density from 3 to 28 mW/m^2. Ideal dosage of thionine was found to be 500 mM, producing 420 mV of maximum voltage and 700 mA/m^2 of maximum current density values.

Yeast extract can also act as mediator for *S. cerevisiae*-based MFC (Sayed et al. 2015). Using plain carbon and gold-plated carbon paper as anode, current density amplified from 94 mA/m^2 in case of control MFC to 190 and 300 mA/m^2, respectively. The power density value increased from 12.9 to 32.6 mW/m^2 and 2 to 70 mW/m^2, with the yeast extract addition for plain and the gold-plated anodes, respectively. The part of the yeast removed as electron exchanging mediator was further affirmed by using gold-plated carbon paper, which further emphasises the role of electron mediators for electricity generation.

Apart from improvement in electron transfer kinematics, the usage of exogenous mediators could be an effective way to deal with cell catabolic mechanism. Exploring the impact of exogenous mediators with varying potentials on the *Melibiosica*-based biofuel cell (Babanova et al. 2011a), it had been witnessed that mediators could expressively impact on the overall power yield of MFC by compelling the living cells to start different catabolic pathways, redirecting electrons from various energetic levels.

Fuel cell design and selection of final electron acceptor (TEA) also influence the performance of yeast-based bio-fuel (Venkata Mohan et al. 2008). Different catholytes *viz.* potassium permanganate, hydrogen peroxide, potassium ferricyanide, ferriine (tris(1,10-phenanthroline)Fe(III)), and ammonium vanadate were investigated for higher power recovery from MFC so far (Haslett et al. 2011, Wei et al. 2012).

Yeast-based MFC – Without Any Exogenous Mediator

The utilization of exogenous mediators is not always suitable for practical applications on the ground that the mediator increases the operational expenses and may further pollute the environment. Due to these, recently the experiments are done mainly on the advancement of bio-electrochemical systems without the use of any artificial mediators, named mediator-less yeast-based MFC. The power acquired ranges between 20 and 70 mW/m^2, depending upon the construction of MFC, anode material used (Raghavulu et al. 2011), the internal resistance (Babanova et al. 2011b), catholyte composition, separator, the anolyte used (Prasad et al. 2007, Hubenova and Mitov 2010) and its conductivity (Hubenova et al. 2011), the applied polarization mode (Hubenova et al. 2014), and also the extent of anaerobicity maintained in the anodic chamber of the system. The factors for each trial do not permit an immediate assessment between maximal power values, as different yeast species have been utilized in different studies; however for the comparison, the power density reported using different yeast species in MFCs are tabulated in Table 9.2.

S. cerevisiae has respiratory chain complexes on its surface. Polysaccharides-made cell wall and non-homogeneous cell plasma membrane structure further hamper electron transport inside the yeast (Cabib et al. 2001). Extracellular exchange of electrons is reliant upon the surface charge of cell and controlled by the cell wall polymers (*viz.* amino groups, etc.). Variations in surface electrical charges of the cell wall depend upon the ion content and pH value (Volkov 2015). Electrical charge on the surface is proportional to voltage difference from 0 to 100 mV. However, the proton pumps can change the membrane potential within seconds. Estimation of membrane potential using microelectrode method was done on *S. cerevisiae*-based MFC and it was found to be in the range of −70 and −45 mV (Borst-Pauwels 1981). Overall, most of the reported experiments of mediator-less yeast-based MFCs can be divided into two broad categories, such as electron transfer by endogenous mediators and by anodic modifications.

Electron transfer by endogenous mediators

C. melibiosica produce electrochemically energetic soluble compounds, which could be used as endogenous mediator (Hubenova et al. 2011). Secreted redox molecules were confirmed by the presence of peaks at anodic (+225 mV vs. Ag/AgCl) and cathodic side (−45 mV vs. Ag/AgCl) by cyclic voltammetry (CV) study of the yeast suspension and its supernatant (Hubenova et al. 2011). The presence of electroactive compounds in anolyte relies upon the method of electron transfer and biofuel cell activity. However in case of polarization, the produced compound at the earlier stages of development added to high steady state current values obtained (Hubenova et al. 2014). The electron transfer in case of mediator-less yeast biofuel cells is assumed to be accomplished via redox molecules secreted by it, because most of the redox molecules are located intercellularly and a rigid wall with 100–200 nm thickness surrounds the cell membrane protein.

Table 9.2. Performance of the yeast-based biofuel cells reported without using any artificial electron transfer mediators

Biocatalysts (Bioagents)	Carbon source	Anode material	Max. power density (mW/m²)	Reference
C. melibiosa	Non-modified carbon felt	Fructose in acetate	45	(Hubenova et al. 2011)
S. cerevisiae	Synthetic wastewater	Graphite plate	25.51	(Raghavulu et al. 2011)
S. cerevisiae	Dextrose in YP	Non-catalyzed graphite plates	30	(Raghavulu et al. 2011)
S. cerevisiae	Lactose	Graphite plate/MWCNT	2.7	(Gal et al. 2016)
	D-glucose		2.8	
	Lactose		33.0	
C. melibiosa	Non-modified carbon felt	Fructose in YP	20	(Babanova et al. 2011)
Hydrangea anomala	Plain graphite	Glucose in YEPD/malt extract	700 (mW/m³)	(Prasad et al. 2007)
S. cerevisiae	Glucose	Graphite plates	28	(Rahimnejad et al. 2012)
C. melibiosa	Non-modified carbon felt	Fructose in YP	27	(Hubenova et al. 2011)
S. cerevisiae	Glucose	Reticulated vitreous carbon	40 (mW/m³)	(Gunawardena et al. 2008)
C. melibiosa	Graphite rods	Fructose in YP	60 (mW/m³)	(Hubenova and Mitov 2010)
A. adeninivorans	Carbon cloth fiber	Glucose in YEPD + $K_3[Fe(CN)_6]$	70	(Haslett et al. 2011)
S. cerevisiae	Glucose	Carbon paper	3.0	(Sayed et al. 2012)

Improvement of the electron transfer by anodic modification

The carbon-based anodes, regardless of whether highly optimized for bacterial attachment, still do not have as high electrical conductivity as of metals for exchanging long distance electron transfer. Thus, scientists have designed new anode materials incorporating metal particles; however, the metal must be carefully selected to have bio-compatibility to evade corrosion and cytotoxicity of living cells (Hubenova et al. 2011, Chen et al. 2012).

Modified carbon paper with thin layers of transition metals *viz.* Co and Ag was investigated to be used as anode catalyst material for improving the performance of air cathode MFCs with *S. cerevisiae* as biocatalyst (Ganguli and Dunn 2012). Schaetzle et al. (2008) showed enhanced power output from *S. cerevisiae* yeast-based biofuel cell by restraining *S. cerevisiae* on carbon nanotube to use as anode catalyst. Anode coated with polyaniline and Pt as catalyst enhanced the power output of *S. cerevisiae* biofuel cell and attained maximum power output of 2.9 W/m^3, which was almost four times higher than the power density achieved by MFC using plain graphite electrodes (Prasad et al. 2007).

Prospects of Yeast-based MFCs

Different yeast strains exhibit different electron transfer mechanisms. In case of *S. cerevisiae* and *H. anomala*, the surface-confined species help in electron transfer, whereas, in case of *C. melibiosica*, *A. adeninivorans*, and *H. polymorpha*, the redox molecules secretes by direct transfer of electrons from the yeast cells to the surface of anode. Anode modification and external mediator can drastically improve the performance of MFC as well. *Kluyberomyces marxianus* is considered one of the most favorable strains of yeast, having a capability to metabolize complex organic materials effectively and demonstrates higher power yield even at higher temperature. Hence, it can be a good alternative for waste remediation with fluctuating temperature. The exo-electrogenicity of certain yeast species could be helpful in developing biofuel cell-based technology for wastewater treatment by utilizing them as biocatalysts.

Algae as Biocatalyst in Microbial Fuel Cells

High oxidation potential and clean reduction product (water) makes oxygen the most common electron acceptor used in MFC technology. On the other hand, supply of oxygen in cathodic side is, however, energy intensive mainly due to the cost of aeration associated with energy required for external mechanical aeration. MCC addresses this challenge by using the algal potential to produce oxygen with an added advantage of using algae for carbon capture, as well as to generate feedstock for emerging biodiesel production. Further, microalgae can also provide an effective solution for nutrient removal from wastewater, thus enlightening the application of MFC in treating wastewater having high nutrient content (Fig. 9.1). Further the algal biomass harvested from cathode can be used for oil extraction and the solid residues can be fed back to the anodic chamber for degradation by microbial consortium (Khandelwal et al. 2018). Hence, MCC emerges as a sustainable technology that uses oxygen produced by algal biomass as electron acceptor for accomplishing concurrent electricity generation, CO_2 sequestration, wastewater treatment, and algal biomass production.

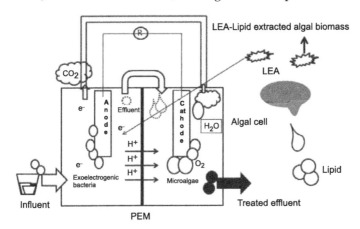

Figure 9.1. Schematic diagram of a microbial carbon-capture cell.

Factors Effecting the Performance of MCC

Algal Bio-cathode

a. *Algal species and concentration:* Algae enhance the performance of the electrochemical system by providing enough oxygen for the reduction reaction to occur. Oxygen generation by algal cells is a result of the photosynthesis process occurring in the cell thylakoid. Hence, type and concentration of algae plays a major role in the overall performance of the system. The photosynthesis rate as well as the carbon-capturing ability varies with algal species, whereas the light and nutrient availability for cell growth is highly dependent on the overall cell concentration in the system. Selection of proper strains is thus one of the vital parameter to be considered in MCC design and operation (Philippart 1995, Shigesada and Okubo 1981, Ugwu and Aoyagi 2008). With an increase in cell concentration due to cell growth, the light reaching the cells inside the chamber drastically decreases due to the self-shading effect of the cells near the surface. As a result, at a higher cell concentration a light inhibition zone is formed within the system, which affects the photosynthesis and hence the system performance. Initial biomass concentration is another important parameter that affects the growth and biomass production of micro-algal species (Li et al. 2013, Ugwu and Aoyagi 2008, Zamfir and Goldberg 2000, Wijffels et al. 1994). A study by

Wang et al. (2013) reported 0.8 g/L of initial biomass concentration as optimal initial concentration, which resulted in highest productivity of green micro-algae *Haematococcus pluvialis*. Similarly a cell concentration of 1–2 g/L was found to be most suitable for cyanobacterium *Synechocystis aquatilis* (Zhang et al. 2001). Selection of strain and maintaining optimal conditions for its growth is thus an important challenge that requires a platform of basic information about algal origin, genetic makeup, and growth characteristics necessary to make it a good candidate for algae-assisted MFC (Table 9.3).

b. *Light energy*: Photosynthesis is the process by which certain species utilize light energy to produce carbohydrate from carbon dioxide and water. Therefore, light is a crucial parameter for algae, in capturing carbon dioxide to carryout photosynthesis and consequently producing oxygen. Hence, light should indeed be provided at the appropriate intensity, duration, and wavelength. Several researchers have observed a proportional increase in biomass productivity with light intensity until the light saturation is reached (Perner-Nochta and Posten 2007, Tilzer 1987, Qiang and Richmond 1996, Jeon et al. 2006). In addition, Li and his group showed that dependence of the response by algal species to varying light intensity varied with algal strains (Li et al. 2012). Hence, it is necessary to find the optimum light intensity required for species under consideration to avoid photo oxidation in presence of excessive light intensity and growth inhibition as a result of less intensity. Recently, highly efficient light-emitting diodes, which scatter less heat energy, have been developed and used in photo-bioreactors (Lee and Palsson 1994, Melnicki et al. 2013, Zhao et al. 2015, Hu and Sato 2017). The part of incident light that isn't utilized by algal cells gets changed into heat energy; wavelengths corresponding to blue light showcased a better performance in terms of growth and productivity than red and green light wavelengths (Blair et al. 2014). Keeping in mind the end goal to improve the micro-algal growth, for a particular algal strain, optimization of light intensity, wavelength and frequency is of utmost importance.

c. *Supply of carbon dioxide:* Impact of global warming is seen all around the world, CO_2 being a noteworthy supporter of this. Apart from burning of fossil fuels, such as oil, natural gas, diesel, petrol, etc., CO_2 is also generated during the anaerobic treatment of wastewater. Microalgae are generally perceived to offer encouraging solution for both bio-fuel production and capturing of discharged CO_2. As discussed earlier, during photosynthesis carbon in its inorganic form (CO_2) gets reduced to organic compounds in chloroplasts and it highly affects the rate of photosynthesis electron transport, and thus the rate of oxygen production (Mukherjee and Moroney 2011). Aside from the above mentioned autotrophic growth mode utilizing the inorganic carbon source, some algae can also grow heterotrophically utilizing carbon sources in the media. Additionally,

Table 9.3. Comparison of the performance of MCC using different algal cultures

Algal species (Cathodic chamber)	Anodic volume (mL)	Power output (mW/m²)	Wastewater treatment efficiency	Details	Reference
Mixed microalgal culture	750	57 (spring), 1.1 (summer)	ND	Synergistic effect between bacteria (anode) & microalgae (cathode)	Venkata Mohan et al. 2014
Anabaena sp.	225	57.80	ND	Varying CO₂–air mixture concentration in sparged gas.	Pandit et al. 2014
Chlorella vulgaris	100	187 (light) and 21 (dark)	ND	MCC with self-sustainable photocathode	Liu et al. 2015
C. vulgaris	800	14.40	ND	Synthetic fruit juice wastewater treatment	Gonzalez et al. 2013
Scenedismus obliquus	250	30	ND	Optimization of cathode material	Kakarla and Min 2014
C. vulgaris	50	62.7	80%	Optimization of light intensity	Gouveia et al. 2014
C. vulgaris	220	4.1 - 5.6 W/m³	ND	Anodic off gas sequestration	Wang et al. 2010
Spirulina platensis	16	1.64 (dark) and 0.132 (light)	ND	Optimization of light intensity and algal concentration	Fu el al. 2010
C. vulgaris	–	42.96	ND	Synthetic fruit juice wastewater treatment	Gonzalez et al. 2014
C. vulgaris	400	2.26 W/m³	81.4%	Study on immobilization of algal cells	Zhou et al. 2012
Chlamydomona sreinhardtii	500	12.95	ND	Effect of light wavelength	Lan et al. 2013
C. pyrenoidosa,	250	6.36 W/m³	87%	Effect of two different algal species	Jadhav et al. 2017
A. ambigua.		4.26 W/m³	82%		

*ND-Not defined

the two growth modes may be joined in a mixotrophic growth, with photosynthesis and respiration occurring simultaneously. Thus, MCC can treat wastewater at anodic side alongside sequester the anodic off gases coming towards cathodic side and remove a part of organic carbon available in the catholyte (Wang et al. 2010). As in case of light energy, different algal species have diverse tolerance limit of confinement to carbon dioxide fixation. Chlorella species, which have high resistance for CO_2, showed a maximum growth at 10% CO_2 (Sung et al. 1998); whereas, *Scenedesmus obliquus* obtained a maximum biomass yield at 15% CO_2 concentration. From various studies, it can be inferred that increase in CO_2 concentration in cathodic chamber containing microalgae species enhances the algal biomass production, lipid content, overall power generation, as well as wastewater treatment efficiency of MCC; however the optimal concentration of CO_2 varies and depends on algal species used (Singh and Singh 2014). As discussed earlier, if the anodic off gas is considered to be the source of CO_2, the amount of CO_2 generated depends on anodic constraints and operating parameters. Hence carbon dioxide delivery to algae is one of the major limitations in the practical application of MCC.

d. *Oxygen production and removal*: The light-driven splitting of water into its elemental constituents (O_2 and H^+) occurs during photosynthesis, releasing molecular oxygen and hydrogen ion. Hydrogen ion is used to convert carbon dioxide into the organic molecules, whereas oxygen gets accumulated in the media. In closed photo bioreactors, dissolved oxygen levels become equivalent to or sometimes more than the air saturation, which causes inhibition of photosynthesis due to photorespiration. Oxygen removal from the algal chamber/closed photo bioreactor is of extreme significance and recently this area is gaining much research interest. Oxygen is normally expelled by using degasser, which not only consumes energy, but also reduces the overall productivity. Hence growing algae in the cathodic chamber of MCC can be an energy efficient remedy to curb the problems associated with photorespiration. Hence, in MCC, quenching of oxygen via oxygen reduction reaction (ORR) in the cathodic chamber helps in lowering the ill-effects of oxygen in photosynthesis process. Oxygen is one of the parameters that determines the power generation in MCC and a concentration level of about 6 mg/L is considered suitable for generation of ambient power in several studies (Khandelwal et al. 2018, Wang et al. 2010).

e. *Nutrient enrichment*: Light, carbon dioxide, and oxygen are some of the constraining factors responsible for algal productivity, followed by nutrient supply. Despite the fact that algal growth is dependent on nitrogen and phosphorus, other nutrients including carbon, silica, and other micronutrients are equally fundamental for better biomass yield. Nutrient sources can be both synthetic media namely BBM, BG11, 3NBBM, etc. (Nahidian et al. 2018), as well as natural nutrient sources, including estuaries (Human et al. 2018), industrial and domestic wastewaters (Wang et al. 2017), etc. Composition of the media, in which algae is cultured, significantly affects the cellular content of algal cell (Hong et al. 2016) and interestingly, different microalgae respond differently to changes made in nutrient media composition. In most of the cases, nitrogen and phosphorus concentrations have a positive correlation with biomass yield, but these conditions are unfavorable for enhancement of lipid content in the cells (Adams et al. 2013). This clarifies the need of optimized nutrient concentration to obtain both good growth rates of biomass as well as high lipid yields so that the scope of power generation and subsequent biodiesel production from harvested biomass in MCC is not compromised.

Anodic constraints and operating conditions

Until now our discussion has been limited to the key parameters affecting the performance of MCC with respect to the processes in cathodic compartment, whereas the limitations based on processes in anodic compartment and operating conditions have not been considered so far. The rate of CO_2 supply is essential for algal photosynthesis, as CO_2 is a prerequisite for the carboxylation of ribulose 1, 5-bisphosphate (RuBP) during the photosynthesis process (Onoda et al. 2005). In MCC, the rate of CO_2 generation depends on the anodic constraints and operating conditions. As discussed above, CO_2 can be supplied by utilizing industrial waste gases or admitting anodic off gas into the cathodic chamber of MCC. In the latter case, CO_2 generated from anodic side is a function of anodic conditions that supports bacterial activity (Logan 2009). For flawless COD removal, CO_2 production, and power generation, operating conditions, such as organic loading rate (OLR) (Hoyos-Santillan et al. 2016), hydraulic retention time (HRT) (Sharma and Li 2010), inoculums (Wang et al. 2010), substrate (Pant et al. 2010), anodic environment and anolyte pH (Jung et al. 2011) ought to be rightly chosen.

Substrate fed in MFC or MCC can be of different COD concentrations depending on the complexity of substrate ranging from complex molecules of starch to simple acetate (Pant et al. 2010). Coulombic efficiency increases with a decrease in complexity of substrate provided (Liu et al. 2009), as it provides ease for bacteria to act upon substrate. The HRT to be provided also depends mainly on the complexity or degradability of the substrate. A better control over bacterial quantity and quality can be achieved by ensuring conditions favorable for bacterial cell sustenance and growth. Focusing on pH, the consortia best performs in slightly alkaline condition (Zhuang et al. 2010, Behera and Ghangrekar 2009). The reason for this is differently addressed in various studies. Certain work demonstrated the higher internal resistance at lower pH (Behera and Ghangrekar 2009), whereas Yuan et al. explained this by dependence of electrocatalytic activity of anodic biofilms on the anolyte pH (Yuan et al. 2011), alongside Zhuang et al. (2010) discussed upon the more negative anodic potential as a result of alkaline pH. To come to a conclusion from the existing knowledge on anodic constraints and operating conditions, MCC can give a better output in terms of power, algal biomass, wastewater treatment, etc. when all the above constraints support the bacterial activity at anodic side.

Design Parameters

Design parameters that affect the performance of MCC include the reactor layout, electrode spacing, electrode material, membrane thickness and area, mixing, and size of chambers. The reactor layout is a critical parameter that influences the power generation and algal production in MCC, which can take two configurations, (a) algal chamber externally connected to the MCC (Gajda et al. 2015, Jiang et al. 2013), and (b) the algal chamber incorporated within MCC. Among the diverse reactor designs, the dual chambered MCCs are most preferred, where anolyte and catholyte are separated by a membrane (proton exchange membrane, PEM), electrodes are linked by an external circuit and CO_2 produced at anodic chamber is transferred to the cathodic chamber (Khandelwal et al. 2018). In single-chambered MCC, where the electrodes were placed in a single chamber without separators, algal bacterial symbiosis occurs (Fu et al. 2010). Aside from this, an airlift MCC system (Hu et al. 2015) and sediment MCC (Neethu and Ghangrekar 2017) achieved high level of CO_2 fixation and wastewater/sediment remediation.

Materials used in fabrication of MCC should be biocompatible, especially the electrode material, as in the case of carbon-based material, such as graphite rod, carbon felt, carbon cloth, etc. Among these electrode materials, the surface area and texture of graphite felt that assists in uniform biofilm formation and colonization of bacteria makes it one of the most suitable electrode material. Although, utilization of appropriate material can diminish the activation losses (Zhou et al. 2011). Reducing the spacing between anode and cathode can bring down the ohmic losses (Doherty et al. 2015), thus increasing the power generation (Table 9.4).

Maintaining anaerobicity in anodic chamber is a prerequisite for the growth of exo-electrogenic bacteria in MCC and hence, separator plays a major role in design of MCC. An ideal separator should have higher proton conductivity, ion transport number, ion exchange capacity, water absorption and minimal oxygen diffusion, resistance, acetate cross-over, biodegradability, etc. (Tanaka 2015). Among different membranes used in MCC, Nafion (Choi et al. 2005) and its modifications including Nafion-silica composite membrane (Xi et al. 2007), Nafion-graphene oxide composite membranes (Peng et al. 2016), bipolar membranes (Karimi et al. 2012), chitosan-graphene oxide mixed-matrix membranes (Holder et al. 2017), glass wool (Xu et al. 2018), ceramic membranes (Yousefi et al. 2017), clayware separator (Jadhav et al. 2016), and coconut shell separator (Neethu et al. 2018) were used successfully in MFC and MCC. The significance of membrane when it comes to scaling up of MCC, considering both stability and cost, has

Table 9.4. Effect of various design parameters on the performance on MCC

Anode material	Cathode material	Reactor configuration	Substrate	Algal biocathode/ catholyte	Power density	Reference
Carbon cloth	Carbon cloth	Dual chamber	SWW	*Spirulina*	0.85 W/m^2	Colombo et al. 2017
Carbon fiber	Carbon paper	Dual chamber	SWW	*Microcystis aeruginosa*	58.4 mW/m^3	Cai et al. 2013
Carbon felt	Carbon felt	Dual chamber	UASB (inoculum) + SWW	*Chlorella*	158.2 ± 15.1 mW/m^2	Zheng et al. 2017
	Stainless steel	MBR-MFC	SWW	*Chlorella pyrenoidosa*	1.2 W/m^3	Yang et al. 2017
Graphite rod	Graphite rod	Dual chamber	SWW	*Chlorella vulgaris*	0.6 mW/m^2	Mitra and Hill 2012
Carbon cloth	Carbon cloth	-	-	*Scenedesmu* ssp.	-	Angelaalincy et al. 2017
Graphite felt	Graphite felt	Dual chamber	SWW	*Desmodesmu* ssp.	64.2 mW/m^2	Wu et al. 2014
Carbon felt	Carbon felt	Dual chamber	SWW	*Chlorellasorokiniana*	3.2 W/m^3	Neethu et al. 2018
Carbon paper	Carbon paper	Dual chamber	*Chlorella pyrenoidosa*	*C. pyrenoidosa*	2.5 mW/m^2	Xu et al. 2015
Carbon brush	Carbon felt	Dual chamber	SWW	*C. vulgaris*	187 mW/m^2	Liu et al. 2015
Graphite felt	Graphite felt	Dual chamber	SWW	*C. vulgaris*	2.7 W/m^3	Khandelwal et al. 2018
Carbon felt	Carbon felt	Dual chamber	Sediment (inoculums) + SWW	*Chlorella*	202.9 ± 18.1 mW/m^2	Zheng et al. 2017

*UASB - Up-flow anaerobic sludge blanket reactor; SWW - Synthetic wastewater

prompted tremendous attempts in the advancement of a suitable material that can viably fit in as a low-cost PEM, and the researchers are still working on it.

Applications of MCC

Wastewater treatment

Huge volume of low strength as well as high-strength wastewaters are produced from domestic, industrial, and agricultural activities throughout the year. Even though various energy-consuming costly treatment techniques have been employed for wastewater remediation, MFC has proven its application in treating wastewater with different strengths and composition as evident from the research on remediation of domestic wastewater (Lefebvre et al. 2011), swine wastewater (Ding et al. 2017), cow-waste slurry (Yokoyama et al. 2006), industrial wastewater including distillery (Sonawane et al. 2014), paper and pulp (Huang and Logan 2008), chocolate industry (Patil et al. 2009), etc. Further, integrating microalgae in the cathodic chamber of MFC (i.e., MCC) will provide an added advantage of removing nutrients from wastewater streams along with carbon capture (Neethu and Ghangrekar 2017).

Electricity generation

The need for new clean energy resources and advancements in the field of biotechnology have driven scientists to utilize the ability of microbes to produce electricity, which eventually led to development of MFC and later on MCC (Santoro et al. 2017). For an MFC using oxygen, the maximum theoretical cell potential will be 1.105 V; however, the observed potential will be practically much less than theoretical potential due to several losses associated with this bio-electrochemical system (Logan et al. 2006). Studies were carried out by varying substrates, inoculums, design factors, materials (electrodes, membranes, etc.), operating conditions, etc. MCC uses the algal potential to produce oxygen with an added advantage of using algae for carbon capture as well as generating feedstock for emerging biodiesel production. Recently, stability of power generation was found to be improved on addition of acetate to domestic wastewater, where a CE of 26% and power density of 252 mW/m^2 was obtained from MCC (Stager et al. 2017). Enhancement of power generation using various low-cost catalysts to accelerate the reaction kinetics is a trending area of research nowadays. The performance of the system as a whole depends on the electrochemical reactions that occur between substrate oxidation to the final electron acceptor.

Carbon capture and biomass production

Microalgae are the most productive biological systems that produce biomass by sequestering carbon dioxide. Incorporating microalgae in cathodic chamber of MCC will serve as bio catalyst as well as an effective candidate for carbon capture. The CO_2 fixation rate varies with species, as evident in the case of *Chlorella vulgaris* with a carbon fixation rate of 6.17 mg/L.h (Bhola et al. 2011) unlike *Scenedesmus* species having the optimal CO_2 consumption rate of 59.19 mg/L.h (Ho et al. 2012). Microalgae can potentially be exploited for CO_2 sequestration; however, the carbon capture for individual algae depends upon the growth kinetics, nutrient availability for growth, operating conditions, and light intensity.

Bottlenecks and Perspectives

Integrating microalgae with MFC is quite challenging as the system is a combination of different units in various phases consolidated together and hence, the overall performance rely upon several operating conditions and design parameters. Response of microalgae to above mentioned parameters varies with species/strains, and as a result, all the essential parameters have to be optimized to a specific species, which is a big challenge. The next challenge faced by this system will be concentration of the algal biomass, increase of which should be balanced with additional requirements of light, nutrient, and CO_2, to avoid deficiency of the same. As discussed in the previous sections, CO_2 is one of the limiting parameters that effects the algal growth and hence the performance of MCC. In cases where CO_2 is supplied from the anodic chamber, it has to be ensured whether algae get sufficient CO_2 essential to support its metabolic activity. Hence, there is a need to study the factors affecting algal biomass production at cathode in terms of CO_2 supply from anode, CO_2 capture rate, O_2 removal rate and thermodynamic cycling, which will involve gas monitoring and flow control of anodic off gases. Another bottleneck will be the nutrient media required by algae, as using costly synthetic media is not always economically viable all the time and additionally growing algae in wastewater will be quite challenging while dealing with pure culture.

Difficulties that would be confronted while integrating algae in MFC are discussed, aside from this MFC itself have several bottlenecks pertaining to the anodic constraints, operating conditions, design parameters, cathodic limitations, etc. The amount of CO_2 generated in anodic chamber is quite important as it has an effect on the algal yield in the cathodic chamber. Hence, it is necessary to provide favorable conditions for the anodic bacterial consortium, which are the sole population responsible for CO_2 generation. Temperature is one such factor on which bacterial activity relies. Exo-electrogenic bacterial activity as well as algal activity change with change in temperature depending upon the strains used, and hence it is a challenge to maintain optimum temperature throughout the operating period of MCC.

Another important concern is the design parameters, which include the configuration and materials employed in fabrication of MCC. Configuration of MCC should be such that it allows easy flow of CO_2 from anodic chamber towards the cathodic chamber with minimal loss. Similarly, a low-cost membrane separator with minimal oxygen and acetate diffusion along with high proton conduction is a prerequisite for efficient performance of MCC, which is yet to be synthesised. In spite of all these, the voltage generated has a positive correlation with the dissolved oxygen concentration of catholyte (Neethu et al. 2018). During the light phase, photosynthetic release of oxygen occurs, whereas during dark phase, respiration process needs oxygen. As oxygen is the key electron acceptor in MCC, its variation will be reflected in the produced voltage, which can be a major challenge while considering the field scale application of MCC. In order to compete realistically with other prevailing power generation technologies and feedstock yield for biodiesel production, MCC should turn out to be more aggressive from the viewpoint of both effectiveness and cost. Consequently, there is a need to optimize various parameters to make MCC a great source of power and establish higher yield of algal biomass for biodiesel production, along with offering efficient treatment to the wastewater and sequestration of CO_2.

Conclusions

Biocatalyst application in MFC greatly enhances its power output. Exo-electrogens are capable of oxidising organic matter into a viable form of energy, electrical energy, because of the syntrophic interactions between the microorganisms. Exchange of electrons between microbes and electrode is the result of different electron transfer modes, either with the presence of exo- or endogenous secreted mediators. However, a better understanding of this electron transfer phenomenon at a molecular level is yet to be modeled in MFC, so that genetically engineered exo-electrogens can be developed for further enhancement of performance of MFC. Yeast, used as biocatalyst in MFC, exhibits different electron transfer mechanisms according to the strains used. Apart from having complicated analysis procedures, proper electron transfer mechanism performed by yeasts is still unknown, which restricts optimized usage of different yeast strains in MFC. In the cathodic part of MFC, oxygenic photosynthesis by microalgae can maintain higher DO level, eliminating the need of mechanical aeration and energy requirement, thus reducing operational cost of MFC. Still there is a need for development of an automated system, based on the optimized design parameters of MCC along with its feasibility evaluation for field scale applications. Hence, if these limitations are addressed precisely by utilization of biocatalysts, such as yeast and algae, which can take forward the MCC technology to an advanced level for its real life futuristic applications.

Acknowledgement

The grant received from The Department of Biotechnology, Government of India (BT/EB/PAN IIT/2012) to undertake this work is duly acknowledged.

REFERENCES

Adams, C., V. Godfrey, B. Wahlen, L. Seefeldt and B. Bugbee. 2013. Understanding precision nitrogen stress to optimize the growth and lipid content tradeoff in oleaginous green microalgae. Bioresour. Technol. 131: 188-194.

Angelaalincy, M., N. Senthilkumar, R. Karpagam, G.G. Kumar, B. Ashokkumar and P. Varalakshmi. 2017. Enhanced extracellular polysaccharide production and self-sustainable electricity generation for PAMFCS by Scenedesmus sp. SB1. ACS Omega 2(7): 3754-3765.

Babanova, S., Y. Hubenova and M. Mitov. 2011a. Influence of artificial mediators on yeast-based fuel cell performance. J. Biosci. Bioeng. 112(4): 379-387.

Babanova, S., Y. Hubenova, M. Mitov and P. Mandjukov. 2011b. Uncertainties of yeast-based biofuel cell operational characteristics. Fuel Cells 11(6): 824-837.

Behera, M. and M.M. Ghangrekar. 2009. Performance of microbial fuel cell in response to change in sludge loading rate at different anodic feed pH. Bioresour. Technol. 100(21): 5114-5121.

Bennetto, H.P. 1990. Electricity generation by microorganisms. Biotechnol. Educ. 1(4): 163-168.

Bennetto, H.P., J.L. Stirling, K. Tanaka and C.A. Vega. 1983. Anodic reactions in microbial fuel cells. Biotechnol. Bioeng. 25(2): 559-568.

Bhola, V., R. Desikan, S.K. Santosh, K. Subburamu, E. Sanniyasi and F. Bux. 2011. Effects of parameters affecting biomass yield and thermal behaviour of Chlorella vulgaris. J. Biosci. Bioeng. 111(3): 377-382.

Blair, M.F., B. Kokabian and V.G. Gude. 2014. Light and growth medium effect on Chlorella vulgaris biomass production. J. Environ. Chem. Eng. 2(1): 665-674.

Borst-Pauwels, G.W.F.H. 1981. Ion transport in yeast. BBA - Rev. Biomembr. 650(2-3): 88-127.

Cabib, E., D.H. Roh, M. Schmidt, L.B. Crotti and A. Varma. 2001. The yeast cell wall and septum as paradigms of cell growth and morphogenesis. J. Biol. Chem. 276(23): 19679-19682.

Cai, P.J., X. Xiao, Y.R. He, W.W. Li, G.L. Zang, G.P. Sheng, M. Hon-Wah Lam, L. Yu and H.Q. Yu. 2013. Reactive oxygen species (ROS) generated by cyanobacteria act as an electron acceptor in the biocathode of a bio-electrochemical system. Biosens. Bioelectron. 39(1): 306-310.

Chaudhuri, S.K. and D.R. Lovley. 2003. Electricity generation by direct oxidation of glucose in mediatorless microbial fuel cells. Nature Biotechnology 21(10): 1229-1232.

Chen, S., G. He, Q. Liu, F. Harnisch, Y. Zhou, Y. Chen, M. Hanif, S. Wang, X. Peng, H. Hou and U. Schröder. 2012. Layered corrugated electrode macrostructures boost microbial bioelectrocatalysis. Energy Environ. Sci. 5(12): 9769.

Choi, P., N.H. Jalani and R. Datta. 2005. Thermodynamics and proton transport in Nafion. J. Electrochem. Soc. 152(8): a1548.

Colombo, A., S. Marzorati, G. Lucchini, P. Cristiani, D. Pant and A. Schievano. 2017. Assisting cultivation of photosynthetic microorganisms by microbial fuel cells to enhance nutrients recovery from wastewater. Bioresour. Technol. 237: 240-248.

Ding, W., S. Cheng, L. Yu and H. Huang. 2017. Effective swine wastewater treatment by combining microbial fuel cells with flocculation. Chemosphere 182: 567-573.

Doherty, L., X. Zhao, Y. Zhao and W. Wang. 2015. The effects of electrode spacing and flow direction on the performance of microbial fuel cell-constructed wetland. Ecol. Eng. 79: 8-14.

Fu, C.C., T.C. Hung, W.T. Wu, T.C. Wen and C.H. Su. 2010. Current and voltage responses in instant photosynthetic microbial cells with Spirulina platensis. Biochem. Eng. J. 52(2-3): 175-180.

Gajda, I., J. Greenman, C. Melhuish and I. Ieropoulos. 2015. Self-sustainable electricity production from algae grown in a microbial fuel cell system. Biomass and Bioenergy 82: 1-7.

Gal, I., O. Schlesinger, L. Amir and L. Alfonta. 2016. Yeast surface display of dehydrogenases in microbial fuel-cells. Bioelectrochemistry 112: 53-60.

Ganguli, R. and B. Dunn. 2012. Electrically conductive, immobilized bioanodes for microbial fuel cells. Nanotechnology 23(29): 294013.

Ganguli, R. and B.S. Dunn. 2009. Kinetics of anode reactions for a yeast-catalysed microbial fuel cell. Fuel Cells 9(1): 44-52.

González, A., P. Cañizares, M.A. Rodrigo, F.J. Fernández, and J. Lobato. 2013. Microbial fuel cell with an algae-assisted cathode: a preliminary assessment. Journal of Power Sources 242: 638-645. http://dx.doi.org/10.1016/j.jpowsour.2013.05.110.

Gonzalez del Campo, A., J.F. Perez, P. Cañizares, M.A. Rodrigo, F.J. Fernandez and J. Lobato. 2014. Study of a photosynthetic MFC for energy recovery from synthetic industrial fruit juice wastewater. International Journal of Hydrogen Energy 39: 21828-21836.

Gouveia, L., C. Neves, D. Sebastião, B.P. Nobre and C.T. Matos. 2014. Effect of light on the production of bioelectricity and added-value microalgae biomass in a photosynthetic alga microbial fuel cell. Bioresource Technology 154: 171-177.

Gunawardena, A., S. Fernando and F. To. 2008. Performance of a yeast-mediated biological fuel cell. Int. J. Mol. Sci. 9(10): 1893-1907.

Hasan, K., M. Grattieri, T. Wang, R.D. Milton, and S.D. Minteer. 2017. Enhanced Bioelectrocatalysis of Shewanella Oneidensis MR-1 by a Naphthoquinone Redox Polymer. ACS Energy Letters 2, no. 9: 1947–1951.

Haslett, N.D., F.J. Rawson, F. Barriëre, G. Kunze, N. Pasco, R. Gooneratne and K.H.R. Baronian. 2011. Characterisation of yeast microbial fuel cell with the yeast Arxula adeninivorans as the biocatalyst. Biosens. Bioelectron. 26(9): 3742-3747.

Ho, S.H., C.Y. Chen and J.S. Chang. 2012. Effect of light intensity and nitrogen starvation on CO_2 fixation and lipid/carbohydrate production of an indigenous microalga Scenedesmus obliquus CNW-n. Bioresour. Technol. 113: 244-252.

Holder, S.L., C.H. Lee and S.R. Popuri. 2017. Simultaneous wastewater treatment and bioelectricity production in microbial fuel cells using cross-linked chitosan-graphene oxide mixed-matrix membranes. Environ. Sci. Pollut. Res. 24(15): 13782-13796.

Hong, M.E., Y.Y. Choi and S.J. Sim. 2016. Effect of red cyst cell inoculation and iron(ii) supplementation on autotrophic astaxanthin production by Haematococcus pluvialis under outdoor summer conditions. J. Biotechnol. 218: 25-33.

Hoyos-Santillan, J., B.H. Lomax, D. Large, B.L. Turner, A. Boom, O.R. Lopez and S. Sjögersten. 2016. Quality not quantity: organic matter composition controls of CO_2 and CH_4 fluxes in neotropical peat profiles. Soil Biol. Biochem. 103: 86-96.

Hu, J.Y. and T. Sato. 2017. A photobioreactor for microalgae cultivation with internal illumination considering flashing light effect and optimized light-source arrangement. Energy Convers. Manag. 133: 558-565.

Hu, X., B. Liu, J. Zhou, R. Jin, S. Qiao and G. Liu. 2015. CO_2 fixation, lipid production, and power generation by a novel air-lift-type microbial carbon capture cell system. Environ. Sci. Technol. 49(17): 10710-10717.

Huang, L. and B.E. Logan. 2008. Electricity generation and treatment of paper recycling wastewater using a microbial fuel cell. Applied Microbiology and Biotechnology 80(2): 349-355.

Hubenova, Y., D. Georgiev and M. Mitov. 2014. Enhanced phytate dephosphorylation by using Candida melibiosica yeast-based biofuel cell. Biotechnol. Lett. 36(10): 1993-1997.

Hubenova, Y. and M. Mitov. 2010. Potential application of Candida melibiosica in biofuel cells. Bioelectrochemistry 78(1): 57-61.

Hubenova, Y.V., R.S. Rashkov, V.D. Buchvarov, M.H. Arnaudova, S.M. Babanova and M.Y. Mitov. 2011. Improvement of yeast-biofuel cell output by electrode modifications. Ind. Eng. Chem. Res. 50(2): 557-564.

Human, L.R.D., M.L. Magoro, T. Dalu, R. Perissinotto, A.K. Whitfield, J.B. Adams, S.H.P. Deyzel and G.M. Rishworth. 2018. Natural nutrient enrichment and algal responses in near pristine micro-estuaries and micro-outlets. Sci. Total Environ. 624: 945-954.

Jadhav, D.A., S.C. Jain and M.M. Ghangrekar. 2016. Cow's urine as a yellow gold for bioelectricity generation in low cost clayware microbial fuel cell. Energy 113: 76-84.

Jadhav, D.A., S.C. Jain and M.M. Ghangrekar. 2017. Simultaneous wastewater treatment, algal biomass production and electricity generation in clayware microbial carbon capture cells. Appl. Biochem. Biotechnol. 183(3): 1076-1092.

Jeon, Y.C., C.W. Cho and Y.S. Yun. 2006. Combined effects of light intensity and acetate concentration on the growth of unicellular microalga Haematococcus pluvialis. Enzyme Microb. Technol. 39(3): 490-495.

Jiang, H.M., S.J. Luo, X.S. Shi, M. Dai and R.B. Guo. 2013. A system combining microbial fuel cell with photobioreactor for continuous domestic wastewater treatment and bioelectricity generation. J. Cent. South Univ. 20(2): 488-494.

Jung, S., M.M. Mench and J.M. Regan. 2011. Impedance characteristics and polarization behavior of a microbial fuel cell in response to short-term changes in medium pH. Environ. Sci. Technol. 45(20): 9069-9074.

Kakarla, R. and B. Min. 2014. Evaluation of microbial fuel cell operation using algae as an oxygen supplier: carbon paper cathode vs. carbon brush cathode. Bioprocess and Biosystems Engineering 37(12): 2453-2461.

Kaneshiro, H., K. Takano, Y. Takada, T. Wakisaka, T. Tachibana and M. Azuma. 2013. A milliliter-scale yeast-based fuel cell with high performance. Biochem. Eng. J. 83: 90-96.

Karimi, S., N. Fraser, B. Roberts and F.R. Foulkes. 2012. A review of metallic bipolar plates for proton exchange membrane fuel cells: materials and fabrication methods. Adv. Mater. Sci. Eng. 1155: 1-22.

Kasem, E.T., T. Tsujiguchi and N. Nakagawa. 2013. Effect of metal modification to carbon paper anodes on the performance of yeast-based microbial fuel cells. Part II: in the case with exogenous mediator, methylene blue. Key Eng. Mater. 534: 82-87.

Khandelwal, A., A. Vijay, A. Dixit and M. Chhabra. 2018. Microbial fuel cell powered by lipid extracted algae: a promising system for algal lipids and power generation. Bioresour. Technol. 247: 520-527.

Kim, H.J., H.S. Park, M.S. Hyun, I.S. Chang, M. Kim and B.H. Kim. 2002. A mediator-less microbial fuel cell using a metal reducing bacterium, Shewanella putrefaciens. Enzyme and Microbial Technology 30(2): 145-152.

Lan, J.C.W., K. Raman, C.M. Huang and C.M. Chang. 2013. The impact of monochromatic blue and red led light upon performance of photo microbial fuel cells (PMFCs) using chlamydomonas reinhardtii transformation F5 as biocatalyst. Biochemical Engineering Journal 78: 39-43. http://dx.doi.org/10.1016/j.bej.2013.02.007.

Lee, C.-G. and B. Palsson. 1994. High-density algal photobioreactors using light-emitting diodes. Biotechnol. Bioeng. 44(10): 1161-1167.

Lefebvre, O., A. Uzabiaga, I.S. Chang, B.H. Kim and H.Y. Ng. 2011. Microbial fuel cells for energy self-sufficient domestic wastewater treatment – a review and discussion from energetic consideration. Appl. Microbiol. Biotechnol. 89(2): 259-270.

Li, L., J. Weiner, D. Zhou, Y. Huang and L. Sheng. 2013. Initial density affects biomass-density and allometric relationships in self-thinning populations of Fagopyrum esculentum. J. Ecol. 101(2): 475-483.

Li, Y., W. Zhou, B. Hu, M. Min, P. Chen and R.R. Ruan. 2012. Effect of light intensity on algal biomass accumulation and biodiesel production for mixotrophic strains Chlorella kessleri and Chlorella protothecoide cultivated in highly concentrated municipal wastewater. Biotechnol. Bioeng. 109(9): 2222-2229.

Liu, T., L. Rao, Y. Yuan and L. Zhuang. 2015. Bioelectricity generation in a microbial fuel cell with a self-sustainable photocathode. Sci. World J. 2015: 864568.

Liu, Z., J. Liu, S. Zhang and Z. Su. 2009. Study of operational performance and electrical response on mediator-less microbial fuel cells fed with carbon- and protein-rich substrates. Biochem. Eng. J. 45(3): 185-191.

Logan, B.E. 2009. Exoelectrogenic bacteria that power microbial fuel cells. Nat. Rev. Microbiol. 7(5): 375-381.

Logan, B.E., B. Hamelers, R. Rozendal, U. Schröder, J. Keller, S. Freguia, P. Aelterman, W. Verstraete and K. Rabaey. 2006. Microbial fuel cells: methodology and technology. Environ. Sci. Technol. 40(17): 5181-5192.

Melnicki, M.R., G.E. Pinchuk, E.A. Hill, L.A. Kucek, S.M. Stolyar, J.K. Fredrickson, A.E. Konopka and A.S. Beliaev. 2013. Feedback-controlled led photobioreactor for photophysiological studies of cyanobacteria. Bioresour. Technol. 134: 127-133.

Mitra, P. and G.A. Hill. 2012. Continuous microbial fuel cell using a photoautotrophic cathode and a fermentative anode. Can. J. Chem. Eng. 90(4): 1006-1010.

Mukherjee, B. and J.V. Moroney. 2011. Algal carbon dioxide concentrating mechanisms. eLS, John Wiley & Sons Ltd., Chichester.

Nahidian, B., F. Ghanati, M. Shahbazi and N. Soltani. 2018. Effect of nutrients on the growth and physiological features of newly isolated Haematococcus pluvialis TMU1. Bioresour. Technol. 255: 229-237.

Neethu, B., G.D. Bhowmick and M.M. Ghangrekar. 2018. Enhancement of bioelectricity generation and algal productivity in microbial fuel cell using low cost coconut shell as membrane separator. Biochemical Engineering Journal 133: 205-213.

Neethu, B. and M.M. Ghangrekar. 2017. Electricity generation through a photo sediment microbial fuel cell using algae at the cathode. Water Sci. Technol. 76(12): 3269-3277.

Onoda, Y., K. Hikosaka and T. Hirose. 2005. The balance between RuBP carboxylation and RuBP regeneration: a mechanism underlying the interspecific variation in acclimation of photosynthesis to seasonal change in temperature. Funct. Plant Biol. 32(10): 903-910.

Pandit, S., B.K. Nayak and D. Das. 2012. Microbial carbon capture cell using cyanobacteria for simultaneous power generation, carbon dioxide sequestration and wastewater treatment. Bioresource Technology 107: 97-102.

Pant, D., G. Van Bogaert, L. Diels and K. Vanbroekhoven. 2010. A review of the substrates used in microbial fuel cells (MFCs) for sustainable energy production. Bioresour. Technol. 101(6): 1533-1543.

Patil, S.A., V.P. Surakasi, S. Koul, S. Ijmulwar, A. Vivek, Y.S. Shouche and B.P. Kapadnis. 2009. Electricity generation using chocolate industry wastewater and its treatment in activated sludge based microbial fuel cell and analysis of developed microbial community in the anode chamber. Bioresour. Technol. 100(21): 5132-5139.

Peng, K.-J., J.-Y. Lai and Y.-L. Liu. 2016. Nanohybrids of graphene oxide chemically-bonded with Nafion: preparation and application for proton exchange membrane fuel cells. J. Memb. Sci. 514: 86-94.

Permana, D., D. Rosdianti, S. Ishmayana, S.D. Rachman, H.E. Putra, D. Rahayuningwulan and H.R. Hariyadi. 2015. Preliminary investigation of electricity production using dual chamber microbial fuel cell (dcMFC) with Saccharomyces cerevisiae as biocatalyst and methylene blue as an electron mediator. Procedia Chem. 17: 36-43.

Perner-Nochta, I. and C. Posten. 2007. Simulations of light intensity variation in photobioreactors. J. Biotechnol. 131(3): 276-285.

Philippart, C.J.M. 1995. Effects of shading on growth, biomass and population maintenance of the intertidal seagrass Zostera noltii hornem. in the Dutch Wadden Sea. J. Exp. Mar. Bio. Ecol. 188(2): 199-213.

Potter, M.C. 1911. Electrical effects accompanying the decomposition of organic compounds. Proc. R. Soc. B Biol. Sci. 84(571): 260-276.

Prasad, D., S. Arun, M. Murugesan, S. Padmanaban, R.S. Satyanarayanan, S. Berchmans and V. Yegnaraman. 2007. Direct electron transfer with yeast cells and construction of a mediatorless microbial fuel cell. Biosens. Bioelectron. 22(11): 2604-2610.

Qiang, H. and A. Richmond. 1996. Productivity and photosynthetic efficiency of Spirulina platensis as affected by light intensity, algal density and rate of mixing in a flat plate photobioreactor. J. Appl. Phycol. 8(2): 139-145.

Raghavulu, S.V., R.K. Goud, P.N. Sarma and S.V. Mohan. 2011. Saccharomyces cerevisiae as anodic biocatalyst for power generation in biofuel cell: influence of redox condition and substrate load. Bioresour. Technol. 102(3): 2751-2757.

Rahimnejad, M., G.D. Najafpour, A.A. Ghoreyshi, F. Talebnia, G.C. Premier, G. Bakeri, J.R. Kim and S.E. Oh. 2012. Thionine increases electricity generation from microbial fuel cell using Saccharomyces cerevisiae and exoelectrogenic mixed culture. J. Microbiol. 50(4): 575-580.

Rawson, F.J., A.J. Gross, D.J. Garrett, A.J. Downard and K.H.R. Baronian. 2012. Mediated electrochemical detection of electron transfer from the outer surface of the cell wall of Saccharomyces cerevisiae. Electrochem. Commun. 15(1): 85-87.

Santoro, C., C. Arbizzani, B. Erable and I. Ieropoulos. 2017. Microbial fuel cells: from fundamentals to applications: a review. J. Power Sources 356: 225-244.

Sayed, E.T., N.A.M. Barakat, M.A. Abdelkareem, H. Fouad and N. Nakagawa. 2015. Yeast extract as an effective and safe mediator for the baker's-yeast-based microbial fuel cell. Ind. Eng. Chem. Res. 54(12): 3116-3122.

Sayed, E.T., T. Tsujiguchi and N. Nakagawa. 2012. Catalytic activity of baker's yeast in a mediatorless microbial fuel cell. Bioelectrochemistry 86: 97-101.

Schaefer, A.L., E.P. Greenberg, C.M. Oliver, Y. Oda, J.J. Huang, G. Bittan-Banin, C.M. Peres, S. Schmidt, K. Juhaszova, J.R. Sufrin and C.S. Harwood. 2008. A new class of homoserine lactone quorum-sensing signals. Nature 454(7204): 595-599.

Schaetzle, O., F. Barrière and K. Baronian. 2008. Bacteria and yeasts as catalysts in microbial fuel cells: electron transfer from micro-organisms to electrodes for green electricity. Energy Environ. Sci. 1(6): 607.

Sharma, Y. and B. Li. 2010. Optimizing energy harvest in wastewater treatment by combining anaerobic hydrogen producing biofermentor (HPB) and microbial fuel cell (MFC). Int. J. Hydrogen Energy 35(8): 3789-3797.

Shigesada, N. and A. Okubo. 1981. Analysis of the self-shading effect on algal vertical distribution in natural waters. J. Math. Biol. 12(3): 311-326.

Shkil, H., A. Schulte, D.A. Guschin and W. Schuhmann. 2011. Electron transfer between genetically modified Hansenula polymorpha yeast cells and electrode surfaces via OS-complex modified redox polymers. ChemPhysChem 12(4): 806-813.

Singh, S.P. and P. Singh. 2014. Effect of CO_2 concentration on algal growth: a review. Renew. Sustain. Energy Rev. 38: 172-179.

Sonawane, J.M., E. Marsili and P. Chandra Ghosh. 2014. Treatment of domestic and distillery wastewater in high surface microbial fuel cells. Int. J. Hydrogen Energy 39(36) ??.

Srikanth, S. and S. Venkata Mohan. 2012. Change in electrogenic activity of the microbial fuel cell (MFC) with the function of biocathode microenvironment as terminal electron accepting condition: influence on overpotentials and bio-electro kinetics. Bioresour. Technol. 119: 241-251.

Stager, J.L., X. Zhang and B.E. Logan. 2017. Addition of acetate improves stability of power generation using microbial fuel cells treating domestic wastewater. Bioelectrochemistry 118: 154-160.

Sung, K.D., J.S. Lee, C.S. Shin and S.C. Park. 1998. Isolation of a new highly CO_2 tolerant fresh water Microalga chlorella sp. KR-1. Renew. Energy 16(1-4): 1019-1022.

Tanaka, Y. 2015. Ion Exchange Membranes: Fundamentals and Applications. Second Edition. Elsevier.

Tilzer, M.M. 1987. Light-dependence of photosynthesis and growth in cyanobacteria: implications for their dominance in eutrophic lakes. New Zeal. J. Mar. Freshw. Res. 21(3): 401-412.

Ugwu, C.U. and H. Aoyagi. 2008. Influence of shading inclined tubular photobioreactor surfaces on biomass productivity of C. sorokiniana. Photosynthetica 46(2): 283-285.

Venkata Mohan, S., R. Saravanan, S.V. Raghavulu, G. Mohanakrishna and P.N. Sarma. 2008. Bioelectricity production from wastewater treatment in dual chambered microbial fuel cell (MFC) using selectively enriched mixed microflora: effect of catholyte. Bioresour. Technol. 99(3): 596-603.

Venkata Mohan, S., S. Srikanth, P. Chiranjeevi, S. Arora and R. Chandra. 2014. Algal biocathode for in situ terminal electron acceptor (TEA) production: synergetic association of bacteria-microalgae metabolism for the functioning of biofuel cell. Bioresource Technology 166: 566-574.

Volkov, V. 2015. Quantitative description of ion transport via plasma membrane of yeast and small cells. Front. Plant Sci. 6: 425.

Walker, A.L. and C.W. Walker. 2006. Biological fuel cell and an application as a reserve power source. J. Power Sources 160(1): 123-129.

Wang, A., D. Sun, N. Ren, C. Liu, W. Liu, B.E. Logan and W.M. Wu. 2010. A rapid selection strategy for an anodophilic consortium for microbial fuel cells. Bioresour. Technol. 101(14): 5733-5735.

Wang, J., D. Han, M.R. Sommerfeld, C. Lu and Q. Hu. 2013. Effect of initial biomass density on growth and astaxanthin production of Haematococcus pluvialis in an outdoor photobioreactor. J. Appl. Phycol. 25(1): 253-260.

Wang, J.H., T.Y. Zhang, G.H. Dao, X.Q. Xu, X.X. Wang and H.Y. Hu. 2017. Microalgae-based advanced municipal wastewater treatment for reuse in water bodies. Applied Microbiology and Biotechnology 101(7): 2059-2675.

Wei, L., H. Han and J. Shen. 2012. Effects of cathodic electron acceptors and potassium ferricyanide concentrations on the performance of microbial fuel cell. Int. J. Hydrogen Energy 37(17): 12980-12986.

Wijffels, R.H., A.W. Schepers, M. Smit, C.D. de Gooijer and J. Tramper. 1994. Effect of initial biomass concentration on the growth of immobilized Nitrosomonas europaea. Appl. Microbiol. Biotechnol. 42(1): 153-157.

Wilkinson, S., J. Klar and S. Applecart. 2006. Optimizing biofuel cell performance using a targeted mixed mediator combination. Electroanalysis 18(19-20): 2001-2007.

Wu, Y. Cheng, Z. jie Wang, Y. Zheng, Y. Xiao, Z. hui Yang and F. Zhao. 2014. Light intensity affects the performance of photo microbial fuel cells with Desmodesmus sp. A8 as cathodic microorganism. Appl. Energy 116: 86-90.

Xi, J., Z. Wu, X. Qiu and L. Chen. 2007. Nafion/SiO_2 hybrid membrane for vanadium redox flow battery. J. Power Sources 166(2): 531-536.

Xu, C., K. Poon, M.M.F. Choi and R. Wang. 2015. Using live algae at the anode of a microbial fuel cell to generate electricity. Environ. Sci. Pollut. Res. ??

Xu, L., Y. Zhao, C. Tang and L. Doherty. 2018. Influence of glass wool as separator on bioelectricity generation in a constructed wetland-microbial fuel cell. J. Environ. Manage. 207: 116-123.

Yang, Q., Y. Lin, L. Liu and F. Yang. 2017. A bio-electrochemical membrane system for more sustainable wastewater treatment with MnO_2/PANI modified stainless steel cathode and photosynthetic provision of dissolved oxygen by algae. Water Sci. Technol. 76(7): 1907-1914.

Yokoyama, H., H. Ohmori, M. Ishida, M. Waki and Y. Tanaka. 2006. Treatment of cow-waste slurry by a microbial fuel cell and the properties of the treated slurry as a liquid manure. Anim. Sci. J. 77(6): 634-638.

Yousefi, V., D. Mohebbi-Kalhori and A. Samimi. 2017. Ceramic-based microbial fuel cells (MFCs): a review. Int. J. Hydrogen Energy 42(3): 1672-1690.

Yuan, Y., B. Zhao, S. Zhou, S. Zhong and L. Zhuang. 2011. Electrocatalytic activity of anodic biofilm responses to pH changes in microbial fuel cells. Bioresour. Technol. 102(13): 6887-6891.

Zamfir, M. and D.E. Goldberg. 2000. The effect of initial density on interactions between bryophytes at individual and community levels. J. Ecol. 88(2): 243-255.

Zhang, K., S. Miyachi and N. Kurano. 2001. Evaluation of a vertical flat-plate photobioreactor for outdoor biomass production and carbon dioxide bio-fixation: effects of reactor dimensions, irradiation and cell concentration on the biomass productivity and irradiation utilization efficiency. Appl. Microbiol. Biotechnol. 55(4): 428-433.

Zhao, Y., S. Sun, C. Hu, H. Zhang, J. Xu and L. Ping. 2015. Performance of three microalgal strains in biogas slurry purification and biogas upgrade in response to various mixed light-emitting diode light wavelengths. Bioresour. Technol. 187: 338-345.

Zheng, W., T. Cai, M. Huang and D. Chen. 2017. Comparison of electrochemical performances and microbial community structures of two photosynthetic microbial fuel cells. J. Biosci. Bioeng. 124(5): 551-558.

Zhou, M., H. He, T. Jin and H. Wang. 2012. Power generation enhancement in novel microbial carbon capture cells with immobilized Chlorella vulgaris. Journal of Power Sources 214: 216–219. http://dx.doi.org/10.1016/j.jpowsour.2012.04.043.

Zhou, M., M. Chi, J. Luo, H. He, and T. Jin. 2011. An Overview of Electrode Materials in Microbial Fuel Cells. Journal of Power Sources 196(10: 4427-4435

Zhuang, L., S. Zhou, Y. Li and Y. Yuan. 2010. Enhanced performance of air-cathode two-chamber microbial fuel cells with high-pH anode and low-pH cathode. Bioresour. Technol. 101(10): 3514-3519.

Cost-effective Carbon Catalysts and Related Electrode Designs for the Air Cathode of Microbial Fuel Cells

Wei Yang, Jun Li[1,2,3*], Qian Fu[1,2,3], Liang Zhang[1,2,3], Xun Zhu[1,2,3] and Qiang Liao[1,2,3]

[1] Key Laboratory of Low-grade Energy Utilization Technologies and Systems, Ministry of Education, Chongqing 400030, China
[2] Institute of Engineering Thermophysics, School of Energy and Power Engineering, Chongqing University, Chongqing 400030, China
[3] No. 174, Shazhengjie, Shapingba, Chongqing, Chongiqng 400044, China

Introduction

The growing demand for energy and continuous concerns over environmental issues have encouraged human beings to explore alternative energy sources to reduce the dependency on fossil fuels and to develop new technologies to alleviate environmental pollution. Currently, microbial fuel cell (MFC) is regarded as an emerging renewable energy technology due to its capability of simultaneous electricity generation and wastewater treatment (Logan et al. 2006). MFC is a device that uses electrochemical active bacteria to catalyze the oxidation of organic pollutants in wastewater. The produced electrons and protons transfer to the cathode, where the cathodic electron acceptor (such as oxygen) was reduced by combining the electrons and protons. A schematic diagram of a single chamber MFC (SC-MFC) is shown in Fig. 10.1 (Liu et al. 2004, Zhang et al. 2010a). Due to the advantages, such as mild operation conditions and substrate (fuel) versatility, MFC has been demonstrated as a promising technology to recover electricity from wastewater.

At the early stage of MFC development, many soluble cathodic final electron acceptors, e.g., ferricyanide (Zhang et al. 2013a), persulfate (Li et al. 2009), permanganate (You et al. 2006), and tri-iodine (Li et al. 2010) have been used to sustain the overall reaction. For the future practical application of MFCs, the use of these chemicals is not suitable due to the need for regular replenishment during the operation, thus increasing the

*Corresponding author: lijun@cqu.edu.cn

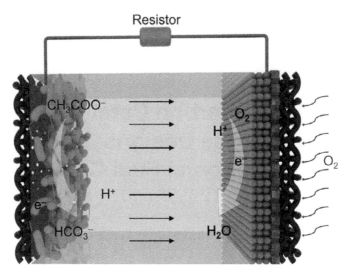

Figure 10.1. Schematic diagram of a SC-MFC.

operational cost. Alternatively, an air cathode MFC was considered as the most promising configuration due to the freely available oxygen in the air and the decreased operational costs (Wang et al. 2013). But the power output of an air cathode was usually limited by the sluggish oxygen reduction reaction (ORR) (Watson et al. 2013b). To reduce the overpotential for ORR, platinum (Pt) was commonly applied as the catalyst for the air cathode. However, the high cost and decreasing catalytic activity over time, resulting from chemical poisoning and biological fouling, would conversely influence its practical utilization (Zhang et al. 2014f). Recently, carbon-based catalysts, such as activated carbon (AC), graphene, and carbon nanotube (CNT), were considered efficient alternatives to Pt for cathode ORR (Cheng and Wu 2013, Feng et al. 2011a, Feng et al. 2011b). Exploring nanomaterials for ORR is at the forefront of material research. It was reported that heteroatom doped graphene can promote the ORR activity and a four-electron (4e$^-$) pathway in both alkaline and acidic media (Chen et al. 2015). Although with a relatively lower catalytic activity than the carbon nanomaterials, commercial AC and biomass-derived carbon have also attracted increasing attention for their application in ORR catalysts, because of the low cost and ease of preparation. Watson et al. (2013a) reported that the P_{max} of 2,450 ± 43 mW/m^2 can be achieved using ammonia-gas-treated AC as the ORR catalyst in MFCs. Similarly, carbon catalysts derived from biomass, including cellulose phosphate, cornstalk, and luffa sponge, achieved a higher P_{max} than Pt/C (Li et al. 2014b, Liu et al. 2015a, Sun et al. 2016). These studies validated the potential application of these materials in MFCs.

A typical air cathode for MFCs usually consists of a catalyst layer (CL), a support/current collector, and a gas diffusion layer (GDL). The schematic diagram of an air cathode is shown in Fig. 10.2. The ORR process occurs in the CL. A good CL usually allows for maximizing the triple-phase interfaces (TPIs) in which oxygen, protons, and electrons are simultaneously present. The GDL is a multi-functional layer that allows oxygen in the air to diffuse into the CL for ORR and prevents water leakage from MFC reactors. The electrode support/current collector functions as a cathode structural support and conducts the electrons from the external circuit to the CL for ORR. The performance of the air cathode is closely related to the cathode material properties, such as porosity or pores distribution, electrical conductivity, and wettability of the cathode, because these properties significantly affect the oxygen supply, H^+/OH^- transport, and electron transfer in the TPIs of the CL. Different electrodes and structures have variable physical and chemical properties, which have crucial impacts on the overall performance of MFCs. The development of suitable electrode materials and electrode structure design is therefore an effective approach to improve the power density and to enable the practical application of MFCs.

Unlike an anode that is usually immersed in the bulk solution, an air cathode is usually placed with one side facing the air for oxygen supply and the other side facing the solution for H^+/OH^- transport, leading to a more complex electrode configuration. The diagram of a common air cathode configuration is shown in Fig. 10.3. The commonly used air cathode has a multi-layered structure consisting of a CL and GDL. A CL, where ORR occurs, is placed on the water-facing side of GDL and is in direct contact with the electrolytes in MFCs. The major body of the GDL is usually a macroporous teflonized carbon paper/cloth with a high hydrophobicity. They are used as the structure supports for the cathode and to conduct the electrons from the circuit to the CL. The teflonized carbon paper/cloth additionally functions

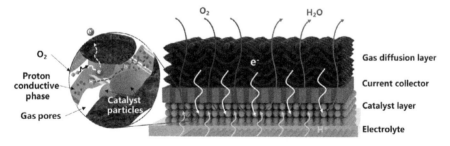

Figure 10.2. Schematic diagram of an air cathode in MFCs.

Figure 10.3. Diagram of a typical air cathode configuration.

as a waterproof layer to prevent the leakage of water from the MFC reactor, while allowing the oxygen supply from the air into the CL. To reduce ohmic resistance between the CL and the macroporous support and to provide non-permeable support during catalyst deposition, a carbon layer consisting of carbon black and polytetrafluoroethylene (PTFE), namely, a micro porous layer (MPL), is usually added between the support and the CL. To improve oxygen supply and to prevent water leakage from the air cathode MFCs, an MPL or a carbon-based layer consisting of carbon black and PTFE is applied on the air-facing side of the MFC cathode. In many studies, several hydrophobic polymer layers (HPLs) were additionally added by spraying the PTFE suspension on the MPL to strengthen the resistance of the air cathode to a high water pressure. Compared to the traditional cathode with Pt/C, cathodes fabricated using a carbon catalyst would face the challenge that the CL is usually constructed with a high catalyst-loading rate (usually > 20 mg/cm² for AC cathodes, compared with 0.5 mg/cm² for Pt cathode), which would cause a thicker CL, thus resulting in a longer transport distance and a larger mass transfer resistance for H^+ or OH^-. Due to the formation of biofilm and salt precipitation after long-term operation, the mass transport would further deteriorate.

This chapter discusses the progress and reviews the recent development of carbon materials that have been used for the air cathodes of MFCs over the past decades. This chapter (1) reviews in detail commercial ACs, carbon nanomaterials, and biomass-derived carbon catalysts for cathodes; (2) provides different cathode designs using these carbon materials; and (3) discusses the critical challenges for the practical application of the air cathode MFCs, such as poison tolerance, long-term stability, and mass transport limitations in cathodes

Carbon-based ORR Catalysts

Activated Carbon

The tailored porosity and pore-size distribution of AC catalysts have widened the utilization of AC for various demanding applications, such as catalysis/electrocatalysis, separation of multisized molecules, energy storage in capacitors, electrodes, CO_2 capture, and H_2 storage (Jain et al. 2016). Recently, as an efficient ORR catalyst, AC was widely applied for the ORR catalysts in air cathodes, as shown in Table 10.1.

Compared to Pt/C catalysts, the MFCs with AC catalysts have achieved a comparable or even higher performance. For example, Zhang et al. reported an MFC using AC as the air cathode catalyst and Ni mesh as the current collector. The as prepared MFC delivered a maximum power density (P_{max}) of 1,220 mW/m², which was higher than that (1,060 mW/m²) using a Pt-based cathode (Zhang et al. 2009a).

In general, commercial AC, such as AC powder and granular AC, is obtained from various precursors (Fig. 10.4). Watson et al. compared the

Table 10.1. Comparison of Pt/C and different carbon catalysts in air cathode of MFCs

Carbon catalyst	Binder	Current collector	Anode	Substrate	Power density (mW/m²)	
Pt/C	Nafion	Carbon cloth	Carbon fiber brush	Acetate	1060	Zhang et al. 2009a
AC	PTFE	Nickel foam	Carbon fiber brush	Acetate	1190 ± 50	Cheng and Wu 2013
AC	PTFE	SSM	Carbon mesh	Acetate	802	Dong et al. 2012b
Quaternary ammonium modified AC	PTFE	SSM	Carbon fiber brush	Acetate	2781 ± 36	Wang et al. 2014b
Carnation-like MnO_2 modified AC	PTFE	SSM	Carbon felt	Acetate	1554	Zhang et al. 2014d
Ammonia gas treated AC	PTFE	SSM	Carbon fiber brush	Acetate	2450 ± 40	Watson et al. 2013a
N-type Cu_2O modified AC	PTFE	SSM	Carbon felt	Acetate	1390 ± 76	Zhang et al. 2015b
Silver modified AC	PTFE	SSM	Not mentioned	Acetate	1080 ± 60	Pu et al. 2014
$Cu_{0.92}Co_{2.08}O_4$ modified AC	PTFE	SSM	Carbon felt	Wastewater	1895	Wang et al. 2017
Nano urchin-like $NiCo_2O_4$ modified AC	PTFE	SSM	Carbon felt	Wastewater	1730 ± 14	Ge et al. 2016
Phosphorus-doped AC	PTFE	SSM	Carbon felt	Acetate	1096 ± 33	Chen et al. 2014
Three-dimensional Cu_xO modified AC	PTFE	SSM	Carbon felt	Wastewater	1550 ± 47	Liu et al. 2015d

(Contd.)

Nitrogen doped CNTs	—	Carbon cloth	Carbon fiber brush	Acetate	1600 ±50	Feng et al. 2011b
MnO$_2$ modified CNTs	Nafion	Carbon paper	Graphite felt	Glucose	210	Zhang et al. 2011b
Manganese and polypyrrole modified CNTs	Nafion	Carbon cloth	Carbon cloth	Acetate	213	Lu et al. 2013b
Iron phthaleincyanide modified CNTs	Nafion	Carbon cloth	Carbon cloth	Glucose	601	Yuan et al. 2011
Activated nitrogen-doped carbon nanofibers	Nafion	Carbon paper	Carbon granules	Acetate	1377 ± 46	Yang et al. 2014b
Nitrogen doped graphene	Nafion	Carbon cloth	Carbon fiber brush	Acetate	1350 ± 15	Feng et al. 2011a
Iron and nitrogen doped graphene	Nafion	Carbon paper	Carbon felt	Acetate	1149.8	Li et al. 2012
Nanotubular MnO$_2$ modified graphene	PTFE	Carbon cloth	Carbon cloth	Glucose	3359	Awan et al. 2014
Nitrogen doped graphene	—	Nickel mesh	Graphite fiber brush	Acetate	1470 ±80	Wang et al. 2016
Cornstalk-derived carbon	PTFE	SSM	Carbon fiber brush	Glucose	1122	Sun et al. 2016
Plant moss derived biocarbon	Nafion	Carbon cloth	Carbon cloth	Acetate	703 ± 16	Zhou et al. 2016b

(Contd.)

Table 10.1. (*Contd.*)

Carbon catalyst	Binder	Current collector	Anode	substrate	Power density (mW/m²)	
Nitrogen and phosphorus doped bamboo charcoal	PTFE	SSM	Carbon fiber brush	Acetate	1719 ± 82	Yang et al. 2017a
Chlorella pyrenoidosa	PTFE	SSM	Carbon fiber brush	Acetate	2068 ± 30	Fan et al. 2017
Sewage sludge biochar	Nafion	Carbon cloth	Carbon cloth	Acetate	500 ± 17	Yuan et al. 2013a
Livestock sewage sludge derived carbon	Not mentioned	Carbon cloth	Graphite fiber brush	Acetate	1273 ± 3	Deng et al. 2016b

Figure 10.4. AC catalyst used for the air cathode in MFCs. (A) AC powder,
(B) and (C) granular AC.

catalytic activity of the catalysts derived from different precursors (e.g., bituminous coal, coconut shell, and hardwood), and concluded that the catalysts derived from bituminous coal had a higher P_{max} (1,620 ± 10 mW/ m²) than other catalysts in MFCs. They attributed these differences to the variable functional groups and the pore structure of ACs, which highly depends on the raw materials, carbonization conditions, and activation processes (Wang et al. 2013, Watson et al. 2013b). Dong et al. (2012a) reported that the electron transfer number of the AC catalyst towards ORR was better linearly related to the micropore area rather than the mesopore area, demonstrating that the micropores in AC catalysts probably contribute to providing additional ORR sites, and therefore resulted in an improved ORR activity. Notably, this study also showed that the P_{max} had a close correlation with the microporosity by normalizing the power density to the surface area of micropores, indicating that increasing the micropore was beneficial for improving the performance (Dong et al. 2012a). Based on this principle, Wang et al. treated the AC catalysts by submerging catalyst powders into 3 M KOH solutions and heating at 85°C for 6 hours to increase the micropores by 3.3–8.7%. This resulted in an enhanced ORR activity of the AC catalysts and improved the P_{max} by 16% from 804 ± 70 to 957 ± 31 mW/m² in the MFCs (Wang et al. 2013). These results suggest that the increased microporosity contributed to the enhancement of TPIs, leading to a better utilization of the AC catalysts and therefore improved MFC performance (Li et al. 2014a). However, from the viewpoint of mass transfer, larger pores (e.g., macropores and mesopores) could work as channels to the transfer of the reactant, e.g., the supply of H⁺, OH⁻, and O_2 for ORR, and thus are beneficial for reducing the overpotential caused by the mass transfer. Li et al. increased the porosity of the CL generated by the AC by introducing NH_4HCO_3 as the pore former, and observed that an optimal macro and mesoporosity led to the improvement of power generation in the MFCs. However, too many larger pores would conversely affect the electrical conductivity of the cathode, and thus lead to a lower cathode performance (Li et al. 2014a).

Several researchers have also reported the improvement of ORR activity by modifying the surface chemistry of ACs by nitrogen-containing functional groups. For example, Watson et al. suggested that treating ACs using ammonia gas at a temperature of 700°C could enhance its ORR activity in neutral pH solution. The MFC using the treated AC delivered a P_{max} of $2,450 \pm 43$ mW/m^2, which was a 28% increase in power compared to the MFC using the untreated AC ($1,910 \pm 188$ mW/m^2) (Watson et al. 2013a). Similarly, Zhang et al. reported a nitrogen-doped AC obtained by using cyanamide as nitrogen precursor. An MFC using the nitrogen-doped AC delivered an ~32% higher P_{max} than that using bare AC (0.31 ± 0.08 W/m^2) (Zhang et al. 2014a).

Carbon Nanomaterials

Nanocarbons, such as graphene, CNTs, and carbon nanofibers are also widely used for the ORR catalysts in the air cathode MFCs, due to their high electrical conductivity, large specific surface area, and tunable heteroatom doping (Table 10.1). Gong et al. (2009) reported that vertically aligned nitrogen-containing CNTs exhibited a four-electron pathway with a superb performance toward ORR. Feng et al. 2011b) demonstrated that the MFC using nitrogen-CNTs as the ORR catalyst exhibited a P_{max} up to $1,600 \pm 50$ mW/m^2 (Fig. 10.5 A), which was ~13% higher than that ($1,393 \pm 35$ mW/m^2) of the MFC using Pt catalysts. Quantum mechanics calculations suggested that nitrogen dopants increased the positive charge density of carbon atoms and induced charge delocalization, which could enhance parallel diatomic adsorption of O_2 and weaken O–O bonds during ORR (Yuan et al. 2016). Apart from using nitrogen doping, CNTs are usually mixed with conductive polymers (PANI or PPy) or/and metal oxides (MnO_2, NiO) to improve the ORR activity (Ghasemi et al. 2016, Huang et al. 2015, Jiang et al. 2014, Liew et al. 2015). Liew et al. (2015) reported the MFC using a manganese oxide/

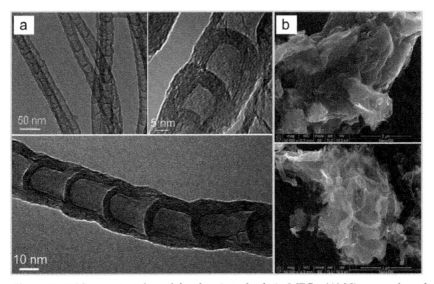

Figure 10.5. Nanomaterial used for the air cathode in MFCs. (A) Nitrogen-doped CNTs (Feng et al. 2011b), (B) Iron- and nitrogen-functionalized graphene (Li et al. 2012).

functionalized CNT nanocomposite as the ORR catalyst obtained a higher P_{max} (520 mW/m²) compared to that without manganese oxide (440 mW/m²). It was reported that graphene mixed with CNTs can serve as a composite catalyst for ORR. Chen et al. prepared a nitrogen-doped graphene/carbon nanotube composite using a hydrothermal process. They observed that a positive shift of 32 and 43 mV in the onset potential for ORR can be obtained by the composite catalyst, compared to nitrogen-doped graphene and CNTs, respectively (Chen et al. 2013). Carbon nanofiber possesses a mesoporous structure, excellent porous interconnectivity, and high surface area, and therefore it can be a viable alternative for the ORR catalyst. Yang et al. proposed a nitrogen-doped carbon nanofiber with hierarchical pores as the efficient ORR catalyst. The results showed that the catalyst achieved a P_{max} of $1{,}377 \pm 46$ mW/m² in MFCs, which was slightly higher than Pt/C ($1{,}307 \pm 43$ mW/m²) (Yang et al. 2014b). Graphene, obtained from the exfoliation of graphite, was regarded as a better candidate than other nanocarbons, not only because it has similar properties of high electrical conductivity, large specific surface area, and high mechanical strength to other nanocarbons, but also because it is free from any contaminants. Similar to other nanocarbons, graphene was usually doped with heteroatom or metal oxides to improve its catalytic activity. An MFC using Fe- and N-doped graphene as the ORR catalysts can deliver a P_{max} of 1,149.8 mW/m², exhibiting ~2.1 times of that using a Pt/C catalyst (561.1 mW/m²) (Fig. 10.5B) (Li et al. 2012).

Biomass-derived Carbons

Although heteroatom-doped nanocarbons exhibit a comparable or even higher ORR performance than Pt, they are still rather expensive for wastewater treatment because the synthesis of nanocarbons involves expensive and sophisticated instruments with high running costs and long production time. In addition, to enhance the ORR performance, expensive and toxic chemical agents are usually required as the sources of heteroatom dopant. All these factors weigh heavily against the real application of the carbon nanomaterials in MFCs. In contrast, carbon materials derived from biomass are preferable for MFC application because they are cheap, naturally abundant, cyclically sustainable, and readily available. In addition, biomass contains numerous heteroatoms (e.g., N and P) and therefore can be used as the potential precursors for the preparation of the heteroatom-doped carbon catalysts with highly specific surface areas. The sources of biomass used for the synthesis of ORR catalysts are diverse and can be classified into plant-derived, animal-derived, and other biomass.

Plant-derived Biomass

Different natural biomass have been widely used as the precursors of the heteroatom-doped carbon ORR catalysts. In general, most of the carbon content in green plants is from CO_2 fixation through photosynthesis by leaves or other green parts, and the content of heteroatom (e.g., nitrogen and phosphorus) is related to enzymes participating in photosynthesis. During the preparation process, the natural heteroatom can be used as dopant, and the self-doped catalyst can be directly obtained from the interaction between the heteroatom and carbon contained in the natural biomass. As reported, several types of biomass, including okara, ginkgo leaves, cornstalk, soy protein, pulse flour, soybean shells, amaranthus, plant moss, and water hyacinth have demonstrated an efficient ORR catalytic activity in MFCs (Alatalo et al. 2016, Gao et al. 2015, Gokhale et al. 2014, Liu et al. 2015c, Pan et al. 2014, Sun et al. 2016, Wang et al. 2015a, Zhou et al. 2016b). For example, Zhou et al. (2016b) synthesized a plant moss derived carbon nanoparticle-coated porous biocarbon as an ORR catalyst and demonstrated a 703 ± 16 mW/m^2 P$_{max}$ in MFCs (Fig. 10.6A). Li et al. (2018) prepared an onion-derived AC as a carbon catalyst to facilitate ORR and obtained a P$_{max}$ of 742 ± 17 mW/m^2 in MFCs, which was similar to that of a Pt/C catalyst (763 ± 19 mW/m^2) (Fig. 10.6B). Recently, Yang et al. (2017a) also showed that the bamboo could be directly pyrolyzed to prepare an ORR catalyst using its inherent N- and P-content as self-dopant, a P$_{max}$ of $1,056 \pm 38$ mW/m^2 was produced with the catalyst (Fig. 10.6C). However, the nitrogen content in the dry matter of green plant is low (0.5–1.5% in roots and stems and 3–5% in leaves) (Antolini 2016). This would lead to a limited number of the active site for ORR. Therefore, to produce a sufficient amount of active sites for ORR, the use of additional heteroatom dopant or nitrogen-enriched plants as the

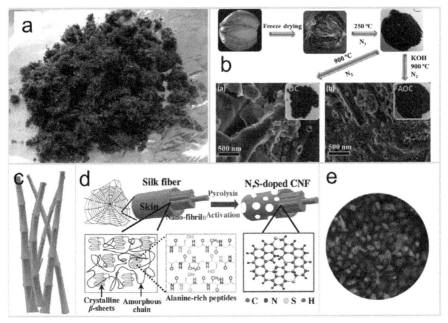

Figure 10.6. Different biomass used as precursors of carbon catalysts. (A) Plant moss (Zhou et al. 2016b), (B) onion (Li et al. 2018), (C) bamboo branches (Yang et al. 2017a), (D) spider silk (Zhou et al. 2016a), (E) *Chlorella pyrenoidosa* (Fan et al. 2017).

precursors is favorable. For example, Gao et al. (2014) prepared nitrogen-enriched carbon from bamboo fungus containing high protein and realized a superior ORR activity. Alatalo et al. (2016) and Zhu et al. (2012) reported that the soy protein and the soy milk derived from soybean could be directly used for the synthesis of ORR catalysts without additional dopants because of its high N content (38-45%).

Animal-derived Biomass

Animal-derived biomass contains a higher nitrogen content than plant biomass, and thus is regarded as a suitable alternative for ORR catalyst precursors. Biomass derived from animals, such as eggs, blood, bones, and animal liver had been reported as precursors for the ORR catalysts. (Guo et al. 2014, Wang et al. 2014a, Wang et al. 2015b, Wu et al. 2016, Zhang et al. 2015a, Zhang et al. 2014b, Zhang et al. 2014c). Among the different biomass derived from animals, eggs were considered a promising precursor for preparing an ORR catalyst because eggs contain abundant proteins, cholesterols, and lecithins. Carbon contained in these substances can be utilized as the precursor for carbon frameworks, while N and P can serve as the intrinsic, homogeneous doping of carbon framework to achieve catalytic activity, waiving the need of the extrinsic doping sources. For example, Wang et al. (2014a) synthesized a non-precious tremella-like mesoporous carbon as the ORR electrocatalyst using carbonized egg white as carbon source. Wu et al. (2016) also reported an ORR catalyst based on N, P, and Fe codoped mesoporous carbon microspheres from eggs without the introduction of extrinsic dopants. In addition, some nitrogen-containing biomass, e.g., gelatin, honey, sucrose, spider silk, and DNA of animals, was proposed for the preparation of nitrogen-doped carbonaceous catalysts (Lu et al. 2013a, Nam et al. 2014, Nunes et al. 2015, Wang et al. 2012, Zhou et al. 2016a). A previous study by Zhou et al. (2016a) reported natural carbon nanofibers derived from spider silk as ORR catalysts in MFCs; an MFC using the as-prepared catalyst presented a P_{max} of 1,800 mW/m^2, which was 1.56 times higher than that of the MFC with Pt/C catalyst (Fig. 10.6D).

Other Biomass

In addition to plant- and animal-derived biomass, several studies have reported that biomass, such as algae and sludge, can be used as the sources for ORR catalyst preparation. Algae can be classified as unicellular microalgae (such as Chlorella and Nitzschia) and multicellular macroalgae (also known as seaweeds). The main components of algal biomass include carbohydrates, proteins, lipids, and nucleic acids. The nitrogen content of algae changes during different growth periods; it is usually from 4% to 8% of the dry weight (Antolini 2016). Due to its high protein content, abundance on earth, and low cost, algae can be used as a promising precursor for the synthesis of N-doped carbon catalysts toward ORR. For example, Liu et al. (2016a) prepared N-doped carbon nanoparticles catalyst by using spiral seaweeds as a precursor and achieved a comparable performance compared to that of Pt/C in RDE tests. A study by Fan et al. (2017) applied *Chlorella pyrenoidosa* as a precursor to prepare the ORR catalysts (Fig. 10.6 E); the as-prepared catalyst delivered a P_{max} of 2,068 ± 30 mW/m², which was 13% higher than that with commercial 20 wt.% Pt/C (1,826 ± 37 mW/m²) in MFCs.

In addition, sludge is a kind of waste biomass that is chiefly produced during wastewater treatment, and usually has a high content of carbon and nitrogen, as well as metals. Due to its high N and metal content, sludge is considered to have potential serving as the precursors for ORR catalysts. For example, Yuan et al. (2013a) reported a sewage-sludge-derived ORR catalyst; an MFC using the prepared catalyst achieved a comparable P_{max} of 500 ± 17 mW/m² to that of the Pt cathode. Recently, Deng et al. synthesized hierarchical carbon catalysts with interconnected honeycomb-like macromesoporous carbon frameworks with N, P, and Fe doping by direct carbonizating the sewage sludge at N_2 atmosphere. The results showed that the MFCs with the prepared ORR catalysts delivered a comparable P_{max} of 1,273 ± 3 mW/m² to that of commercial Pt/C (1,294 ± 2 mW/m²), confirming its feasibility as ORR catalysts for air cathodes (Deng et al. 2016a).

Carbon-based Cathode Designs

Air cathode MFCs are considered one of the most promising architecture for scale-up because the cathode is directly exposed to the air and it does not require additional energy to supply oxygen. In addition to a high activity toward ORR, a good air cathode should be porous to allow a sufficient oxygen supply to the CLs. The cathode should also be partly hydrophobic to permit the penetration of liquid electrolyte into the CLs to create a large amount of TPIs. Furthermore, water leakage from MFCs must be avoided, and the materials must be good electrical conductors to minimize the ohmic loss. Obviously, the design of the air cathode considerably influences the physical and chemical properties of these components and therefore greatly affects the performance of the air cathode. The typical fabrication procedure for an air cathode with an HPL and MPL is shown in Fig. 10.7. The cathode preparation usually follows several steps: (1) the preparation of the MPL by coating a mixture of carbon black and PTFE suspension onto a side of teflonized carbon cloth/paper and heat treating at 370~400°C, (2) the HPL (PTFE layer) preparation by brushing four layers of PTFE suspension onto the carbon layer and heat treating at 370~400°C, (3) the preparation of an MPL by coating a layer of the mixture of carbon black and PTFE suspension onto the other side of the carbon cloth/paper, and (4) the CL preparation by dipping/brushing/spraying a layer of a mixture of the catalyst and PTFE or Nafion binder on the MPL. Based on the structure and preparation approaches, the air cathodes for MFCs can be classified into carbon-cloth/ paper-based, metal mesh, monolithic/free-standing, and packed-bed cathode (Cheng et al. 2006a, Yang et al. 2016, Yang et al. 2017b, Zhang et al. 2009a, Zhang et al. 2013b).

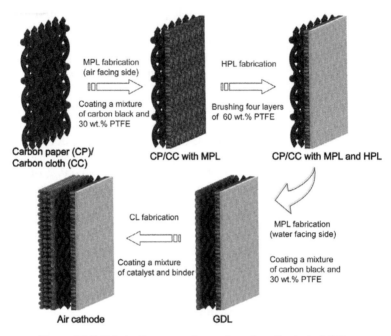

Figure 10.7. Fabrication procedure of an air cathode in MFCs.

Carbon Cloth/Paper-based Cathode

GDL and MPL

Carbon paper and carbon cloth were widely employed as the electrode supports in the fabrication of air cathodes due to their high gas permeability and electrical conductivity. Both carbon cloth and carbon paper are carbon-fiber-based porous materials: carbon paper is non-woven, while carbon cloth is woven fabric; thus, no binder is needed for the use of these two materials. For the preparation of GDLs, carbon cloth/paper or cloth is immersed into 10-20% PTFE suspension, followed by heat treating at 300-400°C. The MPL is usually prepared by spraying or spreading a mixture of carbon black, solvent (mixture of water and isopropyl alcohol), and 30 wt.% PTFE suspension onto the GDL, and sintering at 300-400°C for 1 hour. For example, Cheng et al. (2006b) prepared the MPL by coating a mixture (20 µl/mg of carbon powder) of carbon black and 30 wt.% PTFE suspension onto the carbon cloth, drying for 2 hours at room temperature, and followed by heat treating at 370°C for 30 minutes. Santoro et al. (2011) reported a three-layer cathode by inserting an MPL between the CL and GDL. The results indicated that an MFC with a three-layer cathode exhibited a lower ohmic resistance (R_{ohmic}: 26–34 Ω), a lower water loss (0.57 g/d/cm²), and a higher COD removal efficiency (71–78%) than that without an MPL (Ohmic resistance: 105 Ω, water flux: 0.75 g/d/cm², COD removal efficiency: 49–68%). Similarly, Papaharalabos et al. (2013) applied an MPL, which comprised a mixture of nano-sized carbon black particles (Vulcan XC-72R) with a non-ionic surfactant (Triton X100), deionised water, and PTFE (60% emulsion), onto the GDL to prepare the air cathode. MFCs using the as-prepared air cathodes achieved a 31% increase in P_{max} compared to that without an MPL.

CL

The CL is usually prepared by dipping or brushing methods on the top of the GDL or MPL. For instance, Cao et al. (2017) fabricated the CL by brushing the mixture of a graphene-based catalyst, water, isopropanol, and Nafion binder onto the teflonized carbon cloth and dried it in air for 24 hours. Similarly, Wang et al. (2018) fabricated the CL by brushing a paste-like mixture, which comprised a catalyst, water, isopropanol, and binder, on to the water-facing side of the carbon cloth. However, the CL fabrication using the brushing method would cause catalyst loss, leading to an inaccurate catalyst-loading rate. To decrease the catalyst loss and obtain a more homogeneous surface coating, Ge et al. (2016) proposed a dipping method by applying the AC catalyst onto the GDL. In detail, the carbon catalyst and PTFE binder were first dispersed in ethanol, and the obtained liquid mixture was dipped onto the carbon cloth, dried in air followed by heat treating at 350°C for 30 minutes (Ge and He 2015). To obtain an air cathode with an acceptable performance, a high catalyst loading is required for the CL (more than 10 mg/cm²). This may result in a labor-consuming process and the detachment of the catalyst powder from the CL. In a later study, Santoro et al. prepared the CL by pressing a mixture of commercial AC and 20% PTFE onto a teflonized carbon cloth. The results indicated that a hot-press condition of 9.6 MPa and 150–200°C were the optimized parameters for the air cathode fabrication based on the carbon cloth (Santoro et al. 2014).

HPL

To reduce water loss, to tune the oxygen diffusion, and to strengthen the tolerance of the air cathode to a high water pressure, the HPL is usually fabricated using commercial polymer membrane or using hydrophobic polymers, e.g., PTFE, polydimethylsiloxane (PDMS), polyvinylidene fluoride (PVDF), etc. For example, Gong et al. (2014) prepared a four-layer HPL by directly brushing 60% Teflon® emulsion on the air cathode exposed to the air. Cheng et al. (2006a) added an additional PTFE hydrophobic layer on the air-facing side of the conductive carbon base layer; the results showed a 117 mV increase in the cathode potential and a coulombic efficiency (CE) increase of 171% with four hydrophobic layers (Cheng et al. 2006a). Later research by Zhang et al. (2016) reported that an MFC with the HPL composed of carbon black and PTFE had a higher P_{max} (1610 ± 90 mW/m²) than that without the use of carbon black (1110 ±20 mW/m²) due to the very low resistivity of 0.17 Ω cm and a high oxygen mass transfer coefficient of $1.07 \pm 0.08 \times 10^{-3}$ cm/s compared to the HPL without any carbon black (1990 Ω cm, $1.37 \pm 0.04 \times 10^{-3}$ cm/s). Therefore, the advantage of carbon-free HPL would be overwhelmed by the decrease in cathode performance to some degree.

Metal Mesh/Foam-based Cathode

Electrode supports

Metal mesh/foam, such as nickel mesh, nickel foam, and stainless-steel mesh, is commonly used as air cathode supports and current collectors in MFCs. Zhang et al. (2009a) reported the first air cathode using metallic substrate (nickel mesh) as the cathode support and current collector due to the corrosion resistance and the low electrical resistance of nickel. Later research by Cheng and Wu also applied nickel foam as the cathode support and current collector due to its high porosity and electrical conductivity. The cathode was prepared by mixing AC catalysts with PTFE binders to form slurry, which was filled into the three-dimensional framework of Ni foam. The results showed that the cathode had a low ohmic resistance and a high catalyst utilization (Cheng and Wu 2013). Although metallic nickel possesses high corrosion resistance and electrical conductivity, its cost remains high for wastewater treatment. Dong et al. (2012b) reported using stainless-steel mesh (SSM) as an alternative to nickel foam to further reduce the cost. Thus, the stainless-steel mesh was one of the most widely used cathode supports for the air cathode of MFCs due to its corrosion resistance, low electrical resistance, and low cost (Fig. 10.8).

GDL

Usually, the GDL for metal-mesh-based cathodes can be classified into two types: the first one is prepared using a mixture of carbon black and PTFE suspension, which is similar to that in carbon-cloth/paper-based cathode, and the second is carbon-free GDL prepared using hydrophobic polymer materials. For example, Cheng and Wu (2013) prepared the GDL by pasting a mixture of carbon black (6 mg/cm²) and PTFE (15 mg/cm²) onto one side of the nickel foam (Cheng and Wu 2013). To reduce the fabrication cost and develop an accurate and labor-saving method, Dong et al. reported a rolling-press method to fabricate the GDL. In this method, the GDL was prepared by dispersing 40 wt.% carbon black and 60 wt.% PTFE suspension into ethanol

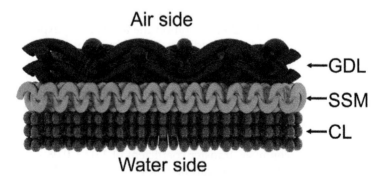

Figure 10.8. Diagram of an air cathode based on stainless-steel mesh.

to obtain a dough-like paste, which was then rolled to form a 0.35-mm thick layer and was rolling-pressed onto a side of the SSM to produce a sheet electrode of 0.4-mm thickness. To form the three-dimensional microchannels for air supply, the GDL was finally sintered for 25 minutes at 340°C (Dong et al. 2012b). The rolling-press method provided a scalable, highly reproducible, and low-cost approach to fabricate the GDL for MFCs. Although great improvements have been achieved, a sintering temperature ranging from 340°C to 370°C is still needed during the GDL fabrication process to achieve a uniform PTFE distribution. Therefore, other fluoropolymers with a low melting point, such as PVDF (melting point 177°C), was used for the GDL preparation.

CL

The fabrication of the cathode CL with spraying or brushing was usually regarded as a coarse and labor-consuming method for commercialization. To solve the problem, a CL based on a metallic substrate is usually prepared by the pressing or rolling-press method. Zhang et al. prepared a CL by using the pressing method. In detail, a mixture of AC and PTFE binder was first prepared to produce a paste and then pressed onto a side of a nickel foam at 15 MPa to form a CL with a 0.45-mm thickness (total weight of CL and nickel mesh ~1 kg/m^2) (Zhang et al. 2009a). However, the uniformity of catalyst distribution in the metal mesh or foam was uncontrollable using this method. To optimize the CL preparation and to extend the TPIs, Dong et al. reported a rolling-press approach to fabricate air cathodes using SSM as the current collector. The difference between this method and the pressing method is that the catalyst paste prepared from the mixture of AC and PTFE is first rolled to a uniform CL and then rolled onto one side of the SSM. Similar to the GDL preparation, the heat treatment of the CL was energy consuming, which would increase the cost of MFC application. To address this problem, Wang et al. proposed a binder-free cathode in which the CL was fabricated using direct growth of graphene onto a nickel mesh through chemical vapor deposition (CVD) without the involvement of a binder, providing a new method for the fabrication of the binder-free CL air cathodes for MFCs; however, the long-term stability of the air cathode has not been verified yet (Wang et al. 2016).

Monolithic or Self-standing Cathode

Although several efforts have been devoted to the optimization of the fabrication of air cathodes, the process remains complex because it usually involves the preparation of CL and GDL, as well as their assembly on the electrode support. Therefore, it is important to explore simple cathode-fabrication methods for the practical application of MFCs. Recently, several researchers have proposed some approaches to prepare self-assembly or free-standing cathodes for MFC applications. For example, Yang et al. fabricated a three-dimensional N- and Fe-doped CNT sponge cathode using the two-stage CVD method; an MFC using this air cathode had a much higher P_{max} of 20.3 W/m^3 than 12.8 W/m^3 for a Pt/C air cathode (Fig. 10.9A) (Yang et al. 2016). Although the prices of nanocarbons, such as CNTs remain expensive for MFC cathode preparation, these methods can provide an opportunity to prepare self-assembly or free-standing cathodes using low-cost carbon materials. Recently, Yang et al. reported that the outer surface teflonized bamboo charcoal tube (BCT) prepared by bamboo tube carbonization was used as a monolithic cathode in MFCs (Fig. 10.9B) (Yang et al. 2017b). In the prepared BCT cathode, the porous carbon can be directly used as the active sites to catalyze ORR avoiding the use of heteroatom doping. In addition, the inherent porous and tubular structures in BCT can be used for the supply channels for oxygen/proton and the mechanical support, separately, an MFC using the BCT cathode showed a comparable P_{max} (40 W/m^3) with the MFC using Pt/C cathode.

Packed-bed Cathode

Packed-bed air cathode was a simple design to avoid the need for binders in the CL and GDL, and thus was beneficial to minimize the ohmic loss. Kalathil et al. (2011) proposed a packed-bed air cathode prepared by packing

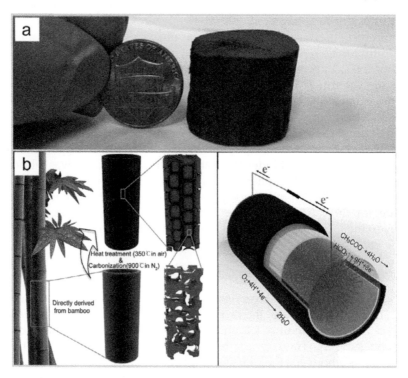

Figure 10.9. Monolithic or self-standing air cathodes. (A) CNT sponge as a self-standing cathode (Yang et al. 2016), (B) bamboo-charcoal-tube-derived air cathode (Yang et al. 2017b).

granular AC in stainless-steel cages or near the current collector (metal mesh) in the cathode chamber that is filled with aerated electrolyte. This design is free of binders, but requires additional energy for aeration to the cathode. In addition, because of the lower diffusion coefficient of oxygen in water than in air, the oxygen supply is a limiting step affecting the cathode performance. To solve this problem, Zhang et al. (2013b) reported a packed-bed air cathode fabricated by packing AC particles onto an SSM; the cathode was placed horizontally above anode chamber with an air-breathing design. However, compared to the traditional cathode design, this cathode cannot withstand the water pressure on the cathode. Therefore, the leakage of electrolyte from the cathode can be a limitation for MFC operation because this cathode is constructed without a gas diffusion layer. Furthermore, the water press to the cathode could induce a change in H^+/OH^- and oxygen transport, as well as the TPI distribution in the cathode, thus affecting the cathode performance. In comparison, the latter exhibited superiority for practical application due to its air-breathing design rather than the usage of water aeration. Although the packed-bed air cathode was simply fabricated without the use of a GDL, binder and corresponding brushing/spraying, rolling press or pressing processes, the long-term stability of the cathode has not been reported yet.

Critical Challenges for Practical Application

Poison Tolerance and Stability of ORR Catalysts

Poison tolerance

Pt is vulnerable to poison in MFC-relevant conditions. Recently, several studies have shown that various ions commonly present in wastewater, such as S^{2-} and NH_4^+, decreased the activity of the Pt catalysts in a short time of operation, because the poison species absorbed on the surface of catalysts would cover the active sites, leading to a decrease of the available catalytic active sites (Santoro et al. 2016, Verdaguer-Casadevall et al. 2012). Verdaguer-Casadevall et al. (2012) revealed that the presence of ammonium significantly decreased the ORR activity of a polycrystalline Pt catalyst. The metabolic product of sulfate-rich wastewater, commonly from animal husbandry, food processing, pulp and paper wastewater, etc., such as sulfide, has been demonstrated as a major reason for Pt degradation in MFC systems (Logan et al. 2005, Stams et al. 2006, Zhao et al. 2008). Recently, Santoro et al. (2016) reported that the current loss of Pt/C was four times more than that of carbon catalysts in the presence of 5 mM S^{2-} concentration. In addition, research by Yang et al. (2017a) indicated that bamboo charcoal (BC) catalysts have a superior poison tolerance than Pt/C under the relevant MFC condition, which usually contains not only a substrate but also species, such as metabolites, Cl^-, SO_4^{2-}, S^{2-}, and HPO_4^{2-} at several to hundreds of mM, as shown in Fig. 10.10. Although there is still no report to verify the detailed effect of each ion on the cathode ORR activity, these studies revealed the

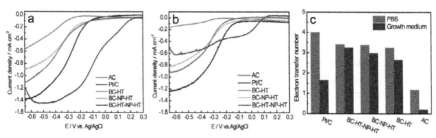

Figure 10.10. Rotating disk electrode tests in phosphate-buffered solution (A), and growth medium (B), and corresponding electron transfer numbers (C) (Yang et al. 2017a).

advantage of carbon-based ORR catalysts over Pt/C in terms of poison tolerance in real wastewater treatment. Additionally, Li et al. reported that AC catalysts exhibited a better stability than Pt/C cathodes during four months of operation. Similar results were obtained by Zhang et al. (2014e); compared to a Pt/C cathode, three different ORR catalysts (raw AC, Fe, and carbon black) showed a superior stability or durability during 17 months of operation (Zhang et al. 2014e).

Cathode biofouling

In a SC-MFC, an air cathode fabricated using a carbon ORR catalyst exhibits a good biocompatibility, which favors the formation of a cathode biofilm. It was reported that the cathode biofilms thickness can be up to 2 ~ 3 mm after a long term operation (Yuan et al. 2013b). A thick biofilm on the surface of the cathode CL can be a diffusion barrier for the ion (e.g., H$^+$, OH$^-$, etc.), which can adversely affect the cathode performance. A previous study reported that the OH$^-$ accumulation resulting from the limited H$^+$/OH$^-$ transfer caused by the cathode biofilm led to a higher local pH near the surface of the cathode CL to 10.0 ± 0.3, compared to the cathode in the absence of biofilm (9.4 ± 0.3). From the Nernst equation, it was easily deduced that the overpotential of the cathode with biofilm was 0.18 ± 0.02 V, which was higher than the cathode in the absence of biofilm (0.14 ± 0.02) (Yuan et al. 2013b). In addition, the extracellular secretions (including proteins and polysaccharides) of the attached biofilm may change the physicochemical properties of the catalysts (including surface functional groups, pore-size distributions, and hydrophilicity), and thus reduce their ORR catalytic activity.

At present, only few studies have reported approaches to inhibit the biofilm growth on air cathodes based on carbonaceous catalysts. Liu et al. (2015b) proposed a method to inhibit the growth of biofilm on air cathodes by incorporating enrofloxacin into the CL; the decrease of P$_{max}$ (21%) was much lower than that of the control cathode (33%) after two months of operation. Similarly, Li et al. (2014c) applied bifunctional quaternary ammonium compounds (QAC) to modify the CL to inhibit the growth of biofilm on the cathode; the results indicated that the modified AC cathode exhibited a lower charge transfer resistance after two months of operation due to the inhibited growth of the biofilm, as shown in Fig. 10.11. In addition, Ag-containing catalysts were prepared as antibacterial ORR catalysts in air cathodes. For example, Dai et al. (2016) reported a bifunctional Ag/Fe/N/C catalyst for biofilm growth inhibition and ORR catalysis. The results indicated that the presence of Ag can alleviate the limitation of H^+/OH^- transport by inhibiting the formation of biofilm, and thus leading to a high performance of air cathodes. Although these methods can effectively inhibit the growth of biofilm and alleviate the accumulation of OH^- in the interface between the bulk solution and CL surface, they can increase the cost and intricacy of cathode fabrication. Moreover, the leakage of the chemicals during the long term operation can be detrimental to the environment, which overweighs the beneficial effect of the MFCs for pollution treatment. Additionally, the operation time for the MFCs using antibacterial chemicals for cathode biofilm inhibition is only 2-3 months (Li et al. 2014c, Liu et al. 2015b). The long-term stability of these air cathodes remain unverified.

Salt precipitation

In addition to the excess growth of biofilm on the surface of the water-facing side of the cathode, salt precipitation is also a crucial factor affecting the performance of air cathodes, because salt precipitation in cathodes would severely block the pore structures in the air cathodes after long-term operation, leading to additional transport resistance to air supply and ion transfer (An et al. 2017). The reason for salt precipitation in air cathodes is still under debate. Zhang et al. (2014e) suggested that salt precipitation formed

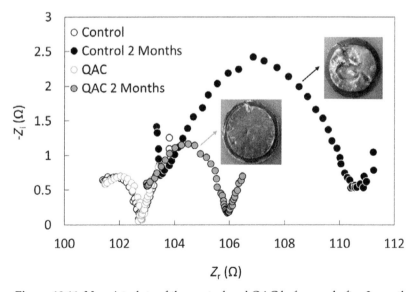

Figure 10.11. Nyquist plots of the control and QAC before and after 2 months of operation (Li et al. 2014c).

on the air facing side of cathode was primarily due to electrolyte evaporation and adsorption of the AC, while An et al. (2017) reported that the electric-field-induced salt precipitation was the major reason for the deterioration of the cathode performance. An et al. (2017) also found that the ion transfer from the anode to the cathode tends to penetrate into micropores and mesopores of the AC catalysts, leading to increased local ion concentration and accelerated salt precipitation. Although the mechanism underlying salt precipitation has been explored, no solution has been reported for this problem. Salt precipitation will be a crucial factor in the design metrics for the practical application of air cathodes.

Mass Transfer in the Air Cathode

H^+/OH^- transport

The CL of carbon-catalyst-based air cathodes is much thicker than that of Pt/C cathodes, due to the high catalyst-loading rate over dozens of milligrams per square centimeter compared to the typical 0.5 mg cm^{-2} catalyst-loading in Pt cathodes (Fan et al. 2017, Yang et al. 2017a). A thicker CL would increase the diffusion resistance of H^+/OH^- to the CL, and thus cause additional performance loss compared to that in Pt/C cathodes. The mass transfer in the CL can be usually affected by several factors: (1) porosity and pore distribution of the catalysts, (2) the hydrophilic or hydrophobic property of the CL, and (3) the surface properties of the catalysts. A previous study by Li et al. (2014a) has reported that the mesopores and macropores can serve as the channels for mass transfer in the CL. With a 38% increase of porosity in a pore range of 0.5 μm to 0.8 μm, the P_{max} was increased by 33%. Similarly, Liu et al. (2016b) confirmed that an increment in mesoporosity can facilitate the mass transfer in the CL and thus enhance the cathode performance. In addition, introducing the hydrophilic binder to the CL can enhance the proton transfer from bulk solution to the catalyst by enhancing the water uptake of the cathode structure. Saito et al. (2011) proposed that by using the polystyrene-*b*-poly (ethylene oxide) polymer as a neutral hydrophilic catalyst binder for MFC cathodes, an ~15% increase of power density can be achieved compared to the control sample. Furthermore, changing physical and chemical properties by chemical modification can facilitate the transfer of H^+ and OH^- in the CL. A study by Wang et al. (2014b) reported that carbon catalysts modified by quaternary ammonium effectively decreased the local pH in the cathode CL and alleviated OH^- accumulation, thus reducing the concentration overpotential.

O_2 transport in cathode

Since the cathode is directly exposed to the ambient atmosphere, and there are no auxiliary fans/compressors in the air cathode MFCs, diffusion due to the oxygen concentration difference is the primary transport mechanism for delivering oxygen to the cathode CL. Therefore, the stable operation of the air cathode MFCs requires minimized overall mass transfer resistance of oxygen through the layered cathode structure so that the voltage loss due to the oxygen concentration polarization can be reduced.

It was found that the properties of the GDL, such as porosity, thickness and materials, and preparation methods are crucial factors influencing the oxygen transport. Zhang et al. (2011a) reported that the GDL with a 30% porosity exhibited a higher oxygen transfer coefficient compared to 70% porosity due to the blockage of oxygen supply resulting from water flooding at a higher porosity. Cheng et al. (2006a) investigated the effect of HPL thickness on the cathode performance in MFCs. It was observed that the oxygen transfer coefficient decreased with the increase in HPL thickness, indicating that HPL restricted the oxygen flux into the system (Cheng et al. 2006a). Similar results were observed by Qiu et al. and Yang et al. They applied PVDF as the alternative to PTFE to prepare the HPL for the air cathode, and found that the dissolved oxygen concentration in the anode chamber decreased with the successive addition of PVDF layers due to the increased oxygen diffusion resistance of the air cathodes (Qiu et al. 2015, Yang et al. 2014a, Yang et al. 2015). In addition, Zhang et al. compared GDLs synthesized from different materials (Fig. 10.12) and found that the GDL comprised carbon black/PTFE and a porous PTFE layer exhibited a higher mass transfer coefficient than that with the wipe-based GDL with PDMS and a PTFE layer, indicating that the properties of the GDL significantly affected the oxygen supply to the cathode and therefore the cathode performance (Zhang et al. 2016).

Several studies have demonstrated that the pore distribution in the CL significantly influenced the cathode performance. A previous study by Li et al. (2014a) suggested that the oxygen transfer coefficient of the cathode increased from 0.7×10^{-4} to 2.1×10^{-4} cm/s with an increase in CL porosity from 8.18% to 12.4%, indicating that the increase in CL porosity provided more interlaced channels for oxygen supply. As the mean free path of gas molecules ($10^{-7} \sim 10^{-8}$ m) is much higher than the diameter of micropores ($<10^{-9}$ m), the Knudsen number of gas molecules in micropores is higher than 10; thus, the gas transport in the micropores of CL was considered to be controlled by Knudsen diffusion (Fig. 10.13), suggesting that the dominated collisions between gas molecules and the porous electrode would consume the most gas kinetic energy, with only a small amount of gas traveling through the cathode (Shi and Huang 2015). Therefore, micropores tend to deteriorate the oxygen transport across the cathode and lead to a high oxygen transfer resistance and a lower oxygen transfer coefficient. Previous research by Liu et al. (2016b) additionally confirmed that mesopore and macropore structures favor the oxygen gas transfer in the cathode compared to micropore structures. Although, with a lower oxygen transfer ability, the micropores benefit the improved specific pore area, and therefore extend the TPIs, providing active sites to facilitate catalytic activity. Therefore, a proper ratio of micropores to mesopores and macropores is important for the

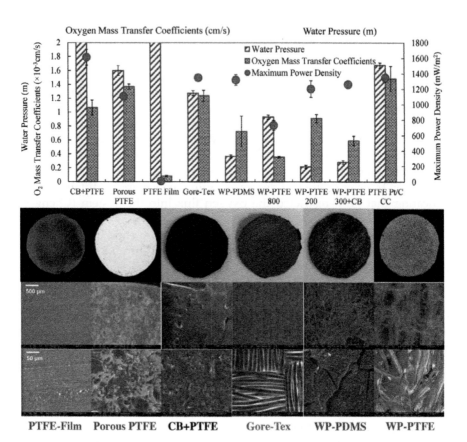

Figure 10.12. P_{max}, oxygen diffusion coefficient, and water pressure tolerance of cathodes with different GDLs (Zhang et al. 2016).

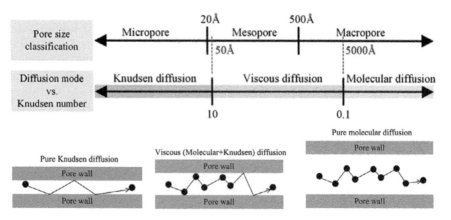

Figure 10.13. Schematic diagram of gas diffusion in porous media (Shi and Huang 2015).

cathode ORR because a proper ratio can simultaneously satisfy an effective oxygen transfer and sufficient active sites for ORR.

Although sufficient oxygen supply in the CL is desired to ensure a robust ORR process, oxygen leakage from the cathode to the anode chamber can induce thick biofilm establishment on the surface of the cathode CL, not only causing a lower columbic efficiency due to the aerobic respiration by bacteria, but also resulting in the blockage of H^+/OH^- transfer to the cathode and the reduced ORR activity of the carbon catalysts. The complex relationship between oxygen supply and the properties of the air cathode causes the MFC performance to behave in a complex manner. To increase the CE value of the system while inhibiting the cathodic biofouling, inserting separators between the anode and air cathode was a common method to limit the diffusion of oxygen into the anode chamber (Fan et al. 2007, Liu and Logan 2004, Watson et al. 2011). The oxygen transfer efficiency of different separators is shown in Table 10.2.

Previous research by Kondaveeti et al. (2014) reported that membranes, such as Nafion membranes, cation exchange membranes, and CMI-7000, exhibited a better performance in limiting the oxygen leakage from the cathode into the anode than polypropylene (PP80 or PP100), polyphenylene sulfide (PPS), and sulfonated polyphenylene sulfide (S-PPS). Fan et al. (2007) also used J-cloth as the separator in an air cathode MFC; the CE value significantly increased from 35% to 71%, due to the reduced oxygen diffusion flux. However, the employment of the separator also limited the proton transfer from the anode to the cathode and increased the internal ohmic resistance of the system. Zhang et al. used nylon and glass fiber filter separators with different pore sizes in the air-cathode MFCs. The results indicated that CEs were inversely associated with power production, due to the increased internal resistance (Zhang et al. 2010b).

Conclusion and Perspective

The purpose of MFC technologies is to simultaneously achieve high-efficient wastewater treatment and energy recovery. Accordingly, the cathode should be easy to prepare, long-term stable, and cost-effective. In this regard, carbon materials have been demonstrated as promising candidates for cathode fabrication. Many pioneering studies on different carbon materials have prompted great progress in this field. Herein, the performance and application of different carbon materials in MFC cathodes were summarized. An advantage of carbon-based catalysts and electrodes is the low cost, stable chemical/physical properties, and tunable heteroatom doping, allowing an easy preparation approach for MFC electrodes. Using the property of tunable heteroatom doping, the ORR performance of carbon catalyts can be further enhanced by heteroatom doping, such as nitrogen and phosphorus doping. In addition, the high poison tolerance of carbon as ORR catalysts ensures a good stability and durability of the catalysts under relevant MFC conditions.

Table 10.2. Characteristics of different separators used in MFCs

Separators	Thickness (mm)	K_o (cm s^{-1})	D_o (cm^2 s^{-1})	References
Nafion® 117	0.19	2.05×10^{-4}	3.89×10^{-6}	Hou et al. 2011
Microfiltration membrane	0.13	1.4×10^{-4}	2.7×10^{-6}	Tang et al. 2010
Ultrafiltration membrane	0.04	2.11×10^{-4}	8.44×10^{-6}	Hou et al. 2011
J-Cloth	0.3	29×10^{-4}	8.69×10^{-5}	Zhang et al. 2009b
Glass fiber	1	5.0×10^{-5}	5.0×10^{-6}	Zhang et al. 2009b
Cation exchange membrane (CMI-7000)	0.46	9.4×10^{-5}	4.3×10^{-6}	Kim et al. 2007
Anion exchange membrane (AMI-7001)	0.46	9.4×10^{-5}	4.3×10^{-6}	Kim et al. 2007
Chitosan-functionalized multiwall CNT	0.26	2.66×10^{-4}	6.38×10^{-6}	Venkatesan and Dharmalingam 2013
Sulphonated polyether ether ketone	—	2.4×10^{-6}	4.8×10^{-8}	Ayyaru and Dharmalingam 2011
Sulfonated polystyrene–ethylene–butylene–polystyrene	0.18	3.59×10^{-5}	6.46×10^{-7}	Ayyaru et al. 2012
PP80	—	3.0×10^{-3}	3.0×10^{-6}	Kondaveeti et al. 2014
PP100	—	2.0×10^{-3}	9.0×10^{-6}	Kondaveeti et al. 2014
PPS	—	5.0×10^{-3}	2.6×10^{-6}	Kondaveeti et al. 2014
S-PPS	—	3.4×10^{-3}	1.8×10^{-5}	Kondaveeti et al. 2014

These advantages make it possible to use carbon catalysts in large-scale application.

For electrode fabrication, air-cathode designs, such as pressing/rolling press have been proposed as alternatives to the traditional dipping or brushing methods. These methods usually use high mechanical and electrical conductive materials, such as SSM, nickel mesh, and nickel foam as structure supports and current collectors to replace the carbon cloth/paper. These designs provide an accurate and labor-saving method and reduce the cost of air cathode fabrication. Notably, self-assembly and monolithic electrodes are easy to fabricate without the use of binders, favoring a decrease in costs and improvement of cathode performance. The packed-bed air cathode exhibited a simple cathode design to avoid the use of binders, GDL, or even water aeration. These cathode designs provide the potential for large-scale production and fabrication of cathodes for MFC application.

Although great improvements have been achieved in the past decades, the power output of MFCs remain insufficient for practical applications. In an air-cathode, despite the abundant pore structure and high surface area of carbon catalysts, the mass transport of reaction species, such as H^+, OH^-, and O_2, still needs to be improved due to the rather thick CL caused by high catalyst-loading (usually >20 mg/cm^2). In addition, the good biocompatibility of the carbon-based air cathode has the disadvantage of favoring the growth of cathode biofilms and salt precipitation after a long-term operation, which is an obvious drawback for practical applications. In addition, the metal-mesh-based cathodes, self-assembly and monolithic cathodes, and packed-bed air cathode were studied on a lab scale only; there have been few reports on the understanding of the effect of scaling up and the corresponding fabrication on the air-cathode performance. Therefore, a considerable effort is still required to explore high performance cathode designs based on cost-effective carbon catalysts in the future, in terms of the cathode-fabrication process, feasibility of scale-up, and mass transport properties.

Acknowledgements

This work was supported by National Natural Science Funds for Outstanding Young Scholar (No. 51622602), the National Science Foundation for Young Scientists of China (No. 51506017), and a project supported by the Natural Science Foundation of Chongqing, China (Grant No: cstc2015jcyjA90017).

REFERENCES

Alatalo, S.M., K. Qiu, K. Preuss, A. Marinovic, M. Sevilla, M. Sillanpää, X. Guo and M.-M. Titirici. 2016. Soy protein directed hydrothermal synthesis of porous carbon aerogels for electrocatalytic oxygen reduction. Carbon 96: 622-630.

An, J., N. Li, L. Wan, L. Zhou, Q. Du, T. Li and X. Wang. 2017. Electric field induced salt precipitation into activated carbon air-cathode causes power decay in microbial fuel cells. Water Res. 123: 369-377.

Antolini, E. 2016. Nitrogen-doped carbons by sustainable N- and C-containing natural resources as nonprecious catalysts and catalyst supports for low temperature fuel cells. Renew. Sust. Energ. Rev. 58: 34-51.

Awan, Z., K.S. Nahm and J.S. Xavier. 2014. Nanotubular MnO_2/graphene oxide composites for the application of open air-breathing cathode microbial fuel cells. Biosens. Bioelectron. 53: 528-534.

Ayyaru, S. and S. Dharmalingam. 2011. Development of MFC using sulphonated polyether ether ketone (SPEEK) membrane for electricity generation from waste water. Bioresour. Technol. 102(24): 11167-11171.

Ayyaru, S., P. Letchoumanane, S. Dharmalingam and A.R. Stanislaus. 2012. Performance of sulfonated polystyrene–ethylene–butylene–polystyrene membrane in microbial fuel cell for bioelectricity production. J. Power Sources 217: 204-208.

Cao, C., L. Wei, G. Wang, J. Liu, Q. Zhai and J. Shen. 2017. A polyaniline-derived iron–nitrogen–carbon nanorod network anchored on graphene as a cost-effective air-cathode electrocatalyst for microbial fuel cells. Inorg. Chem. Front. 4(11): 1930-1938.

Chen, L., R. Du, J. Zhu, Y. Mao, C. Xue, N. Zhang, Y. Hou, J. Zhang and T. Yi. 2015. Three-dimensional nitrogen-doped graphene nanoribbons aerogel as a highly efficient catalyst for the oxygen reduction reaction. Small 11(12): 1423-1429.

Chen, P., T.Y. Xiao, Y.H. Qian, S.S. Li and S.H. Yu. 2013. A nitrogen-doped graphene/carbon nanotube nanocomposite with synergistically enhanced electrochemical activity. Adv. Mater. 25(23): 3192-3196.

Chen, Z., K. Li and L. Pu. 2014. The performance of phosphorus (P)-doped activated carbon as a catalyst in air-cathode microbial fuel cells. Bioresour. Technol. 170: 379-384.

Cheng, S., H. Liu and B.E. Logan. 2006a. Increased performance of single-chamber microbial fuel cells using an improved cathode structure. Electrochem. Commun. 8(3): 489-494.

Cheng, S., H. Liu and B.E. Logan. 2006b. Power densities using different cathode catalysts (Pt and CoTMPP) and polymer binders (Nafion and PTFE) in single chamber microbial fuel cells. Environ. Sci. Technol. 40(1): 364-369.

Cheng, S. and J. Wu. 2013. Air-cathode preparation with activated carbon as catalyst, PTFE as binder and nickel foam as current collector for microbial fuel cells. Bioelectrochemistry 92: 22-26.

Dai, Y., Y. Chan, B. Jiang, L. Wang, J. Zou, K. Pan and H. Fu. (2016). Bifunctional Ag/Fe/N/C catalysts for enhancing oxygen reduction via cathodic biofilm inhibition in microbial fuel cells. ACS Applied Materials & Interfaces 8(11): 6992-7002.

Deng, L., H. Yuan, X. Cai, Y. Ruan, S. Zhou, Y. Chen and Y. Yuan. 2016a. Honeycomb-like hierarchical carbon derived from livestock sewage sludge as oxygen reduction reaction catalysts in microbial fuel cells. Int. J. Hydrogen Energ. 41(47): 22328-22336.

Dong, H., H. Yu and X. Wang. 2012a. Catalysis kinetics and porous analysis of rolling activated carbon-PTFE air-cathode in microbial fuel cells. Environ. Sci. Technol. 46(23): 13009-13015.

Dong, H., H. Yu, X. Wang, Q. Zhou and J. Feng. 2012b. A novel structure of scalable air-cathode without Nafion and Pt by rolling activated carbon and PTFE as catalyst layer in microbial fuel cells. Water Res. 46(17): 5777.

Fan, Y., H. Hu and H. Liu. 2007. Enhanced Coulombic efficiency and power density of air-cathode microbial fuel cells with an improved cell configuration. J. Power Sources 171(2): 348-354.

Fan, Z., J. Li, Y. Zhou, Q. Fu, W. Yang, X. Zhu and Q. Liao. 2017. A green, cheap, high-performance carbonaceous catalyst derived from Chlorella pyrenoidosa for oxygen reduction reaction in microbial fuel cells. Int. J. Hydrogen Energ. 42(45): 27657-27665.

Feng, L., Y. Chen and L. Chen. 2011a. Easy-to-operate and low-temperature synthesis of gram-scale nitrogen-doped graphene and its application as cathode catalyst in microbial fuel cells. ACS Nano 5(12): 9611-9618.

Feng, L., Y. Yan, Y. Chen and L. Wang. 2011b. Nitrogen-doped carbon nanotubes as efficient and durable metal-free cathodic catalysts for oxygen reduction in microbial fuel cells. Energ. Environ. Sci. 4(5): 1892-1899.

Gao, S., H. Fan and S. Zhang. 2014. Nitrogen-enriched carbon from bamboo fungus with superior oxygen reduction reaction activity. J. Mater. Chem. A 2(43): 18263-18270.

Gao, S., K. Geng, H. Liu, X. Wei, M. Zhang, P. Wang and J. Wang. 2015. Transforming organic-rich amaranthus waste into nitrogen-doped carbon with superior performance of the oxygen reduction reaction. Energ. Environ. Sci. 8(1): 221-229.

Ge, B., K. Li, Z. Fu, L. Pu, X. Zhang, Z. Liu and K. Huang. 2016. The performance of nano urchin-like $NiCo_2O_4$ modified activated carbon as air cathode for microbial fuel cell. J. Power Sources 303: 325-332.

Ge, Z. and Z. He. 2015. An effective dipping method for coating activated carbon catalyst on the cathode electrodes of microbial fuel cells. RSC Advances 5(46): 36933-36937.

Ghasemi, M., W.R.W. Daud, S.H. Hassan, T. Jafary, M. Rahimnejad, A. Ahmad and M.H. Yazdi. 2016. Carbon nanotube/polypyrrole nanocomposite as a novel cathode catalyst and proper alternative for Pt in microbial fuel cell. Int. J. Hydrogen Energ. 41(8): 4872-4878.

Gokhale, R., S.M. Unni, D. Puthusseri, S. Kurungot and S. Ogale. 2014. Synthesis of an efficient heteroatom-doped carbon electro-catalyst for oxygen reduction reaction by pyrolysis of protein-rich pulse flour cooked with SiO_2 nanoparticles. Phys. Chem. Chem. Phys. 16(9): 4251-4259.

Gong, K., F. Du, Z. Xia, M. Durstock and L. Dai. 2009. Nitrogen-doped carbon nanotube arrays with high electrocatalytic activity for oxygen reduction. Science 323(5915): 760-764.

Gong, X.-B., S.-J. You, X.-H. Wang, J.-N. Zhang, Y. Gan and N.-Q. Ren. 2014. A novel stainless steel mesh/cobalt oxide hybrid electrode for efficient catalysis of oxygen reduction in a microbial fuel cell. Biosens. Bioelectron. 55: 237-241.

Guo, C.-Z., C.-G. Chen and Z.-L. Luo. 2014. A novel nitrogen-containing electrocatalyst for oxygen reduction reaction from blood protein pyrolysis. J. Power Sources 245: 841-845.

Hou, B., J. Sun and Y.-y. Hu. 2011. Simultaneous Congo red decolorization and electricity generation in air-cathode single-chamber microbial fuel cell with different microfiltration, ultrafiltration and proton exchange membranes. Bioresour. Technol. 102(6): 4433-4438.

Huang, J., N. Zhu, T. Yang, T. Zhang, P. Wu and Z. Dang. 2015. Nickel oxide and carbon nanotube composite (NiO/CNT) as a novel cathode non-precious metal catalyst in microbial fuel cells. Biosens. Bioelectron. 72: 332-339.

Jain, A., R. Balasubramanian and M. Srinivasan. 2016. Hydrothermal conversion of biomass waste to activated carbon with high porosity: a review. Chem. Eng. J. 283: 789-805.

Jiang, Y., Y. Xu, Q. Yang, Y. Chen, S. Zhu and S. Shen. 2014. Power generation using polyaniline/multi-walled carbon nanotubes as an alternative cathode catalyst in microbial fuel cells. Int. J. Energ. Res. 38(11): 1416-1423.

Kalathil, S., J. Lee and M.H. Cho. 2011. Granular activated carbon based microbial fuel cell for simultaneous decolorization of real dye wastewater and electricity generation. New Biotechnol. 29(1): 32-37.

Kim, J.R., S. Cheng, S.-E. Oh and B.E. Logan. 2007. Power generation using different cation, anion, and ultrafiltration membranes in microbial fuel cells. Environ. Sci. Technol. 41(3): 1004-1009.

Kondaveeti, S., J. Lee, R. Kakarla, H.S. Kim and B. Min. 2014. Low-cost separators for enhanced power production and field application of microbial fuel cells (MFCs). Electrochim. Acta 132: 434-440.

Li, D., L. Deng, H. Yuan, G. Dong, J. Chen, X. Zhang, Y. Chen and Y. Yuan. 2018. N, P-doped mesoporous carbon from onion as trifunctional metal-free electrode modifier for enhanced power performance and capacitive manner of microbial fuel cells. Electrochim. Acta 262: 297-305.

Li, D., Y. Qu, J. Liu, W. He, H. Wang and Y. Feng. 2014a. Using ammonium bicarbonate as pore former in activated carbon catalyst layer to enhance performance of air cathode microbial fuel cell. J. Power Sources 272: 909-914.

Li, J., Q. Fu, Q. Liao, X. Zhu, D.-d. Ye and X. Tian. 2009. Persulfate: a self-activated cathodic electron acceptor for microbial fuel cells. J. Power Sources 194(1): 269-274.

Li, J., Fu, Q., Zhu, X., Liao, Q., Zhang, L., Wang, H. 2010. A solar regenerable cathodic electron acceptor for microbial fuel cells. Electrochim. Acta, 55(7): 2332-2337.

Li, J., S. Wang, Y. Ren, Z. Ren, Y. Qiu and J. Yu. 2014b. Nitrogen-doped activated carbon with micrometer-scale channels derived from luffa sponge fibers as electrocatalysts for oxygen reduction reaction with high stability in acidic media. Electrochim. Acta 149: 56-64.

Li, N., Y. Liu, J. An, C. Feng and X. Wang. 2014c. Bifunctional quaternary ammonium compounds to inhibit biofilm growth and enhance performance for activated carbon air-cathode in microbial fuel cells. J. Power Sources 272: 895-899.

Li, S., Y. Hu, Q. Xu, J. Sun, B. Hou and Y. Zhang. 2012. Iron- and nitrogen-functionalized graphene as a non-precious metal catalyst for enhanced oxygen reduction in an air-cathode microbial fuel cell. J. Power Sources 213: 265-269.

Liew, K.B., W.R.W. Daud, M. Ghasemi, K.S. Loh, M. Ismail, S.S. Lim and J.X. Leong. 2015. Manganese oxide/functionalised carbon nanotubes nanocomposite as catalyst for oxygen reduction reaction in microbial fuel cell. Int. J. Hydrogen Energ. 40(35): 11625-11632.

Liu, F., L. Liu, X. Li, J. Zeng, L. Du and S. Liao. 2016a. Nitrogen self-doped carbon nanoparticles derived from spiral seaweeds for oxygen reduction reaction. RSC Advances 6(33): 27535-27541.

Liu, H. and B.E. Logan. 2004. Electricity generation using an air-cathode single chamber microbial fuel cell in the presence and absence of a proton exchange membrane. Environ. Sci. Technol. 38(14): 4040-4046.

Liu, H., R. Ramnarayanan and B.E. Logan. 2004. Production of electricity during wastewater treatment using a single chamber microbial fuel cell. Environ. Sci. Technol. 38(7): 2281-2285.

Liu, Q., Y. Zhou, S. Chen, Z. Wang, H. Hou and F. Zhao. 2015a. Cellulose-derived nitrogen and phosphorus dual-doped carbon as high performance oxygen reduction catalyst in microbial fuel cell. J. Power Sources 273: 1189-1193.

Liu, W., S. Cheng, D. Sun, H. Huang, J. Chen and K. Cen. 2015b. Inhibition of microbial growth on air cathodes of single chamber microbial fuel cells by incorporating enrofloxacin into the catalyst layer. Biosens. Bioelectron. 72: 44-50.

Liu, X., Y. Zhou, W. Zhou, L. Li, S. Huang and S. Chen. 2015c. Biomass-derived nitrogen self-doped porous carbon as effective metal-free catalysts for oxygen reduction reaction. Nanoscale 7(14): 6136-6142.

Liu, Y., K. Li, B. Ge, L. Pu and Z. Liu. 2016b. Influence of micropore and mesoporous in activated carbon air-cathode catalysts on oxygen reduction reaction in microbial fuel cells. Electrochim. Acta 214: 110-118.

Liu, Z., K. Li, X. Zhang, B. Ge and L. Pu. 2015d. Influence of different morphology of three-dimensional Cu_xO with mixed facets modified air–cathodes on microbial fuel cell. Bioresour. Technol. 195: 154-161.

Logan, B.E., B. Hamelers, R. Rozendal, U. Schröder, J. Keller, S. Freguia, P. Aelterman, W. Verstraete and K. Rabaey. 2006. Microbial fuel cells: methodology and technology. Environ. Sci. Technol. 40(17): 5181-5192.

Logan, B.E., C. Murano, K. Scott, N.D. Gray and I.M. Head. 2005. Electricity generation from cysteine in a microbial fuel cell. Water Res. 39(5): 942-952.

Lu, J., X. Bo, H. Wang and L. Guo. 2013a. Nitrogen-doped ordered mesoporous carbons synthesized from honey as metal-free catalyst for oxygen reduction reaction. Electrochim. Acta 108: 10-16.

Lu, M., L. Guo, S. Kharkwal, H.Y. Ng and S.F.Y. Li. 2013b. Manganese–polypyrrole–carbon nanotube, a new oxygen reduction catalyst for air-cathode microbial fuel cells. J. Power Sources 221: 381-386.

Nam, G., J. Park, S.T. Kim, D.-b. Shin, N. Park, Y. Kim, J.-S. Lee and J. Cho. 2014. Metal-free Ketjenblack incorporated nitrogen-doped carbon sheets derived from gelatin as oxygen reduction catalysts. Nano Lett. 14(4): 1870-1876.

Nunes, M., I.M. Rocha, D.M. Fernandes, A.S. Mestre, C.N. Moura, A.P. Carvalho, M.F. Pereira and C. Freire. 2015. Sucrose-derived activated carbons: electron transfer properties and application as oxygen reduction electrocatalysts. RSC Advances 5(124): 102919-102931.

Pan, F., Z. Cao, Q. Zhao, H. Liang and J. Zhang. 2014. Nitrogen-doped porous carbon nanosheets made from biomass as highly active electrocatalyst for oxygen reduction reaction. J. Power Sources 272: 8-15.

Papaharalabos, G., J. Greenman, C. Melhuish, C. Santoro, P. Cristiani, B. Li and I. Ieropoulos. 2013. Increased power output from micro porous layer (MPL) cathode microbial fuel cells (MFC). Int. J. Hydrogen Energ. 38(26): 11552-11558.

Pu, L., K. Li, Z. Chen, P. Zhang, X. Zhang and Z. Fu. 2014. Silver electrodeposition on the activated carbon air cathode for performance improvement in microbial fuel cells. J. Power Sources 268: 476-481.

Qiu, Z., M. Su, L. Wei, H. Han, Q. Jia and J. Shen. 2015. Improvement of microbial fuel cell cathodes using cost-effective polyvinylidene fluoride. J. Power Sources 273: 566-573.

Saito, T., T.H. Roberts, T.E. Long, B.E. Logan and M.A. Hickner. 2011. Neutral hydrophilic cathode catalyst binders for microbial fuel cells. Energ. Environ. Sci 4(3): 928-934.

Santoro, C., A. Agrios, U. Pasaogullari and B. Li. (2011). Effects of gas diffusion layer (GDL) and micro porous layer (MPL) on cathode performance in microbial fuel cells (MFCs). International Journal of Hydrogen Energy 36(20): 13096-13104.

Santoro, C., K. Artyushkova, S. Babanova, P. Atanassov, I. Ieropoulos, M. Grattieri, P. Cristiani, S. Trasatti, B. Li and A.J. Schuler. 2014. Parameters characterization and optimization of activated carbon (AC) cathodes for microbial fuel cell application. Bioresour. Technol. 163: 54-63.

Santoro, C., A. Serov, L. Stariha, M. Kodali, J. Gordon, S. Babanova, O. Bretschger, K. Artyushkova and P. Atanassov. 2016. Iron based catalysts from novel low-cost organic precursors for enhanced oxygen reduction reaction in neutral media microbial fuel cells. Energ. Environ. Sci 9(7): 2346-2353.

Shi, X. and T. Huang. 2015. Effect of pore-size distribution in cathodic gas diffusion layers on the electricity generation of microbial fuel cells (MFCs). RSC Advances 5(124): 102555-102559.

Stams, A.J., F.A. De Bok, C.M. Plugge, V. Eekert, H. Miriam, J. Dolfing and G. Schraa. 2006. Exocellular electron transfer in anaerobic microbial communities. Environ. Microbiol. 8(3): 371-382.

Sun, Y., Y. Duan, L. Hao, Z. Xing, Y. Dai, R. Li and J. Zou. 2016. Cornstalk-derived nitrogen-doped partly graphitized carbon as efficient metal-free catalyst for oxygen reduction reaction in microbial fuel cells. ACS Appl. Mater. Inter. 8(39): 25923-25932.

Tang, X., K. Guo, H. Li, Z. Du and J. Tian. 2010. Microfiltration membrane performance in two-chamber microbial fuel cells. Biochem. Eng. J. 52(2-3): 194-198.

Venkatesan, P.N. and S. Dharmalingam. 2013. Characterization and performance study on chitosan-functionalized multi walled carbon nano tube as separator in microbial fuel cell. J. Membr. Sci. 435: 92-98.

Verdaguer-Casadevall, A., P. Hernandez-Fernandez, I.E. Stephens, I. Chorkendorff and S. Dahl. 2012. The effect of ammonia upon the electrocatalysis of hydrogen oxidation and oxygen reduction on polycrystalline platinum. J. Power Sources 220: 205-210.

Wang, D., Z. Ma, M. Zhang, N. Zhao and H. Song. 2018. Fe/N-doped graphene with rod-like CNTs as an air-cathode catalyst in microbial fuel cells. RSC Advances 8(3): 1203-1209.

Wang, H., X. Bo, C. Luhana and L. Guo. 2012. Nitrogen doped large mesoporous carbon for oxygen reduction electrocatalyst using DNA as carbon and nitrogen precursor. Electrochem. Commun. 21: 5-8.

Wang, H., K. Wang, J. Key, S. Ji, V. Linkov and R. Wang. 2014a. Egg white derived tremella-like mesoporous carbon as efficient non-precious electrocatalyst for oxygen reduction. J. Electrochem. Soc. 161(10): H637-H642.

Wang, J., K. Li, L. Zhang, B. Ge, Y. Liu, T. Yang and D. Liu. 2017. Temperature-depended Cu 0.92 Co 2.08 O 4 modified activated carbon air cathode improves power output in microbial fuel cell. Int. J. Hydrogen Energ. 42(5): 3316-3324.

Wang, Q., X. Zhang, R. Lv, X. Chen, B. Xue, P. Liang and X. Huang. 2016. Binder-free nitrogen-doped graphene catalyst air-cathodes for microbial fuel cells. J. Mater. Chem. A 4(32): 12387-12391.

Wang, R., H. Wang, T. Zhou, J. Key, Y. Ma, Z. Zhang, Q. Wang and S. Ji. 2015a. The enhanced electrocatalytic activity of okara-derived N-doped mesoporous carbon for oxygen reduction reaction. J. Power Sources 274: 741-747.

Wang, R., K. Wang, Z. Wang, H. Song, H. Wang and S. Ji. 2015b. Pig bones derived N-doped carbon with multi-level pores as electrocatalyst for oxygen reduction. J. Power Sources 297: 295-301.

Wang, X., C. Feng, N. Ding, Q. Zhang, N. Li, X. Li, Y. Zhang and Q. Zhou. 2014b. Accelerated OH–transport in activated carbon air cathode by modification of quaternary ammonium for microbial fuel cells. Environ. Sci. Technol., 48(7): 4191-4198.

Wang, X., N. Gao, Q. Zhou, H. Dong, H. Yu and Y. Feng. 2013. Acidic and alkaline pretreatments of activated carbon and their effects on the performance of air-cathodes in microbial fuel cells. Bioresour. Technol. 144: 632-636.

Watson, V.J., C.N. Delgado and B.E. Logan. 2013a. Improvement of activated carbons as oxygen reduction catalysts in neutral solutions by ammonia gas treatment and their performance in microbial fuel cells. J. Power Sources, 242: 756-761.

Watson, V.J., C. Nieto Delgado and B.E. Logan. 2013b. Influence of chemical and physical properties of activated carbon powders on oxygen reduction and microbial fuel cell performance. Environ. Sci. Technol. 47(12): 6704-6710.

Watson, V.J., T. Saito, M.A. Hickner and B.E. Logan. 2011. Polymer coatings as separator layers for microbial fuel cell cathodes. J. Power Sources 196(6): 3009-3014.

Wu, H., J. Geng, H. Ge, Z. Guo, Y. Wang and G. Zheng. 2016. Egg-derived mesoporous carbon microspheres as bifunctional oxygen evolution and oxygen reduction electrocatalysts. Adv. Energy. Mater. 6(20): 1600794.

Yang, G., C. Erbay, S.-i. Yi, P. de Figueiredo, R. Sadr, A. Han and C. Yu. 2016. Bifunctional nano-sponges serving as non-precious metal catalysts and self-standing cathodes for high performance fuel cell applications. Nano Energy 22: 607-614.

Yang, W., W. He, F. Zhang, M.A. Hickner and B.E. Logan. 2014a. Single-step fabrication using a phase inversion method of poly (vinylidene fluoride) (PVDF) activated carbon air cathodes for microbial fuel cells. Environ. Sci. Tech. Let. 1(10): 416-420.

Yang, W., K.-Y. Kim and B.E. Logan. 2015. Development of carbon free diffusion layer for activated carbon air cathode of microbial fuel cells. Bioresour. Technol. 197: 318-322.

Yang, W., J. Li, D. Ye, X. Zhu and Q. Liao. 2017a. Bamboo charcoal as a cost-effective catalyst for an air-cathode of microbial fuel cells. Electrochim. Acta 224: 585-592.

Yang, W., J. Li, L. Zhang, X. Zhu and Q. Liao. 2017b. A monolithic air cathode derived from bamboo for microbial fuel cells. RSC Advances 7(45): 28469-28475.

Yang, X., W. Zou, Y. Su, Y. Zhu, H. Jiang, J. Shen and C. Li. 2014b. Activated nitrogen-doped carbon nanofibers with hierarchical pore as efficient oxygen reduction reaction catalyst for microbial fuel cells. J. Power Sources 266: 36-42.

You, S., Q. Zhao, J. Zhang, J. Jiang and S. Zhao. 2006. A microbial fuel cell using permanganate as the cathodic electron acceptor. J. Power Sources 162(2): 1409-1415.

Yuan, H., Y. Hou, I.M. Abu-Reesh, J. Chen and Z. He. 2016. Oxygen reduction reaction catalysts used in microbial fuel cells for energy-efficient wastewater treatment: a review. Mater. Horiz. 3(5): 382-401.

Yuan, Y., T. Yuan, D. Wang, J. Tang and S. Zhou. 2013a. Sewage sludge biochar as an efficient catalyst for oxygen reduction reaction in an microbial fuel cell. Bioresour. Technol. 144: 115-120.

Yuan, Y., B. Zhao, Y. Jeon, S. Zhong, S. Zhou and S. Kim. 2011. Iron phthalocyanine supported on amino-functionalized multi-walled carbon nanotube as an alternative cathodic oxygen catalyst in microbial fuel cells. Bioresour. Technol. 102(10): 5849-5854.

Yuan, Y., S. Zhou and J. Tang. 2013b. In situ investigation of cathode and local biofilm microenvironments reveals important roles of OH– and oxygen transport in microbial fuel cells. Environ. Sci. Technol. 47(9): 4911-4917.

Zhang, B., Z. Wen, S. Ci, S. Mao, J. Chen and Z. He. 2014a. Synthesizing nitrogen-doped activated carbon and probing its active sites for oxygen reduction reaction in microbial fuel cells. ACS Appl. Mater. Inter. 6(10): 7464-7470.

Zhang, F., S. Cheng, D. Pant, G. Van Bogaert and B.E. Logan. 2009a. Power generation using an activated carbon and metal mesh cathode in a microbial fuel cell. Electrochem. Commun. 11(11): 2177-2179.

Zhang, F., K.S. Jacobson, P. Torres and Z. He. 2010a. Effects of anolyte recirculation rates and catholytes on electricity generation in a litre-scale upflow microbial fuel cell. Energ. Environ. Sci. 3(9): 1347-1352.

Zhang, F., D. Pant and B.E. Logan. 2011a. Long-term performance of activated carbon air cathodes with different diffusion layer porosities in microbial fuel cells. Biosens. Bioelectron. 30(1): 49-55.

Zhang, J., Q. Li, C. Zhang, L. Mai, M. Pan and S. Mu. 2015a. A N-self-doped carbon catalyst derived from pig blood for oxygen reduction with high activity and stability. Electrochim. Acta 160: 139-144.

Zhang, J., S. Wu, X. Chen, K. Cheng, M. Pan and S. Mu. 2014b. An animal liver derived non-precious metal catalyst for oxygen reduction with high activity and stability. RSC Advances 4(62): 32811-32816.

Zhang, J., S. Wu, X. Chen, M. Pan and S. Mu. 2014c. Egg derived nitrogen-self-doped carbon/carbon nanotube hybrids as noble-metal-free catalysts for oxygen reduction. J. Power Sources 271: 522-529.

Zhang, L., J. Li, X. Zhu, D. Ye and Q. Liao. 2013a. Anodic current distribution in a liter-scale microbial fuel cell with electrode arrays. Chem. Eng. J. 223: 623-631.

Zhang, P., K. Li and X. Liu. 2014d. Carnation-like MnO_2 modified activated carbon air cathode improve power generation in microbial fuel cells. J. Power Sources 264: 248-253.

Zhang, X., S. Cheng, X. Huang and B.E. Logan. 2010b. The use of nylon and glass fiber filter separators with different pore sizes in air-cathode single-chamber microbial fuel cells. Energ. Environ. Sci 3(5): 659-664.

Zhang, X., S. Cheng, X. Wang, X. Huang and B.E. Logan. 2009b. Separator characteristics for increasing performance of microbial fuel cells. Environ. Sci. Technol. 43(21): 8456-8461.

Zhang, X., W. He, W. Yang, J. Liu, Q. Wang, P. Liang, X. Huang and B.E. Logan. 2016. Diffusion layer characteristics for increasing the performance of activated carbon air cathodes in microbial fuel cells. Environmental Science: Water Research & Technology 2(2): 266-273.

Zhang, X., K. Li, P. Yan, Z. Liu and L. Pu. 2015b. N-type Cu_2O doped activated carbon as catalyst for improving power generation of air cathode microbial fuel cells. Bioresour. Technol. 187: 299-304.

Zhang, X., D. Pant, F. Zhang, J. Liu, W. He and B.E. Logan. 2014e. Long-term performance of chemically and physically modified activated carbons in air cathodes of microbial fuel cells. ChemElectroChem 1(11): 1859-1866.

Zhang, X., J. Shi, P. Liang, J. Wei, X. Huang, C. Zhang and B.E. Logan. 2013b. Power generation by packed-bed air-cathode microbial fuel cells. Bioresour. Technol. 142: 109-114.

Zhang, X., X. Xia, I. Ivanov, X. Huang and B.E. Logan. 2014f. Enhanced activated carbon cathode performance for microbial fuel cell by blending carbon black. Environ. Sci. Technol. 48(3): 2075-2081.

Zhang, Y., Y. Hu, S. Li, J. Sun and B. Hou. 2011b. Manganese dioxide-coated carbon nanotubes as an improved cathodic catalyst for oxygen reduction in a microbial fuel cell. J. Power Sources 196(22): 9284-9289.

Zhao, F., N. Rahunen, J.R. Varcoe, A. Chandra, C. Avignone-Rossa, A.E. Thumser and R.C. Slade. 2008. Activated carbon cloth as anode for sulfate removal in a microbial fuel cell. Environ. Sci. Technol. 42(13): 4971-4976.

Zhou, L., P. Fu, X. Cai, S. Zhou and Y. Yuan. 2016a. Naturally derived carbon nanofibers as sustainable electrocatalysts for microbial energy harvesting: a new application of spider silk. Appl. Catal. B-Environ. 188: 31-38.

Zhou, L., P. Fu, D. Wen, Y. Yuan and S. Zhou. 2016b. Self-constructed carbon nanoparticles-coated porous biocarbon from plant moss as advanced oxygen reduction catalysts. Appl. Catal. B-Environ. 181: 635-643.

Zhu, C., J. Zhai and S. Dong. 2012. Bifunctional fluorescent carbon nanodots: green synthesis via soy milk and application as metal-free electrocatalysts for oxygen reduction. Chem. Commun. 48(75): 9367-9369.

Bioelectricity from Municipal Waste

Pranab Jyoti Sarma[1] and Kaustubha Mohanty*[1,2]

[1] Centre for Energy, Indian Institute of Technology Guwahati, Assam - 781039, India
[2] Department of Chemical Engineering, Indian Institute of Technology Guwahati, Assam - 781039, India

Introduction

Solid waste management is a major global challenge that humanity is facing since the last few decades. The dumping of unsegregated municipal wastes in and around us is creating an unhygienic environment, leading to an increase in the size and numbers of landfills. A major portion i.e., more than 60% of the municipal solid wastes in developing countries are organic fraction (Chakhtoura et al. 2014), which constitutes a mixture of carbohydrates, proteins, lipids and fibers (cellulose, hemicellulose, and lignin). It is estimated that 143,449 metric tons per day of urban municipal solid waste is generated, as per the Central Pollution Control Board (CPCB), 2014-15. With an increase in urban growth rate of 3.0%–3.5% per year, the per capita waste generation is also increasing at a rate of 1.3% every year (CPHEEO 2016). However, most of the solid wastes are not properly collected by the authorities in towns and cities. Moreover, uncontrolled disposal and burning up of these wastes is the common practice observed in many areas leading to a significant health risk for the large masses along with rapid degradation of the environment (Varma et al. 2013). At present, the biological management of OFMSW is mainly carried out by either composting or anaerobic digestion. Composting is a widely familiar fundamental process of treatment of biodegradable waste. On the other hand, anaerobic digestion has some advantages over composting; however, certain operational hindrances exist, as methanogenic bacteria are sensitive to many toxic substances present in waste and low energy conversion efficiency compared to other fermentation products.

Microbial electrochemical technologies are rapidly developing and promising novel technologies with no exception to microbial fuel cells (MFCs), wherein anaerobic digestion of OFMSW break down organic matter

*Corresponding author: kmohanty@iitg.ac.in

into carbon dioxide, protons, and electrons. The electrons that are produced during fermentation are transferred onto the anode surface. The electrons flow from anode through the external load before being consumed at the cathode. As a result, the flow of electrons between the anode and the cathode generates a potential difference leading to the generation of electrical power (Fig. 11.1). This organic fraction municipal solid waste (OFMSW) represents an attractive source of biomass energy, particularly in developing countries.

The organic matter present in municipal solid wastes is easily biodegradable, as it has a lesser fraction (<2.4%) of lignin mass (Maroun et al. 2007), which represents a valuable source of feed for microorganisms for electricity generation in MFCs. However, generation of bioelectricity in MFCs using the municipal solid waste has been rarely reported in literature till date.

Table 11.1. Physical composition of municipal solid waste

	Composition (%)							
Year	Biodegradables	Paper	Rubber	Metal	Glass	Rags	Other	Inerts
1996	42.21	3.63	0.60	0.49	0.60	—	—	45.13
2005	47.43	8.13	9.22	0.50	1.01	4.49	4.016	25.16

CPHEEO 2016

Figure 11.1. Schematic representation of MFC setup converting biodegradable wastes to electricity.

Solid Waste Characterization for MFC

Solid wastes are complex heterogeneous mixtures, and hence determination of their compositions is not easy. However, characterization of solid waste is an important criteria to evaluate its physical and chemical composition depending on sources and types of solid wastes. The nature and composition of wastes in MFCs will affect electricity production by virtue of relative proportions of different components present, the moisture content, and the specific nature of the biodegradable element.

Physical composition of solid wastes *viz.*, moisture content, particle size, density, temperature, and pH are important factors affecting their degradation rate and resource recovery process. The moisture content of solid wastes is usually expressed as the weight of moisture per unit weight of wet or dry material. Moisture content for most of the municipal solid wastes varies from 15 to 40%, depending on the composition of the wastes and the humidity and weather conditions. Most microorganisms, including bacteria, require a minimum of approximately 12% moisture for growth under optimum pH range of 6-8.

The moisture content affects the performance of MFCs (Wang et al. 2014), as greater water contents facilitated formation of more electron-mobile solution, thus promoting transfer of electrons to the cathode. A 10% greater content of water in the MFC resulted in 3-fold greater output voltage. Karluvali et al. (2015) studied the effect of temperature on electricity generation from OFMSW in a tubular microbial fuel cell in fed-batch mode. It was found that with the rise of temperature in mesophilic range of 20°C to 35°C, the maximum current density increased from 197.7 mA m^{-2} to 355.4 mA m^{-2} and maximum power density production tripled from 14.8 mW m^{-2} to 47.6 mW m^{-2}.

Densities of the organic fraction of solid wastes are often important to assess the total mass and volume of water present. Unfortunately, due to mixed nature of solid wastes, there is no uniformity in the densities of solid wastes reported in the literature. Also, densities of solid wastes vary with geographic location, season of the year, the length of storage time, etc.

The particle size in OFMSW is an important consideration in the resource recovery process through MFCs, as it affected the microbial degradation rate. It is highly desirable to have homogeneity and increased surface area/volume ratio, and it reduces the potential for preferential liquid flow paths through the waste. The size of the particles also influence waste packing densities, and reduction in particle size could increase surface area available to degradation by bacteria.

If the OFMSW solid wastes are to be used as fuel in a microbial fuel cell, determination of ultimate analysis of its component (APHA.1985) is essential, which typically involves the determination of the percent C (carbon), H (hydrogen), O (oxygen), N (nitrogen), S (sulphur), P (phosphorus), total Kjeldahl nitrogen (TKN), conductivity, total solids (TS), total volatile solids (TVS), and chemical oxygen demand (COD). To determine the amount of organic matter present and to ensure proper mix of waste, these analyses are necessary for achieving suitable C/N ratios for biological conversion processes.

Biological Properties of MSW

OFMSW is composed of various components, which can be classified into various categories *viz.*, water-soluble constituents containing sugars, starches, amino acids, and various organic acids, cellulose, hemicellulose, fats, oils, and waxes, lignin, lignocellulose, proteins, etc. The type and composition of waste will highly define the amount of electricity generation in a MFC. Table 11.2 (Khan et al. 2017) shows the types of organic substrates used in MFC with corresponding current and power densities.

Table 11.2. Comparative analysis of current and power densities from different types of substrate and microbial communities in MFCs.

Type of substrate	Type of strain	Maximum current or power densities
Starch	Pure culture of *Clostridium butyricum*	13,000 mAm^{-2}
Malt extract, yeast extract and glucose (Food and beverage industry)	Pure culture of *E. cloacae*	670 mAm^{-2}
Carboxymethyl cellulose (CMC)	Co-culture of *Clostridium cellulolyticum* and *G. sulfurreducens*	500 mAm^{-2}
Lactate (waste from dairy etc.)	Pure culture of *S. oneidensis* MR-1	50 mAm^{-2}
Cellulose particles (paper industry)	Pure culture of *Enterobacter cloacae*	200 mAm^{-2}
Municipal solid waste	*Geobacter, Bacteroides* and *Clostridium*	355.4 mAm^{-2}
Orange peel waste biomass	*Pseudomonas* was the predominant genera	847 ± 18.4 mAm^{-2}

One of the salient characteristic of MFCs is that almost all of the organic components present in OFMSW are rich sources of food for microorganism, and hence can be easily converted into energy. A mixture of biowastes can actually result in higher extractable current than any single component present in wastes (Barua et al. 2010).

Physical and Biological Transformations of Solid Waste

During the process of electricity generation in MFCs, the solid waste undergoes various transformations owing to microbial activity. The physical and biological transformation processes are highly essential for effective management of solid wastes, as well as resource recovery. This directly affects the development of an effective MFC technology system. The physical transformations that occur during solid waste degradation include (1) separation of components, (2) reduction in volume, and (3) size reduction of component due to degradation process. Physical transformations do not involve a change in phase (e.g., solid to gas), unlike chemical and biological transformation processes.

During the biological transformation process, volume and weight of organic material get reduced due to microbial process and the end product results in a humus-like material that can be used as a soil conditioner. In a mixed type of MSW, a variety of microorganisms apart from bacteria are also involved in the biological transformation of organic wastes *viz.*, fungi, yeasts, and actinomycetes. The success of MFC technology also relies on the development of anaerobic condition in the anodic chamber during biological transformation, where degradation of organic wastes is carried out.

The degradation ability and corresponding electric potential development from different household waste substrates using lab scale MFC was studied (Sanju et al. 2016), and the presence of oxidizing agent like $KMnO_4$ was found, which increases electricity generation capabilities.

Microbial Community Analysis in OFMSW

Municipal solid waste usually contains varieties of different complex substrates; hence, a synergistic consortium of hydrolytic and fermentative microorganisms is usually needed to break down such wastes. Therefore, characterization of bacterial communities of the biofilm in MFCs is highly important and can be analyzed by pyrosequencing of 16S rRNA, as it allows identification of various dominant and rare operational taxonomic units (OTUs) in a community (Chakhtoura et al. 2014). Through pyrosequencing of 16S rRNA gene, the presence of syntrophic interaction between exoelectrogenic *Geobacter* and fermentative *Bacteroides* was seen in a food waste-based MFC. It was also found that different organic loading rate of food wastes affects the power output of MFCs (Jia et al. 2013).

Working Analysis of Municipal Waste-based MFC

In MFCs electricity generation is fundamentally carried out by the metabolic activity of the microbes. The power generation capacity of MFCs is directly dependent on metabolic rate, which indicates that a higher metabolic rate helps in more efficient utilization and transformation of organic matter. This shows that the higher growth rate of microbes helps in greater electricity generation.

MFC technology is a continuous emerging technology with new and innovative research leading to make this technology a larger scale sustainable energy generation process in the future. The MFCs technology runs mainly in two configurations *viz.*, MFC with air cathodes or single chamber MFC, and MFC with liquid feed cathode or dual chamber MFC. In single chamber MFC, both the electrodes are placed in a single chamber with cathode in direct contact with external environment either in presence or absence of a membrane. The power density is usually high in single chambered MFC as compared to dual chamber, and hence single chamber MFCs are gaining more popularity along with the use of biocathode, rather than the use of expensive chemical catalysts (Khan et al. 2007).

Conclusion

Generation of electricity from municipal solid waste-based microbial fuel cell has the advantage of being a self-sustaining technology, as the microorganisms can break down a range of organic waste. MFCs based on solid waste are very good examples of green technology, which provide dual benefit of electricity generation by effective solid waste management, and maintaining a healthy pollution-free environment.

Though the potential of solid waste MFC has not been harnessed to its optimum level and the amount of electricity generation is low, it is very much needed for the future as it doesn't induce competition to any food products for bio-energy generation, such as with bio-ethanol or bio-diesel from corn, soybean, etc. affecting food supply for larger mass of populations. Systems must be scaled up through careful consideration of all important factors *viz.*, the design of MFCs, electrode material, types of waste, etc. to make it a viable technology for our future.

REFERENCES

American Public Health Association (APHA) New York. 1985. Standard Methods for the Examination of Water and Waste Water, 16th Ed.

Barua, P.K. and D. Deka. 2010. Electricity generation from biowaste based microbial fuel cells. Int. J. Energy, Info & Comm. 1: 77-92.

CPHEEO. 2016. Manual on municipal solid waste management. Central Public Health and Environmental Engineering Organization, New Delhi.

Jia, J., Yu. Tang, B. Liu, Di. Wu, N. Ren and D. Xing. 2013. Electricity generation from food wastes and microbial community structure in microbial fuel cells. Biores. Technol. 144: 94-99.

Chakhtoura, J.E., M. El-Fadel, H.A. Rao, D. Li, S. Ghanimeh and P.E. Saikaly. 2014. Electricity generation and microbial community structure of air-cathode microbial fuel cells powered with the organic fraction of municipal solid waste and inoculated with different seeds. Biomass & Bioenergy 67: 24-31.

Karluvali, A., E.O. Koroglu, A. Cetinkaya and B. Ozkaya. 2015. Electricity generation from organic fraction of municipal solid wastes in tubular microbial fuel cell. Sep. Purif. Technol. 156: 502-511.

Khan, M.D., N. Khan, S. Sultana, R. Joshi, S. Ahmed, E. Yu, K. Scott, A. Ahmad and M.Z. Khan. 2017. Bioelectrochemical conversion of waste to energy using microbial fuel cell technology. Process Biochem. 57: 141-158.

Maroun, R. and M. El. Fadel 2007. Start-up of anaerobic digestion of source-sorted organic municipal solid waste in the absence of classical inocula. Environ. Sci. Technol. 41(19): 6808-6814.

Sanju, S. and R. Pawels. 2016. Microbial fuel cell (MFC) technology for household waste reduction and bioenergy production. CiVEJ 3: 119-126.

Varma, S.V. and A.S. Kalamdhad. 2013. Composting of Municipal Solid Waste (MSW) mixed with cattle manure. Int. J. Env. Sci. 3: 2068-2079.

Wang, X., J. Tang, J.Cui, Q. Liu, J.P. Giesy and M. Hecker. 2014. Synergy of electricity generation and waste disposal in solid-state microbial fuel cell (MFC) of cow manure composting. Int. J. Electrochem. Sci. 9: 3144-3157.

Biological Conversion of Food Waste to Value Addition in Microbial Fuel Cell

C. Nagendranatha Reddy[1] and Booki Min*

Department of Environmental Science and Engineering, Kyung Hee University, Seocheon-dong, Yongin-si, Gyeonggi-do 446-701, Republic of Korea

[1] Present Address: Bhuma Shobha Nagireddy Memorial College of Engineering and Technology, Kota Kandukur Metta, Allagadda - 518543, Kurnool (Dist), Andhra Pradesh, India

Introduction

The exponential expansion of global population and rapid urbanization has led to a huge demand for food production. This extensive production of food exceeding demand has consequently generated large amounts of food waste/wastewaters (FW) by domestic and associated food processing industries (Ghosh et al. 2016). Around one-third (~1300 million tons) of the overall food generated for human intake is being lost or wasted, accounting to nearly 680 billion USD in industrialized countries and 310 billion USD in developing countries (http://www.fao.org/save-food/resources/keyfindings/en/). Dumping of waste/wastewaters without prior treatment may affect the integrity and equilibrium of the environment due to the presence of higher organic fraction and nutrients, thereby demanding the advancements in competent wastewater treatment approaches to curtail environmental pollution and develop sustainable society. In several countries, the FW is being landfilled or incinerated or discharged into civil wastewater treatment plants after simple recycling processes, putting public health and environment at a greater risk. At present, landfill is not a feasible option due to inadequate land available and possible contamination of ground and surface waters, while incineration is a highly energy-intensive process with simultaneous production of hazardous gases and ashes (Crowley et al. 2003). In addition, incineration and landfill processes do not recover any energy and resource. Due to the limitations associated with the above-mentioned methods, they are not considered to be economically viable and

*Corresponding author: bkmin@khu.ac.kr

environmentally friendly approaches for safe disposal of FW (Jeswani and Azapagic 2016, Dai et al. 2015). Hence, attaining energy neutral treatment of waste/wastewaters without further emphasizing the detrimental effects on environment and endangering other life forms is necessary (http://unesdoc. unesco.org/images/0024/002471/247153e.pdf). Currently, the research is being focused on the development of bioenergy using waste (biodegradable FW) as feedstock, thereby developing sustainable development research.

Therefore, efforts have been made to treat and explore the energies carried by the FW or its leachate through biological methods for simultaneous reduction of waste quantities and production of value-added chemicals. More recently, renewable energy from biological sources is termed as potential ways to reduce the future needs of fossil fuels and to overcome the global warming crisis. Till date, many technologies involving biological means have been assessed for organic waste metabolism (Puyol et al. 2017, Paritosh et al. 2017, Owusu and Asumadu-Sarkodie 2016). The organic content of FW serves as a potential feedstock to valorize spectrum of bio-based products including biofuels (Coma et al. 2017, Paritosh et al. 2017). To essentially establish a sustainable and effective remediation strategy, microbial fuel cell (MFC) would be feasible to diminish the dependence on the fossil-based feedstock and to limit the impact on the environment. MFC is a hybrid bio-electrochemical device that has the inherent advantages of directly transforming the energy stored in organic substrate to electrical energy mediated by anode respiring/electroactive bacteria as biocatalyst (Franks and Nevin 2010, Li et al. 2014, Goswami and Mishra 2018, Min et al. 2005). This biocatalyzed metabolic process provides a new approach for effective and environment friendly strategies for waste elimination, minimization, and effective utilization to various value-added products. MFCs have unique advantage of developing new electrical infrastructures for remote areas where FW generation is higher and left to decompose. This biological production of bioenergy by utilizing FW as feedstock by MFC technology will rule out the biofuel/bioenergy separation and purification expenditures, which are considered to be a bottleneck for commercial application of bioconversion technologies (Ravindran and Jaiswal 2016a, b, Goud et al. 2011). In this chapter, the emphasis is curbed to the utilization of FW as substrate in MFCs for wastewater treatment and simultaneous electricity harvesting.

Food Waste/Wastewaters as Potential Feedstock

The loss of food in the food supply chain that is produced for human feeding is termed as food waste (Girotto et al. 2015, Parfitt et al. 2010). This is related to retailers' and consumers' behaviors. Ineffective waste management strategies and rapid development leads to FW accumulation. Around the world, approximately 2 million tons of municipal solid waste has been generated every year, of which FW is a major component (25-70%) (Aschemann-Witzel et al. 2015). In United States of America, United

Kingdom, and Japan, approximately 30-40% of the overall food produced is lost as waste. In the European Union, almost 90 million tons of FW was produced in the year 2010, but it is expected to increase 126 million tons by 2020. However, South Africa has the least FW generation of 9 million tons per annum (Li et al. 2013a, Otles et al. 2015).

Source of Food Waste/Wastewaters Generation

FW generation includes disposal activities related to food harvesting, processing, distribution, transportation, consumption, storage, etc. This loss can be attributed by spoilage or contamination during transportation and storage, distribution, remains that are not edible by humans, losses during processing, over preparation, which exceeds the consumption, etc. FW generation is classified into pre-consumer (generated from farming, treating, handling, circulation, etc.) and post-consumer wastes (wastes from improper storage, surplus meal preparation, and ingestion, etc.) (http://www.fao.org/docrep/014/mb060e/mb060e02.pdf). This FW generation at household level is influenced by household socio-demographic characters, consumption behavior, and food patterns. In medium- and high-income countries, great extent of food is wasted even if it is still fit for consumption. However, in low-income countries, comparatively less food is wasted at the consumer level, with maximum during the initial phase of the food supply chain (Olinto et al. 2011, Abdullah et al. 2016).

Studies show that the most perishable products form the highest fraction of FW and they can be grouped as seven commodities *viz.*, cereals, pulses and oil crops, roots and tubers, fruits and vegetables, dairy products, fish and meat. The key sources of FW are being generated from the food and drink manufacturing industries, followed by grocery retailers, catering business, and finally comes the food wasted by humans, known as domestic or household or kitchen waste. North America, Oceania, and Europe top the per capita food loss and waste with ~280 kg/year, while the south and south East Asia with minimum of ~130 kg/year. In USA, the most wasted food items (in wt %) are vegetables or its remains (~23), milk products (~20), grains and fruits (~14), and meat products (~10) (Cuellar and Webber 2010, Li et al. 2013a, Kosseva 2013). The common food items wasted may vary from place to place.

Composition of Food Waste/Wastewaters

Food waste is a reservoir of various organic fractions *viz.*, carbohydrates, proteins, lipids, and traces of inorganic components, which can be further digested into simpler organic compounds *viz.*, glucose, amino acids, fatty acids, etc. (Lin et al. 2013, Dahiya et al. 2018, Nguyen et al. 2017). The organic waste, also called biowaste, is particularly challenging due to its diverse composition and moisture content. The composition of FW varies with location, source, diet plans, and type at that particular point of time. These wastes include fruits and vegetable remains, blood and flesh from meat and fish, wastewaters from various food processing and manufacturing industries, and other discarded items after consumption. The characteristics of food waste/wastewaters include low concentrations of suspended solids (SS), high biological oxygen demand (BOD) and chemical oxygen demand (COD), and variations in other operational parameters.

FW is a combination of various ingredients that are heterogeneous in composition. The general composition of solid FW are carbohydrates (lignin, starch, cellulose and hemicellulose, etc.), proteins, lipids, etc. (Uçkun Kiran et al. 2014). The total amount of carbohydrates accounts to 35-69%, while the protein concentrations vary between 3.9-21.9%. The higher cellulose content hinders the valorization process as the hydrolysis of cellulose is very challenging (Ren et al. 2007). The canteen-based FW contains the remains of rice, vegetables, and meat, while the household has peels of vegetables and fruits in higher quantities, followed by dairy products, bakery items, grains, etc. The total solid (initial) amount of fruits and vegetable waste is ~8-18%, with the overall volatile solids (VS) content of ~87%. The organic portion includes hemicellulose and sugars (75%), cellulose (9%), and lignin (5%) with high moisture content (80-90%). The citrus waste contains high organic material with many carbohydrates (soluble and insoluble) (Kaparaju et al. 2012). Only a portion of the animal or bird is consumed by humans. Therefore, approximately 50-54, 52, 60-62, 68-72 and 78% of each cow, sheep, pig, chicken, and turkey, respectively end up as meat for human consumption with the remaining as waste. The wastewater contains high COD, BOD, SS, total phosphorus, and nitrogen with strong odor. Approximate composition of meat contains 20-80% of water, 9-34% of protein, 1.5-65% of fats, and 1-12% minerals. Seafood processing wastewaters contain high fat, oil and grease content (FOG), nitrogen, salinity, COD, and BOD. Previous reports shows 1-72.5 and 12.5-37.5 kg per ton of product in processing and filleting industries, respectively. The major components of fish meat include proteins (58%), ash (22%), ether extract (19%), lipids, omega-3 fatty acids, and minerals (iron, calcium, phosphorus, selenium, iodine, copper, etc.) with very limited levels of carbohydrates, which varies from species to species, age, season, environment, gender, etc. Vitamins A and D are produced in large quantities from the liver of cod and halibut species (Esteban et al. 2007).

The food and the allied processing industries produce huge magnitudes of wastes worldwide. Some of the industries are olive mills, dairy, meat processing, food and other cereals production and processing, wine and breweries, cassava mills, rice mills, chocolate mills, palm oil mills, etc. The olive mills generate olive cake and olive mill wastewater as the final waste in large quantities. The composition includes huge quantities of polyphenolic compounds, polyalcohols, long chain fatty acids, suspended solids, pectin, volatile acids, colloids, tannins, lipids, and other nitrogen containing compounds (Zagklis et al. 2015). They are characterized by black colored juice with low pH and high sugar content. The cheese production in dairy industries generate cheese-whey as a by-product, which is high in carbohydrates (Mollea et al. 2013, Kosseva et al. 2009). The constituents include high lactose content, soluble proteins, lipids, and mineral salts of 4.5-5, 0.6-0.8, 0.4-0.5, and 8-10% (w/v), respectively with added lactic and citric acids, B-group vitamins, and non-protein nitrogen compounds (urea and uric acid). Lactose, the main constituent of whey wastewater, is responsible for high COD (60-80 g/L) and BOD (30-50 g/L) (Guimaraes et al. 2010). The wastewater from brewing, distillation, and wine manufacturing processes of fermentation industries contains high BOD and COD with varying phenols, tannins, and other organic acids concentrations. The liquid waste from the distillery process contains high concentrations of COD (>100 g O_2/L), BOD (61-70 g/L), salinity (EC 250-300 dS/m), sulfates, metal, and potassium ions (Satyawali and Balakrishnan 2008). Table 12.1 highlights the characteristics of various food-based wastewaters.

Problems Associated with Food Waste/Wastewaters Accumulation

The problems associated with surplus FW generation include deleterious impact on environment, economic, and social levels. The organic matter and the nutrients of FW decays on buildup, providing breeding grounds for disease-causing microorganisms. FW also contributes to emission of green house gases during its disposal in landfills and contaminates the soil and water, thereby disrupting the biogenic cycles. It affects the income of both farmers and consumer due to the high costs associated with food wastage under economic impacts. Social impacts of FW may be attributed to the general perception of worldwide food security (Kosseva 2013). The current legislation prioritizes the prevention of FW and gives least emphasis on its disposal. In industrialized countries, higher quantities of food are lost or wasted due to more production, which exceeds the demand. Hence, the initial steps towards reducing the FW generation are proper communication and cooperation with farmers, which leads to prevention of over production, over supply, thereby balancing the problems associated with surplus and shortage of crops during the period (HLPE 2014).

Current Practices of Food Waste/Wastewaters Management

The consideration of FW as an unexploited resource to generate bioenergy and other platform chemicals leads to treatment of accumulated waste. Numerous biological technologies have been explored to recover biogas, biofuels, bioactive compounds, biodegradable plastics, bio preservatives, etc. as final products (Pham et al. 2015, Ghosh et al. 2016). From the energy security point, the energy recovery from organic fraction of FW is an added advantage, thereby motivating the utilization of FW towards reduction of disposal burdens on environment, evading natural resources depletion, and reducing risks on human health and environment. Animal feed use, composting, incineration, land filling, and anaerobic digestion are the current FW management techniques. Today, the maximum amount of FW generated is being recycled in the form of compost and its use as animal

Table 12.1. Composition of various food waste/wastewaters

Waste/Wastewater type	pH	TSS (g/L)	TS	COD (g/L)	BOD (g/L)	Nitrogen content	References
Canteen waste	6.5	23	36 g/l	330	247.5	-	Goud et al. 2011
Vegetable and fruit waste	5.9-8.7	0.21-4.12	-	1.5-18.7	0.8-9.6	-	Thassitou and Arvanitoyannis 2001
Olive mill processing wastewater	3-5	65	6.39%	100	43	1.2-1.5%	
Meat processing wastewater	4.9-8.1	0.3-6.4	-	0.5-15.9	0.15-4.6	50-841 mg/l	Kosseva 2013
Dairy wastewater	7.1-8.1	0.36-0.92	-	0.980-7.5	0.680-4.5	-	Britz et al. 2006
Cheese wastewater	5.5-9.5	0.5-2.5	-	1-7	0.588-5	102 mg/l	
Cheese\whey wastewater	5.2	0.188-2.330	-	0.377-2.214	0.189-6.219	13-172 mg/l	
Brewery wastewater	4.6-7.3	530-3728	1289-12248 mg/l	1179-5848	1609-3980	0-13.05 mg/l	Enitan et al. 2015
Ice cream wastewater	6.25	265	664 mg/l	11900	524	88.51 mg/l	Qasim and Mane 2013
Sweet-Snacks wastewater	5.64	225	764 mg/l	9720	487	95.2 mg/l	
Coffee wastewater	4.27-4.40	-	-	9270-14800	470-551	32-57 mg/l	Rattan et al. 2015
Spent wash (Distillery)	4.2	4200	42400 mg/l	57164	32300	1254 mg/l	Ansari et al. 2012
Palm oil mill effluent	3.4-5.2	5000-54000	11500-79000 mg/l	15000-100000	10250-43750	180-1400 mg/l	Madaki and Send 2013

feed (Lin et al. 2013). Composting produces valuable soil fertilizer, which has higher economic efficiency and is environmentally friendly, but the high moisture content generates detrimental leachate, with possible microbial contamination (Esteban et al. 2007). In most of the developing countries, FW is either landfilled or incinerated to generate bioenergy (Ngoc and Schnitzer 2009, Othman et al. 2013). The process of anaerobic digestion (AD) serves as an alternative way to produce CO_2 and CH_4 as the final products (Othman et al. 2013). However, the limitations of incomplete recovery, preprocessing and pretreatment, high investment costs for large tanks, dependence on various operational parameters, odor generation, etc. have suggested imminent ways to treat FW along with the recovery of various resources. The technology of MFC aids in overcoming the above mentioned disadvantages due to its inherent ability to valorize complex and recalcitrant wastewaters towards renewable energy generation and efficient recovery (Nagendranatha Reddy et al. 2015, Fornero et al. 2010, Pandey et al. 2016).

Biological Conversion of Food Waste/Wastewaters to Bioelectricity by MFC

MFC is probably the most advanced process for valorization of FW towards the generation of various value-added products. The MFCs essentially produce bioelectricity from the oxidation (anaerobically) of diverse organic and biodegradable compounds into energy, water, and CO_2 by biofilms of electroactive bacteria (Pant et al. 2012). Thus, the organic fraction rich food waste/wastewaters can be considered potential substrates for bioelectricity, biofuels, and platform chemicals generation in MFC operation (Fig. 12.1). The advantages of MFC over conventional bioprocesses are high-energy conversion and recovery, convenient transmission and utilization of bioelectricity, developing new applications based on the requirement, generating diverse spectrum of value-added products, etc. (Rozendal et al. 2008). Hence, the use of low value waste as feedstock aids in confronting critical environmental issues, such as pollution reduction and sustainable and renewable energy sources (Pant et al. 2012). Since the first usage of MFC for bioelectricity generation, various complex and composite wastewaters (varying from domestic to industrial effluents) have been engaged in bioenergy generation with simultaneous waste reduction/treatment (Elmekawy et al. 2015, Pant et al. 2012). Based on the source and characteristics of food waste/wastewaters, they are classified as vegetable and fruit waste, mixed (kitchen and canteen) waste, food processing wastewaters, and fermented food waste/wastewaters.

Vegetable and Fruit Waste

With the presence of higher carbohydrates content, fruit and vegetable waste makes a feasible and potential feedstock for utilization as feedstock. Venkata Mohan et al. (2010) evaluated the treatment of composite vegetable waste as

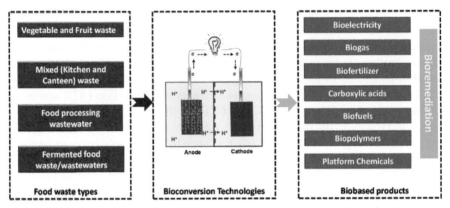

Figure 12.1. Schematic pathway of various products generation from food waste.

potential substrate in single-chambered mediator-less MFC. The MFC reactor operated at three different COD concentrations (2.08, 1.39, and 0.70 kg COD/ m^3-day), showing the highest power generation of 57.38 mW/m^2 with lower substrate load. With vegetable waste as substrate in dual-chambered MFC, influence of ultrasonic pretreatment, organic loadings, and anolyte pH on performance of MFC was evaluated by Tao et al. (2013). Results revealed that higher power generation (10.19 W/m^3) and organic compounds removal efficiency (TCOD; 62.5%) were achieved with ultrasonication pretreatment at >1.0 W/mL, which is almost double when compared to corresponding untreated control samples. Another study using U-shaped MFC design with household vegetable waste extract as substrate depicted the maximum power density (PD), current density (CD), and internal resistance of 88990 mW/ m^2, 314.4 mA/m^2, and 123.23 Ω, respectively, on the 2nd day of operation, suggesting the efficiency of U-shaped MFC design in comparison to dual-chambered MFCs (Javed et al. 2017). And, in order to evaluate the optimum composition of vegetable waste for treatment in MFCs, liquid extracts of potato, tomato, and lettuce along with different edge lengths of potato cubes (3, 5, and 7 mm) were evaluated. The potato extract showed higher CD of 100.2 mA/m^2, followed by tomato (86.7 mA/m^2), and lettuce (72.2 mA/ m^2) with similar COD removal efficacy of ~91.4 + 1.8% by the end of cycle operation. Potato cubes with 3, 5, and 7 mm showed maximum CD of 189.1, 178.9, and 163.3 mA/m^2, respectively with higher retention times. The difference in the performance is attributed to the higher carbohydrates present in potato that plays an imperative role in metabolic activity of microorganisms (Du et al. 2015).

Studies using fruit wastes as substrates in MFC were also evaluated. Miran et al. (2016) evaluated the bioelectricity production from unfiltered, filtered, and powdered orange peel waste (OPW). The maximum power densities of 358.8 ± 15.6, 320.5 ± 22.6, and 277.5 ± 5.3 mW/m^2 were obtained with filtered, unfiltered, and powered waste, respectively, representing the significance of particle size with regard to maximum voltage generation. The pectinase and polygalacturonase enzyme activities were also evaluated for pectin and cellulose, the key components of orange peel waste. Another study demonstrated the new approaches for the generation of bioelectricity from various waste citrus fruits using an MFC, and the results were compared to other reactor configurations. The result showed that MFCs have generated higher power output as compared to their counterpart reactors (Khan and Obaid 2015). The MFC (membrane-less single-chambered) operation with both vegetable and fruit waste at different ratios is evaluated for bioelectricity generation by Logrono et al. (2015). A blend of soil-organic matter was used as microbiological resource in the experiment, which showed maximum voltage generation of 330 mV with fruit to vegetable ratio of 25:75, but the final voltage (184 mV) was higher with 75:25 ratio when compared to 50:50 (118 mV) and 25:75 (100 mV), depicting the preference of microorganisms for easily biodegradable substrates.

Mixed (Kitchen and Canteen) Waste

The common kitchen and canteen waste contains mixture of leftover remains (raw and boiled vegetables, rice, meat, dairy products, etc.). This composite FW can act as a potential substrate for generation of bioelectricity in MFC. Moqsud et al. (2014) evaluated the feasibility of kitchen waste and bamboo waste as substrates in single-chambered MFC, showing the maximum voltage generation of 620 mV and higher nutrient content for soil amendment with kitchen waste. By utilizing the canteen waste, types of the organic compounds present in FW before and after the treatment in MFC were analyzed along with simultaneous bioelectricity generation. A maximum PD of 5.6 W/m^3 was achieved, depicting the easy and ready degradation of aromatic compounds in the hydrophilic and acidic fractions, thereby increasing the carboxylic and alcoholic groups due to hydrolysis and fermentation of composite FW in MFC (Li et al. 2016). Subsequently, the different organic loading rates (OLR) and dilution factors (DF) were evaluated in single-chambered MFC using canteen waste as substrate. OLR 2 (1.74 kg COD/m^3 day) and DF of 15 (3.2 + 0.4 g COD) were optimum for maximum power generation of 5.13 mW/m^2 and 556 mW/m^2 and COD removal efficiencies of 64.8% and 86.4%, respectively depicting the importance of organic loading rates in overall performance of MFC (Goud et al. 2011, Jia et al. 2013). Different inoculum densities (0, 50, 100, 150, 200, and 250 mg/L) of *Chlorella vulgaris* in cathode chamber of MFC showed highest PD (19151 mW/m^3), COD removal efficiency (44%), and total lipid productivity (31%) with optimal initial density of 150 mg/L (Hou et al. 2016).

Food Processing Wastewaters

The effluents from food processing industries constitute high organic content and are available abundantly, making them the most feasible substrates for MFC operation (Katuri et al. 2012). Various wastewaters have been treated by using MFC systems with bioelectricity generation and organic removal efficiencies reaching from 0.8-528 mW/m^2 and 63-95%, respectively. Dairy wastewaters rich in proteins and lipids offer substantial source of bioenergy (Fang and Yu 2000). The long chain fatty acids and release of free ammonia during the proteolysis processes in MFC have minimal impact on the anodic culture for bioelectricity generation (Nimje et al. 2012). Elakkiya and Matheswaran (2013) evaluated the bioelectricity generation from dairy industry wastewater by varying metabolism, pH, and inlet COD concentrations. The results demonstrated the efficiency of anaerobic metabolism for better columbic efficiency with observed PD of 2.7 W/m^3 at pH 7 and inlet COD of 1600 mg/L, which are thus termed as optimal parameters to treat dairy wastewater. In another study conducted by Faria et al. (2017), the higher PD and COD removal efficiencies of 5.1 W/m^3 and 63.5% were observed under continuous mode of operation. In spite of higher PD, the bioreactor depicted lower substrate removal efficiency due to variation in inlet COD concentrations and shorter retention times with continuous operation in comparison to fed batch mode operation. The brewery industry wastewater has high biodegradability due to carbohydrate content with COD concentrations typically ranging from 3-5 g/L. Thus, the wastewater is considered as feedstock for producing high power densities during MFC operation (Wen et al. 2009, Feng et al. 2008, Pant et al. 2010). The bioelectricity generation using single-chambered MFC was evaluated at varying buffering capacities and temperatures to effectively treat brewery wastewater. At 30°C, the full strength raw wastewater showed PD of 205 mW/m^2, which increased by 136% when phosphate buffer was used. Different bioreactor design of hybrid upflow MFC showed the PD of 1.6 W/m^3 (Katuri and Scott 2010). The earlier study of utilizing starch-based wastewater by Kim et al. (2004) showed generation of electricity at the inlet COD of 1700 mg/L. And, the recent study on cassava mill wastewaters as substrates in MFC (30 L; dual-chambered) by Kaewkannetra et al. (2011) demonstrated the high PD of 1771 mW/m^2 at inlet concentrations of 516 g/L. The effluents of the wineries consist of high COD and inhibitory substances (salts and sulfur compounds) that limit the microbial activity in MFC. Hence, the strategies of diluting the effluent or pretreating may enhance the scope for its utilization in MFC (Zhong et al. 2011). The winery effluent generated 41% higher power output in single-chambered MFC when compared to operation with domestic wastewater (Cusick et al. 2010). Cercado-Quezada et al. (2010) evaluated the treatment of different food processing wastewaters *viz.*, apple juice, wine, and yogurt waste at different potentials and inoculum sources. The results suggested the feasibility of yogurt and fruit juice wastewater as potential feedstock by generating the maximum PDs of 54 and 43.7 mW/m^2, respectively, in comparison to wine wastewater (0.8 mW/m^2). Other wastewaters *viz.*, potato processing (Durruty et al. 2012), food processing (Mansoorian et al. 2013), meat processing (Katuri et al. 2012), cheese whey (Antonopoulou et al. 2010), rice mill (Behera et al. 2010), and fish processing (You et al. 2010) were also studied.

Fermented Food Waste/Wastewaters

The possible integration of other technologies with MFC for utilization of hydrolyzed waste for generation of bioenergy, complete utilization of carbon, and effective functioning of bioprocesses have gained considerable attention. Integrating the conventional anaerobic digestion effluents with MFCs can generate bioelectricity by utilizing organic fraction of biogenic wastes *viz.* carbohydrate, lipids, and proteins present in FW, thereby aiding in further removal of carbon load that is discharged into the environment. The treatment of FW in MFC with prior hydrolysis helps in achieving efficient performance (enhanced bioelectrogenesis and COD removal efficiencies) (Sivagurunathan et al. 2017, Kakarla and Min 2014).

Goud and Venkata Mohan (2011) studied the influence of fermented and un-fermented FW on MFC operation at different OLRs, and demonstrated that the fermented FW showed higher performance at the 1.74 kg COD/m^3-day concentration. In a study conducted by Xin et al. (2018), the efficiency of MFCs over anaerobic digesters was illustrated in terms of conversion efficiency (0.245 kWh/ kg FW) by using fungal mash hydrolysate as substrate in cylinder-type air-cathode single-chambered MFCs. The maximum PD of 0.173 W/m^2 was achieved at the initial COD of about 1200 mg/L. The differences in utilizing single and dual-chambered MFC for bioelectrogenesis and treating the feedstock VFAs attained from fermented FW were evaluated by utilizing different microbial sources (inoculum). The maximum PD of 240.3 mW/m^2 was achieved by utilizing the short chain VFAs more rapidly (Choi et al. 2011). In an MFC utilizing FW leachate produced from biohydrogen fermentation, the maximum PD of 1540 mW/m^2 was observed in batch mode, while it was 769.2 mW/m^2 when operated in sequencing batch mode. It also showed the higher energy recovery (CE = 52.6%) achieved with lower organic rates. In a dual-chambered MFC, multi anodes and cathodes arrangement showed enhanced performance (15.14 W/m^3) with optimized substrate concentration (5000 mg/L) (Rikame et al. 2012). With the influence of NaCl (100 mM) on the MFC performance using FW leachate, the maximum PD (1000 mW/m^3) was achieved. By varying the anodic pH gradually from acidic to alkaline conditions (pH 4-9), there was a gradual increase in PD and it reached maximum (9956 mW/m^3) with highest columbic efficiency of 63.4 % at alkaline pH (Li et al. 2013b).

Apart from bioelectricity, other products generated from the MFC, microbial electrolysis cells, electrofermentation, and bioelectrohydrolysis processes utilizing FW as feedstock are volatile fatty acids (VFAs) (Sravan et al. 2018, Pant et al. 2013), simple sugars (Chandrasekhar and Venkata Mohan 2014), biofertilizer (Moqsud et al. 2014, Youn et al. 2015), biohydrogen (Chandrasekhar et al. 2015), bioethanol (Chandrasekhar et al. 2015), biohythane (Sravan et al. 2018), etc. Numerous solid and liquid wastes, such as vegetable and canteen wastes, domestic, industrial, and pretreated wastewaters have been studied for bioelectricity generation. The MFC performance with respect to various wastewater for bioelectricity generation is given in Table 12.2.

Table 12.2. An overview of various food wastes used for the operation of MFCs

Waste/Wastewater type	COD load	Max OCV	Maximum PD/CD	COD removal %	Working volume	Electrodes	Membrane	MFC configuration	Reference
Vegetable waste	2.08, 1.39 and 0.70 kg COD/m³-day	266 mV	57.38 mW/m²	62.86%	430 ml	Non-catalyzed graphite plates	PEM	Single chambered	Venkata Mohan et al. 2010
Vegetable waste	-	566 mV	88990 mW/m²; 314.4 mA/m²	-	-	Graphite rods	Cation exchange membrane	U shaped MFC	Javed et al. 2017
Fruit waste	-	3.43 V	2.92 mW	-	-	Zinc and copper electrodes	Salt bridge	Dual chambered	Khan and Obaid 2015
Vegetable waste	833 and 1828 mg/l	-	189.1 mA/m²	93.2%	0.24 L	Carbon felt	Cation exchange membrane	Dual chambered	Du et al. 2015
Fruit waste	1217 mg/l	0.59 V	358.8 mW/m²; 847 mA/m²	78.3%	200 ml	Graphite cloth and graphite felt	PEM	Dual chambered	Miran et al. 2016
Vegetable and fruit waste	-	330 mV	-	-	-	Carbon fiber	Soil-activated carbon	Single chambered	Logrono et al. 2015

(Contd.)

Table 12.2. (*Contd.*)

Waste/Wastewater Type	COD load	Max OCV	Maximum PD/CD	COD removal %	Working volume	Electrodes	Membrane	MFC configuration	Reference
Canteen waste	1.01, 1.74 and 2.61 kg COD/m³-day	295 mV	390 mA/m²	64.83%	0.43 L	Non-catalyzed graphite plates	PEM	Single chambered	Goud et al. 2011
Canteen waste	4900, 3200 and 2000 mg/l	322 mV	556 mW/m²	86.4%	22 ml	Carbon cloth and brush with graphite fibers and titanium core	–	Single chambered	Jia et al. 2013
Canteen waste	–	463 mV	170.81 mW/m²	76%	500 ml	Non-catalyzed graphite plates	PEM	Single chambered	Venkata Mohan and Chandrasekhar 2011
Kitchen waste	–	620 mV	60 mW/m²	–	–	Carbon fiber	–	Single chambered	Moqsud et al. 2014
Canteen waste	2700 mg/l	0.51 V	5.6 W/m³	87.4%	120 ml	Carbon cloth	–	Single chambered	Li et al. 2016
Canteen waste	220 g/l	260 mV	19151 mW/m³	44%	300 ml	Graphite and sheet copper	PEM	Dual chambered	Hou et al. 2016

Substrate	Concentration/COD	Voltage	Power/current density	Efficiency	Volume	Electrode	Membrane	Chamber	Reference
Fermented canteen waste	1.01, 1.74 and 2.61 kg COD/m³-day	391 mV	530 mA/m²	70%	500 ml	Non-catalyzed graphite plates	PEM	Single chambered	Goud and Venkata Mohan 2011
Fermented canteen waste	3.5 g/L	533 mV	240 mW/m²	57%	200 ml	Carbon felt and paper	Cation-permeable ion exchange	Single and dual chambered	Choi et al. 2011
Fermented canteen waste	1000, 2000, 5000, 10,000 and 20,000 mg/L	1.12 V	15.14 W/m³; 66.75 A/m³	90%	1.2 L	Pure carbon electrodes	PEM	Dual chambered	Rikame et al. 2012
Fermented food waste	1.5, 3.1 and 6.2 g COD/L-d,	0.56 V	1540 mW/m²	96%	24 ml	Graphite brush and carbon cloth	-	Single chambered	Choi and Ahn 2015
Fermented canteen (synthetic) waste	1000 mg/L	-	9956 mW/m³	85.4%	75.6 ml	Carbon felt	Cation exchange membrane	Dual chambered	Li et al. 2013b
Food waste leachate	2860 mg/L	270 mV	1.53 ± 0.11 W/m³; 8.75 ± 0.31 A/m³	95.4%	200 ml	Carbon felt and gas diffusion electrode	PEM	Single chambered	Wang and Lim 2017
Fermented kitchen waste	258, 323, 430, 645 and 1290 mg/L	400 mV	400 mW/m³	76%	300 ml	Graphite and sheet copper	PEM	Dual chambered	Hou et al. 2017

(Contd.)

Table 12.2. (*Contd.*)

Waste/Wastewater Type	COD load	Max OCV	Maximum PD/CD	COD removal %	Working volume	Electrodes	Membrane	MFC configuration	Reference
Fermented canteen waste	1200 mg/L	0.57 V	0.173 W/m²	90%	-	Carbon brush and cloth	-	Single chambered	Xin et al. 2018
Food and cereal-processing wastewater	735, 1670, 3250 and 8920 mg/L	0.29 V	81 mW/m²	95%	28/250 ml	Toray cloth	PEM	Single and dual chambered	Oh and Logan 2005
Apple juice, wine lees and yogurt processing wastewaters	7600/20200 and 14300 mg/L	-	43.7, 0.8 and 53.8 mW/m²	-	50/500 ml	Graphite felt and platinum mesh	-	Single chambered	Cercado-Quezada et al. 2010
Food diary processing wastewater	1562 mg/L	647 mV	0.037 mA/cm²; 0.015 mW/cm²	90%	200 ml	Carbon cloth	-	Single chambered	Nimje et al. 2012
Protein food processing wastewater	1900 mg/L	0.465 V	527 mA/m² and 230 mW/m²	86%	1.5 L	Graphic sheets	PEM	Dual chambered	Mansoorian et al. 2013
Dairy wastewater	1600 mg/L	780 mV	192 mW/m²	91%	0.3 L	Plain graphite plates	PEM	Dual chambered	Elakkiya and Matheswaran 2013
Dairy wastewater	3299 mg/L	576 mV	92.2 mW/m²	63%	350 ml	Carbon toray sheets	PEM	Dual chambered	Faria et al. 2017
Beer brewery wastewater	2250 mg/L	491 mV	528 mW/m²	90%	-	Carbon cloth	-	Single chambered	Feng et al. 2008

Influencing Parameters Affecting the MFC Performance Using Food Waste/Wastewaters

The process of converting the organic waste into value-added products by the MFC technology involves the complex processes of both biological and electrochemical reactions (Kim et al. 2005, Min and Angelidaki 2008). Hence, various parameters involved during the process either directly or indirectly may affect the process, such as FW source and its composition, organic loading rates, electrode chemistry and spacing, hydrolysis of FW, microbial consortia, and other operating parameters.

Source and Composition of Food Waste

The properties of FW and its origin are amongst the important parameters that have significant influence on the overall performance of MFC. The FW originating from the domestic needs and industrial processes are the two major types. The domestic food wastewater is easy to degrade in the MFC, while the industrial processing wastewater are complex and recalcitrant, which might not be favorable feedstock for utilization in MFC as such.

The wastewater containing the blend of composite vegetable and fruit waste in various ratios were evaluated by Logrono et al. (2015), illustrating the complete degradation of carbohydrates in fruit waste, while the cellulose-rich residues have not degraded completely, suggesting the preference for residues easily biodegradable. Another study carried out by Moqsud et al. (2014) depicted the higher bioelectricity generation with kitchen garbage due to the presence of more carbohydrates and sugars in the kitchen waste than the bamboo waste. The MFC studies using FW leachate obtained from prior fermentation processes showed varying bioelectricity generation with the similar COD values (1-1.3 g/L). The initial COD values of 1, 1.2, and 1.3 g/L showed power generation of 657, 0.173, and 1540 mW/m^2, respectively, depicting the differences in the composition of influent waste/wastewater (Li et al. 2013b, Xin et al. 2018, Choi and Ahn 2015).

Organic Loading Rates

Organic loading rates illustrate the quantity of waste in terms of COD that is fed into the MFC per certain time and volume. Literature shows significant influence of organic loading on bioelectricity generation and COD removal in MFCs. For example, the bioelectricity generation from FW leachate at various COD concentrations (1000, 2000, 5000, 10,000, and 20,000 mg/L) showed linear increase till 5000 mg/L, after which it showed gradual decrement with increment in organic loading. This inhibition of performance might be due to increase in VFAs concentration, which further reduce the operating pH, change in microbial community, increase in internal resistance and charge transfer resistance, etc. (Rikame et al. 2012). Higher bioelectricity generation was observed to increase from 188 mV at OLR 1 (1.01 kgCOD/m^3-day) to

295 mV at OLR 2 (1.74 kg COD/m³-day), which then decreased to 245 mV at OLR 3 (2.61 kg COD /m³-day), depicting the substrate mediated inactivation of biocatalyst (Goud et al. 2011). This reveals that optimum organic loading is required for better performance of MFC.

Reactor Configuration

Reactor type and design is the most imperative characteristic that primarily involves substrate-microbe-electrode interactions, thereby governing the efficacy and scalability of an MFC. Various design parameters, such as reactor design (single or dual-chambered), electrode materials type, arrangement and distance between electrodes, separator materials, etc. determine the efficient electron transfer mechanisms and losses associated (Min et al. 2005, Tenca et al. 2013, Rahimnejad et al. 2015, Oh et al. 2004, Ma et al. 2015).

Various reactor configurations with different electrode materials and separator membranes have been evaluated in MFC by utilizing FW as feedstock. The single and dual-chambered MFCs were extensively utilized to treat different FW (Choi et al. 2011, Jia et al. 2013, Hou et al. 2016, Goud et al. 2011). A U-shaped MFC design was evaluated by Javed et al. (2017) for treating household vegetable waste and observed enhanced PD of 88990 mW/m² and lower internal resistance of 123.23 Ω, suggesting the efficiency of U-shaped MFC design in comparison to dual-chambered MFCs. In another study using upflow membraneless MFC, treating high-strength palm oil mill effluent showed maximum PD of 44.6 mW/m² with enhanced mass transfer efficiency.

One of the major challenges in MFC operation is finding low cost, effective, and long lasting electrode materials and membranes. Various conductive anodic materials *viz.*, graphite (granules, rod, brush, etc.), carbon (brush, cloth, paper, etc.), and metal-based (stainless steel, etc.) have been extensively studied (Du et al. 2015, Miran et al. 2016, Jia et al. 2013, Chandrasekhar et al. 2015, Sravan et al. 2018, Hou et al. 2016, Wang and Lim 2017, Cercado-Quezada et al. 2010, Faria et al. 2017). Venkata Mohan and Chandrasekhar (2011) evaluated the influence of variable electrode-membrane assemblies and system buffering capacity by utilizing canteen FW in solid phase microbial fuel cell (SMFC). The configuration (anode placed 5 cm away from cathode) with proton exchange membrane (PEM) showed better power generation of 170.81 mW/m² followed by the without PEM configuration (53.41 mW/m²) showcasing the impact of electrode spacing, membrane, and buffering capacity on the bioelectricity generation. Accounting for higher costs of MFC scalability, efforts have been made to completely eliminate the usage of membranes in MFC operation. Few researchers have also focused on developing new and alternative membrane materials, such as canvas cloths, glass fibers, earthen pots, bipolar membranes, salt bridges, etc. that have superior properties and are cost-effective (Li et al. 2013b, Logrono et al. 2015, Khan and Obaid 2015).

Pretreatment of Food Waste

The composite solid FW and higher complexity of food industry wastewaters necessitate the requirement of pretreatment or hydrolysis for conversion into soluble and monomeric organics (Fig. 12.2). The process of enhancing the bio digestibility of substrate using MFC is called Bioelectrohydrolysis. This increases the availability of enzymes to the carbon-rich materials, thereby improving the overall performance of MFC from wastes. The electrons generated *in situ* or supplied from external source not only generate voltage but also aid in bioremediation or transformation of pollutants (Sreelatha et al. 2015, Nagendranatha Reddy et al. 2016, Nagendranatha Reddy and Venkata Mohan 2016). A bioelectrohydrolysis system was constructed to pretreat the canteen FW for enhancing the biohydrogen (H_2) production efficiency in dark fermentation. The pretreated waste showed higher bioH_2 production and maximum substrate degradation of 29.12 ml/h and 52.2%, respectively in comparison to control (26.75 ml/h; 43.68%) due to functioning of anode for bioconversion of composite organic materials to simpler molecules via the process of catabolism (Chandrasekhar and Venkata Mohan 2014). The electrofermentation of composite canteen FW using –0.6 V applied potential in single-chambered MFC showed higher production of VFAs (4595 mg/L) and biohythane (2.1 L) in comparison to closed circuit (3593 mg/L; 1.41 L) and control (2666 mg/L; 1.61 L) operations, respectively (Sravan et al. 2018).

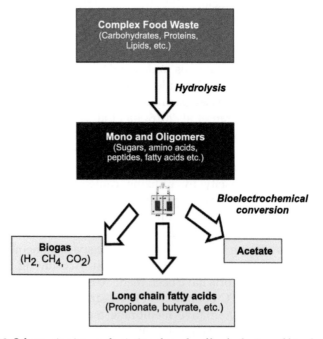

Figure 12.2. Schematic picture depicting the role of hydrolysis and bioelectrochemical reactions in forming value-added products.

Microbial Consortia

The bacterial metabolism involves the process of utilizing carbon-based substrates for the generation of bioelectricity. Hence, the anodic biocatalyst plays an important role in bioconversion and electron transfer mechanisms that occur on the electrodes of the MFC. Synergistic cooperation between multiple species in mixed culture aids in degradation of complex and recalcitrant wastewater, especially industrial wastewaters. Hence, a well-established and balanced microbial community is essential for the wastewater treatment systems, because few bacterial strains consume the electrons rather than generation.

The inoculum source and FW composition are the key ingredients that enrich appropriate bacteria for particular wastewater treatment. In order to curtail the initial bioreactor start-up process and advance the adaptation of particular microorganisms in terms of long term operational stability, indigenous species have to be inoculated into the bioreactor. For example, the performance of MFC utilizing yogurt wastewater was investigated using various inoculum sources. The MFC inoculated with compost-leachate exhibited higher power generation (92 mW/m^2) to the anaerobic sludge inoculated system (54 mW/m^2) (Cercado-Quezada et al. 2010). Similarly, to treat sea food wastewater containing high salinity, marine sediment has to be inoculated in the anode (You et al. 2010). The MFC bioreactor operated with FW hydrolysate showed the dominance of *Moheibacter* (19.8%), *Azospirillum* (18.4%), *Geobacter* (6.6%), *Petrimonas* (4.6%), *Alicycliphilus* (5.2%), *Rhodococcus* (3.8%), and *Pseudomonas* (3.2%). It is reported that the *Moheibacter* and *Petrimonas* are accountable for degradation of carbohydrates present in FW, while the *Pseudomonas*, *Geobacter* and *Alicycliphilus* are known for their exoelectrogenic activity in MFC (Xin et al. 2018, Song et al. 2005). Sravan et al. (2018) portrayed the dominance of *Clostridiaceae*, *Bacteroides*, and *Proteobacteria* families for the production of carboxylic acids, BioH$_2$ and BioCH$_4$ utilizing canteen based FW in electrofermentation unit at closed circuitry and applied potential of –0.6 V. The enrichment of both anaerobic and fermentative bacteria depicts the ability of producing various acids and alcohols utilizing diverse organic materials. Other reports also show the enrichment of *Proteobacteria*, *Bacteroidetes*, and *Firmicutes* in MFCs operated with different food and food processing wastewaters (Jia et al. 2013, Pendyala et al. 2016, Miran et al. 2016). The dominance of *Geobacter*, *Shewanella*, and *Pseudomonas* was observed to be higher in the MFCs operating with FW.

Other Operating Parameters

Various operational parameters *viz.*, pH, temperature, conductivity, dissolved oxygen concentrations, dilutions with sewage, cycle operation time, etc. affect the overall performance of MFC, as all these are closely interlinked (Fig. 12.3). The imbalance in pH affects the functions of extracellular enzymes, thereby limiting the MFC performance. The varying anolyte pH (5, 7, and 9) was evaluated in a 2-chambered MFC treating dairy wastewater illustrating the maximum OCV of 774 mV with pH 7, followed by pH 5 (754 mV) and pH 9 (675 mV) (Elakkiya and Matheswaran 2013). The influence of HRT on the fermentation process was performed at different pH of 5 and 5.5 towards the predominant production of acetate, butyrate, or lactate (Pant et al. 2013). In a study conducted by Feng et al. (2008), the single-chambered MFC treating beer brewery wastewater showed maximum PD of 205 mW/m² at 30°C, which decreased to 170 mW/m² when the temperature was 20°C, depicting the importance of temperature for microbial growth, efficient functioning of enzymes etc. According to Li et al. (2013b), the internal resistance of the MFC system can be reduced by upsurging the ionic conductivity of electrolyte. The NaCl at different concentrations showed the maximum power density of 657.8 + 8.34 mW/m³ at 100 mM concentration, which is almost 2× the value at control operations.

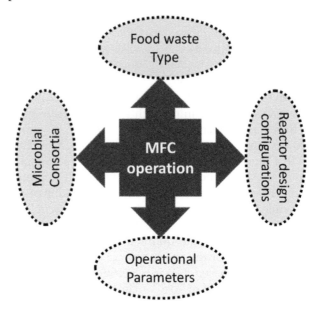

Figure 12.3. Influencing parameters for MFC operation with food waste as substrate.

Current Challenges for Employing Food Waste-based MFCs for Field Application

In spite of MFC being a viable and promising option for treating FW, its efficacy is limited by numerous operational and economic challenges. The primary challenge of MFC operation is scaling up, to meet the daily requirements of waste generation and power generation. Few attempts have been made to establish pilot scale treatment of various wastewaters. Advanced Water Management Center at the University of Queensland have set a 1 m³ reactor to treat wastewater obtained from Foster's brewery, but this MFC has produced maximum currents of 2 A/cell with minimal

PD (0.5 W/m^2) and COD removal (0.2 kg COD/m^3-day) (Logan 2010, Keller and Rabaey 2011). The correlation of power generation and carbon removal is particularly challenging when real wastewater is used because of its composite composition. Other factors that directly affect the scale up process are process control (low current and power densities, etc.), microbial manipulation and system design (cost of electrodes, membranes, installation and operation, etc.), and uneven operational parameters (feed composition, temperature, pH, conductivity, etc.) (Oh and Logan 2007, Logan 2010). Hence, miniaturization of MFC for achieving higher yields is still in its infancy and needs many solutions for practical implementation (Elmekawy et al. 2015). FW containing high content of cellulose and lignin is deliberated as primary challenge for lignocellulosic FW degradation in MFCs (Ren et al. 2007). Finding a common pretreatment method for the FW is also a challenging aspect, as the composition varies with one substrate to another (Elmekawy et al. 2015). Hence, for development of sustainable solutions for MFC operations with FW and fundamental understanding, an integrated and multidisciplinary approach would aid in the development of green and sustainable bioremediation technologies.

Final Thoughts

The MFC conversion process of organic-rich FW is highly promising with efficient energy generation and pollutant treatment, which can foresee many practical applications in the future. Many studies showed that FW with simple composition and higher carbohydrates illustrated high bioelectrogenic activity along with improved COD removal efficiencies. The complex and composite FW/food processing wastewater needs to be pretreated for enhancing degradability and MFC performance.

Acknowledgements

The study was carried out with research grants from National Research Foundation of Korea (2015R1D1A1A09059935; 2018R1A2B6001507) and Korea-India S & T Co-operation Program (2016K1A3A1A19945953).

REFERENCES

Abdullah, N.F., P.S. Teo and L.H. Foo. 2016. Ethnic differences in the food intake patterns and its associated factors of adolescents in Kelantan, Malaysia. Nutrients 8: 1-14.

Ansari, F., A.K. Awasthi and B.P. Srivastava. 2012. Physico-chemical characterization of distillery effluent and its dilution effect at different levels. Arch. Appl. Sci. Res. 4: 1705-1715.

Antonopoulou, G., K. Stamatelatou, S. Bebelis and G. Lyberatos. 2010. Electricity generation from synthetic substrates and cheese whey using a two-chamber microbial fuel cell. Biochem. Eng. J. 50: 10-15.

Aschemann-Witzel, J., I. de Hooge, P. Amani, T. Bech-Larsen and M. Oostindjer. 2015. Consumer-related food waste: causes and potential for action. Sustainability 7: 6457-6477.

Behera, M., P.S. Jana, T.T. More and M.M. Ghangrekar. 2010. Ricemill wastewater treatment in microbial fuel cells fabricated using proton exchange membrane and earthen pot at different pH. Bioelectrochemistry 79: 228-233.

Britz, T.J., C. VanSchalkwyk and Y.T. Hung. 2006. Treatment of dairy processing wastewaters. pp. 1-28. *In*: Wang, L.K., Y.-T. Hung, H.H. Lo and K. Yapijakis (eds.). Waste Treatment in Food Processing Industry. CRC Taylor & Francis Group LLC, Boca Raton, London, New York.

Cercado-Quezada, B., M.L. Delia and A. Bergel. 2010. Testing various food-industry wastes for electricity production in microbial fuel cell. Bioresource Technology 101: 2748-2754.

Chandrasekhar, K. and S. Venkata Mohan. 2014. Bio-electrohydrolysis as a pretreatment strategy to catabolize complex food waste in closed circuitry: function of electron flux to enhance acidogenic biohydrogen production. International Journal of Hydrogen Energy 39: 11411-11422.

Chandrasekhar, K., K. Amulya and S. Venkata Mohan. 2015. Solid phase bio-electrofermentation of food waste to harvest value-added products associated with waste remediation. Waste Management 45: 57-65.

Choi, J., H.N. Chang and J.I. Han. 2011. Performance of microbial fuel cell with volatile fatty acids from food wastes. Biotechnol Lett. 33: 705-714.

Choi, J. and Y. Ahn. 2015. Enhanced bioelectricity harvesting in microbial fuel cells treating food waste leachate produced from biohydrogen fermentation. Bioresource Technology 183: 53-60.

Coma, M., E. Martinez-Hernandez, F. Abeln, S. Raikova, J. Donnelly, T.C. Arnot, M.J. Allen, D.D. Hong and C.J. Chuck. 2017. Organic waste as a sustainable feedstock for platform chemicals. Faraday Discuss 202: 175-195.

Crowley, D., A. Staines, C. Collins, J. Bracken, M. Bruen, J. Fry, V. Hrymak, D. Malone, B. Magette, M. Ryan and C. Thunhurst. 2003. Health and environmental effects of landfilling and incineration of waste – a literature review. Reports 3. http://arrow.dit.ie/schfsehrep/3

Cuellar, A.D. and M.E. Webber. 2010. Wasted food, wasted energy: the embedded energy in food waste in the United States. Environ. Sci. Technol. 44: 6464-6469.

Cusick, R.D., P.D. Kiely and B.E. Logan. 2010. A monetary comparison of energy recovered from microbial fuel cells and microbial electrolysis cells fed winery or domestic wastewaters. Int. J. Hyd. Energy 35: 8855-8861.

Dahiya, S., A.N. Kumar, J.S. Sravan, S. Chatterjee, O. Sarkar and S. Venkata Mohan. 2018. Food waste biorefinery: sustainable strategy for circular bioeconomy. Bioresource Technology 248: 2-12.

Dai, Y.C., M.P.R. Gordon, J.Y. Ye, D.Y. Xu, Z.Y. Lin, N.K.L. Robinson, R. Woodard and M.K. Harder. 2015. Why door stepping can increase household waste recycling. Resources, Conservation and Recycling 102: 9-19.

Du, H., F. Li and C. Feng. 2015. Comparison of the performance of microbial fuel cell for treatment of different vegetable liquids and potato solid with different sizes. Journal of Environmental Engineering Research 52: 379-387.

Durruty, I., P.S. Bonanni, J.F. Gonzalez and J.P. Busalmen. 2012. Evaluation of potato-processing wastewater treatment in a microbial fuel cell. Biores. Technol. 105: 81-87.

Elakkiya, E. and M. Matheswaran. 2013. Comparison of anodic metabolisms in bioelectricity production during treatment of dairy wastewater in microbial fuel cell. Biores. Technol. 136: 407-412.

Elmekawy, A., S. Srikanth, S. Bajracharya, H.M. Hegab, P.S. Nigam, A. Singh, S. Venkata Mohan and D. Pant. 2015. Food and agricultural wastes as substrates for bioelectrochemical system (BES): the synchronized recovery of sustainable energy and waste treatment. Food Research International 73: 213-225.

Enitan, A.M., J. Adiyemo, S. Kumari, F.M. Swalaha and F. Bux. 2015. Characterization of brewery wastewater composition. International Journal of Environmental and Ecological Engineering 9: 1073-1076.

Esteban, M.B., A.J. Garcia, P. Ramos and M.C. Marquez. 2007. Evaluation of fruit, vegetable and fish wastes as alternative feed-stuffs in pig diets. Waste Manag. 27: 193-200.

Fang, H. and H. Yu. 2000. Effect of HRT on mesophilic acidogenesis of dairy wastewater. J. Environ. Eng. 126: 1145-1148. http://www.fao.org/docrep/014/mb060e/mb060e02.pdf (Food and Agriculture Organization of the United Nations) http://www.fao.org/save-food/resources/keyfindings/en/ (Food and Agriculture Organization of the United Nations)

Faria, A., L. Gonclaves, J.M. Peixoto, L. Peixoto, A.G. Brito and G. Martins. 2017. Resources recovery in the dairy industry: bioelectricity production using a continuous microbial fuel cell. Journal of Cleaner Production 140: 971-976.

Feng, Y., X. Wang, B.E. Logan and H. Lee. 2008. Brewery wastewater treatment using air-cathode microbial fuel cells. Appl Microbiol Biotechnol. 78: 873-880.

Fornero, J.J., M. Rosenbaum and L.T. Angenent. 2010. Electric power generation from municipal, food, and animal wastewaters using microbial fuel cells. Electroanalysis 22: 832-843.

Franks, A.E. and K.P. Nevin. 2010. Microbial fuel cells: a current review. Energies 3: 899-919.

Ghosh, P.R., D. Fawcett, S.B. Sharma and G.E.J. Poinern. 2016. Progress towards sustainable utilisation and management of food wastes in the global economy. International Journal of Food Science 2016: 1-22.

Girotto, F., L. Alibardi and R. Cossu. 2015. Food waste generation and industrial uses: a review. Waste Management 45: 32-41.

Goswami, R. and V.K. Mishra. 2018. A review of design, operational conditions and applications of microbial fuel cells. Biofuels 9: 203-220.

Goud, R.K., P.S. Babu and S. Venkata Mohan. 2011. Canteen based composite food waste as potential anodic fuel for bioelectricity generation in single chambered microbial fuel cell (MFC): bio-electrochemical evaluation under increasing substrate loading condition. International Journal of Hydrogen Energy 36: 6210-6218.

Goud, R.K. and S. Venkata Mohan. 2011. Pre-fermentation of waste as a strategy to enhance the performance of single chambered microbial fuel cell (MFC). International Journal of Hydrogen Energy 36: 13753-13762.

Guimaraes, P.M.R., J.A. Teixeira and L. Domingues. 2010. Fermentation of lactose to bio-ethanol by yeasts as part of integrated solutions for the valorization of cheese whey. Biotechnol. Adv. 28: 375-384.

HLPE, 2014. Food losses and waste in the context of sustainable food systems. A report by the High Level Panel of Experts on Food Security and Nutrition of the Committee on World Food Security, Rome.

Hou, Q., H. Pei, W. Hu, L. Jiang and Z. Yu. 2016. Mutual facilitations of food waste treatment, microbial fuel cell bioelectricity generation and *Chlorella vulgaris* lipid production. Biores. Technol. 203: 50-55.

Hou, Q., J. Cheng, C. Nie, H. Pei, L. Jiang, L. Zhang and Z. Yang. 2017. Features of *Golenkinia* sp. and microbial fuel cells used for the treatment of anaerobically digested effluent from kitchen waste at different dilutions. Biores. Technol. 240: 130-136.

Javed, M.M., M.A. Nisar, B. Muneer and M.U. Ahmad. 2017. Production of bioelectricity from vegetable waste extract by designing a U-shaped microbial fuel cell. Pakistan J. Zool., 49: 711-716.

Jeswani, H.K. and A. Azapagic. 2016. Assessing the environmental sustainability of energy recovery from municipal solid waste in the UK. Waste Management 50: 346-363.

Jia, J., Y. Tang, B. Liu, D. Wu, N. Ren and D. Xing. 2013. Electricity generation from food wastes and microbial community structure in microbial fuel cells. Biores. Technol. 144: 94-99.

Kaewkannetra, P., W. Chiwes and T.Y. Chiu. 2011. Treatment of cassava mill wastewater and production of electricity through microbial fuel cell technology. Fuel 90: 2746-2750.

Kakarla, R. and B. Min. 2014. Evaluation of microbial fuel cell operation using algae as an oxygen supplier: carbon paper cathode vs. carbon brush cathode. Bioprocess and Biosystems Engineering 37: 2453-2461.

Kaparaju, P., J. Rintala and A. Oikari. 2012. Agricultural potential of anaerobically digested industrial orange waste with and without aerobic post-treatment. Environ Technol. 33: 85-94.

Katuri, K.P., A.M. Enright, V. O'Flaherty and D. Leech. 2012. Microbial analysis of anodic biofilm in a microbial fuel cell using slaughter-house wastewater. Bioelectrochemistry 87: 164-171.

Katuri, K.P. and K. Scott. 2010. Electricity generation from the treatment of wastewater with a hybrid up-flow microbial fuel cell. Biotechnology and Bioengineering 107: 52-58.

Keller, J. and K. Rabaey. 2011. Experiences from MFC pilot plant operation. http://www.microbialfuelcell.org/Presentations/First%20MFC%20symposium/JK%20 presentation%20MFC%20Pilot%20v3.pdf, dated 11/09/2011

Khan, A.M. and M. Obaid. 2015. Comparative bioelectricity generation from waste citrus fruit using a galvanic cell, fuel cell and microbial fuel cell. Journal of Energy in Southern Africa 26: 90-99.

Kim, B.H., H.S. Park, H.J. Kim, G.T. Kim, I.S. Chang, J. Lee and N.T. Phung. 2004. Enrichment of microbial community generating electricity using a fuel-cell-type electrochemical cell. Appl. Microbiol. Biotechnol. 63: 672-681.

Kim, J.R., B. Min and B.E. Logan. 2005. Evaluation of procedures to acclimate a microbial fuel cell for electricity production. Appl. Microbiol. Biotechnol. 68: 23-30.

Kosseva, M.R., P.S. Panesar, G. Kaur and J.F. Kennedy. 2009. Use of immobilized biocatalysts in the processing of cheese whey. Int. J. Biol. Macromol. 45: 437-447.

Li, H., Y. Tian, W. Zuo, J. Zhang, X. Pan, L. Li and X. Su. 2016. Electricity generation from food wastes and characteristics of organic matters in microbial fuel cell. Biores. Technol. 205: 104-110.

Li, W.W., G.P. Sheng and H.Q. Yu. 2013a. Electricity generation from food industry wastewater using microbial fuel cell technology. pp 249-259. *In*: Kosseva, M.R. and C. Webb (eds.). Food Industry Wastes Assessment and Recuperation of Commodities. Academic Press, Oxford, UK.

Li, W.W., H.Q. Yu and Z. He. 2014. Towards sustainable wastewater treatment by using microbial fuel cells-centered technologies. Energy Environ. Sci. 7: 911-924.

Li, X.M., K.Y. Cheng and J.W.C. Wong. 2013b. Bioelectricity production from food waste leachate using microbial fuel cells: effect of NaCl and pH. Biores. Technol. 149: 452-458.

Lin, C.S.K., L.A. Pfaltzgraff, L. Herrero-Davila, E.B. Mubofu, S. Abderrahim, J.H. Clark, A.A. Koutinas, N. Kopsahelis, K. Stamatelatou and F. Dickson. 2013. FW as a valuable resource for the production of chemicals, materials and fuels: current situation and global perspective. Energ Environ. Sci. 6: 426-464.

Logan, B.E. 2010. Scaling up microbial fuel cells and other bioelectrochemical systems. Appl. Microbiol Biotechnol. 85: 1665-1671.

Logrono, W., G. Ramirez, C. Recalde, M. Echeverria and A. Cunachi. 2015. Bioelectricity generation from vegetables and fruits wastes by using single chamber microbial fuel cells with high Andean soils. Energy Procedia 75: 2009-2014.

Ma, C.Y., C.H. Wu and C.W. Lin. 2015. A novel V-shaped microbial fuel cell for electricity generation in biodegrading rice straw compost. Journal of Advanced Agricultural Technologies 2: 57-62.

Madaki, Y.S. and L. Send. 2013. Palm oil mill effluent (POME) from Malaysia palm oil mills: waste or resource. International Journal of Science and Environment 2: 1138-1155.

Mansoorian, H.J., A.H. Mahvi, A.J. Jafari, M.M. Amin, A. Rajabizadeh and N. Khanjani. 2013. Bioelectricity generation using two-chamber microbial fuel cell treating wastewater from food processing. Enzyme and Microbial Technology 52: 352- 357.

Min, B. and I. Angelidaki. 2008. Innovative microbial fuel cell for electricity production from anaerobic reactors. Journal of Power Sources 180: 641-647.

Min, B., S. Cheng and B.E. Logan. 2005. Electricity generation using membrane and salt bridge microbial fuel cells. Water Res. 39: 1675-1686.

Miran, W., M. Nawaz, J. Jang and D.S. Lee. 2016. Conversion of orange peel waste biomass to bioelectricity using a mediator-less microbial fuel cell. Science of the Total Environment 547: 197-205.

Mollea, C., L. Marmo and F. Bosco. 2013. Valorisation of cheese whey, a by-product from the dairy industry. InTech 549-588.

Moqsud, M.A., K. Omine, N. Yasufuku, Q.S. Bushra, M. Hyodo and Y. Nakata. 2014. Bioelectricity from kitchen and bamboo waste in a microbial fuel cell. Waste Management & Research 32: 124-130.

Nagendranatha Reddy, C., J.A. Annie Modestra, A.N. Kumar and S. Venkata Mohan. 2015. Waste remediation integrating with value addition: biorefinery approach towards sustainable bio-based technologies. pp. 231-256. *In*: V.C. Kalia (ed.). Microbial Factories, Biofuels, Waste Treatment. Vol. 1. Springer India.

Nagendranatha Reddy, C., K. Arunasri, Y.D. Kumar, K.V. Krishna and S. Venkata Mohan. 2016. Qualitative in vitro evaluation of plant growth promoting activity of electrogenic bacteria from biohydrogen producing microbial electrolysis cell towards biofertilizer application. Journal of Energy and Environmental Sustainability 1: 47-51.

Nagendranatha Reddy, C. and S. Venkata Mohan. 2016. Integrated bio-electrogenic process for bioelectricity production and cathodic nutrient recovery from azo dye wastewater. Renewable Energy 98: 188-196.

Ngoc, U.N. and H. Schnitzer. 2009. Sustainable solutions for solid waste management in Southeast Asian countries. Waste Management 29: 1982-1995.

Nguyen, H.T.H., R. Kakarla and B. Min. 2017. Algae cathode microbial fuel cells for electricity generation and nutrient removal from landfill leachate wastewater. International Journal of Hydrogen Energy 42: 29433-29442.

Nimje, V.R., C.Y. Chen, H.R. Chen, C.C. Chen, Y.M. Huang, M.J. Tseng, K.C. Cheng and Y.F. Chang. 2012. Comparative bioelectricity production from various wastewaters in microbial fuel cells using mixed cultures and a pure strain of *Shewanella oneidensis*. Biores. Technol. 104: 315-323.

Oh, S., B. Min and B.E. Logan. 2004. Cathode performance as a factor in electricity generation in microbial fuel cells. Environ Sci. Technol. 38: 4900-4904.

Oh, S.E. and B.E. Logan. 2005. Hydrogen and electricity production from a food processing wastewater using fermentation and microbial fuel cell technologies. Water Res. 39: 4673-4682.

Oh, S.E. and B.E. Logan. 2007. Voltage reversal during microbial fuel cell stack operation. J. Power Sources 167: 11-17.

Olinto, M.T.A., W.C. Willett, D.P. Gigante and C.G. Victora. 2011. Sociodemographic and lifestyle characteristics in relation to dietary patterns among young Brazilian adults. Public Health Nutr. 14: 150-159.

Othman, S.N., Z.Z. Noor, A.H. Abba, R.O. Yusuf and M.A. Abu Hassan. 2013. Review on life cycle assessment of integrated solid waste management in some Asian countries. Journal of Cleaner Production 41: 251-262.

Otles, S., S. Despoudi, C. Bucatariu and C. Kartal. 2015. Food waste management, valorization, and sustainability in the food industry. pp. 3-23. *In*: C.M. Galanakis (ed.). Food Waste Recovery. Academic Press, USA.

Owusu, P.A. and S. Asumadu-Sarkodie. 2016. A review of renewable energy sources, sustainability issues and climate change mitigation. Cogent Engineering 3: 1-14.

Pandey, P., V.N. Shinde, R.L. Deopurkar, S.P. Kale, S.A. Patil and D. Pant. 2016. Recent advances in the use of different substrates in microbial fuel cells toward wastewater treatment and simultaneous energy recovery. Appl. Energy 168: 706-723.

Pant, D., A. Singh, G. Van Bogaert, S.I. Olsen, P.S. Nigam, L. Diels and K. Vanbroekhoven. 2012. Bioelectrochemical systems (BES) for sustainable energy production and product recovery from organic wastes and industrial wastewaters. RSC Adv. 2: 1248-1263.

Pant, D., D. Arslan, G.V. Bogaert, Y.A. Gallego, H.D. Wever, L. Diels and K. Vanbroekhoven. 2013. Integrated conversion of food waste diluted with sewage into volatile fatty acids through fermentation and electricity through a fuel cell. Environmental Technology 34: 1935-1945.

Pant, D., G. VanBogaert, L. Diels and K. Vanbroekhoven. 2010. A review of the substrates used in microbial fuel cells (MFCs) for sustainable energy production. Biores. Technol. 101: 1533-1543.

Parfitt, J., M. Barthe and S. Macnaughton. 2010. Food waste within food supply chains: quantification and potential for change to 2050. Phil. Trans. R. Soc. B365: 3065-3081.

Paritosh, K., S.K. Kushwaha, M. Yadav, N. Pareek, A. Chawade and V. Vivekanand. 2017. Food waste to energy: an overview of sustainable approaches for food waste management and nutrient recycling. Biomed Res Int. 2017: 1-19.

Pendyala, B., S.R. Chaganti, J.A. Lalman and D.D. Heath. 2016. Optimizing the performance of microbial fuel cells fed a combination of different synthetic organic fractions in municipal solid waste. Waste Management 49: 73-82.

Pham, T.P.T., R. Kaushik, G.K. Parshetti, R. Mahmood and R. Balasubramanian. 2015. Food waste-to-energy conversion technologies: current status and future directions. Waste Management 38: 399-408.

Puyol, D., D.J. Batstone, T. Hulsen, S. Astals, M. Peces and J.O. Kromer. 2017. Resource recovery from wastewater by biological technologies: opportunities, challenges, and prospects. Frontiers in Microbiology 7: 1-23.

Qasim, M. and A.V. Mane. 2013. Characterization and treatment of selected food industrial effluents by coagulation and adsorption techniques. Water Resources and Industry 4: 1-12.

Rahimnejad, M., A. Adhami, S. Darvari, A. Zirepour and S.E. Oh. 2015. Microbial fuel cell as new technology for bioelectricity generation: a review. Alexandria Engineering Journal 54: 745-756.

Rattan, S., A.K. Parande, V.D. Nagaraju and G.K. Ghiwari. 2015. A comprehensive review on utilization of wastewater from coffee processing. Environ Sci Pollut Res. 22: 6461-6472.

Ravindran, R. and A.K. Jaiswal. 2016a. Exploitation of food industry waste for high-value products. Trends in Biotechnology 34: 58-69.

Ravindran, R. and A.K. Jaiswal. 2016b. Microbial enzyme production using lignocellulosic food industry wastes as feedstock: a review. Bioengineering 3: 1-22.

Ren, Z., T.E. Ward and J.M. Regan. 2007. Electricity production from cellulose in a microbial fuel cell using a defined binary culture. Environ. Sci. Technol. 41: 4781-4786.

Rikame, S.S., A.A. Mungray and A.K. Mungray. 2012. Electricity generation from acidogenic food waste leachate using dual chamber mediator less microbial fuel cell. International Biodeterioration & Biodegradation 75: 131-137.

Rozendal, R.A., H.V.M. Hamelers, K. Rabaey, J. Keller and C.J.N. Buisman. 2008. Towards practical implementation of bioelectrochemical wastewater treatment. Trends Biotechnol. 26: 450-459.

Satyawali, Y. and M. Balakrishnan. 2008. Wastewater treatment in molasses based alcohol distilleries for COD and color removal: a review. J. Environ. Manage. 86: 481-497.

Sivagurunathan, P., C. Kuppam, A. Mudhoo, G.D. Saratale, A. Kadier, G. Zhen, L. Chatellard, E. Trably and G. Kumar. 2017. A comprehensive review on two-stage integrative schemes for the valorization of dark fermentative effluents. Critical Reviews in Biotechnology 38: 868-882.

Song, J., K. Xu, H. Ma and J. Huang. 2005. Method for producing single cell protein from apple pomace by dual solid state fermentation. Patent no. CN 1673343-A.

Sravan, J.S., S.K. Butti, O. Sarkar, K.V. Krishna and S. Venkata Mohan. 2018. Electrofermentation of food waste-regulating acidogenesis towards enhanced volatile fatty acids production. Chemical Engineering Journal 334: 1709-1718.

Sreelatha, S., G. Velvizhi, C. Nagendranatha Reddy, J.A. Modestra and S. Venkata Mohan. 2015. Solid electron acceptor effect on biocatalyst activity in treating azo dye based wastewater. RSC Adv. 5: 95926-95938.

Tao, K., X. Quan and Y. Quan. 2013. Composite vegetable degradation and electricity generation in microbial fuel cell with ultrasonic pretreatment. Environmental Engineering and Management Journal 12: 1423-1427.

Tenca, A., R.D. Cusick, A. Schievano, R. Oberti and B.E. Logan. 2013. Evaluation of low cost cathode materials for treatment of industrial and food processing wastewater using microbial electrolysis cells. International Journal of Hydrogen Energy 38: 1859-1865.

Thassitou, P.K. and I.S. Arvanitoyannis. 2001. Bioremediation: a novel approach to food waste management. Trends in Food Sci. Tech. 12: 185-196.

Uçkun Kiran, E., A.P. Trzcinski, W.J. Ng and Y. Liu. 2014. Bioconversion of food waste to energy: a review. Fuel 134: 389-399. http://unesdoc.unesco. org/images/0024/002471/247153e.pdf (The United Nations World Water Development Report on "Wastewater: The Untapped resource")

Venkata Mohan, S., G. Mohanakrishna and P.N. Sarma. 2010. Composite vegetable waste as renewable resource for bioelectricity generation through non-catalyzed open-air cathode microbial fuel cell. Biores. Technol. 101: 970-976.

Venkata Mohan, S. and K. Chandrasekhar. 2011. Solid phase microbial fuel cell (SMFC) for harnessing bioelectricity from composite food waste fermentation: influence of electrode assembly and buffering capacity. Biores. Technol. 102: 7077-7085.

Wang, Z.J. and B.S. Lim. 2017. Electric power generation from treatment of food waste leachate using microbial fuel cell. Environ. Eng. Res. 22: 157-161.

Wen, Q., Y. Wu, D. Cao, L. Zhao and Q. Sun. 2009. Electricity generation and modeling of microbial fuel cell from continuous beer brewery wastewater. Biores. Technol. 100: 4171-4175.

Xin, X., Y. Ma and Y. Liu. 2018. Electric energy production from food waste: microbial fuel cells versus anaerobic digestion. Biores. Technol. 255: 281-287.

You, S.J., J.N. Zhang, Y.X. Yuan, N.Q. Ren and X.H. Wang. 2010. Development of microbial fuel cell with anoxic/oxic design for treatment of saline sea food wastewater and biological electricity generation. J. Chem. Technol. Biotechnol. 85: 1077-1083.

Youn, S., J. Yeo, H. Joung and Y. Yang. 2015. Energy harvesting from food waste by inoculation of vermincomposted organic matter into microbial fuel cell (MFC). IEEE Sensors 2015: 1-4.

Zagklis, D.P., A.I. Vavouraki, M.E. Kornaros and C.A. Paraskeva. 2015. Purification of olive mill wastewater phenols through membrane filtration and resin adsorption/desorption. J Haz Mat. 285: 69-76.

Zhong, C., B. Zhang, L. Kong, A. Xue and J. Ni. 2011. Electricity generation from molasses wastewater by an anaerobic baffled stacking microbial fuel cell. J. Chem. Technol. Biotechnol. 86: 406-413.

Optimization of Energy Production and Water Treatment in MFCs by Modeling Tools

V.M. Ortiz-Martínez[1,2]*, M.J. Salar-García[1,2]*, A. de Ramón-Fernández[3], A.P. de los Ríos[2], F.J. Hernández-Fernández[1] and P. Andreo-Martínez[2]

[1] Technical University of Cartagena, Department of Chemical and Environmental Engineering, C/Doctor Fleming s/n, Cartagena, Murcia, Spain, E-30202

[2] University of Murcia, Department of Chemical Engineering, Espinardo, Murcia, Spain, E-30100

[3] University of Alicante, Department of Computer Technology, Alicante, Spain, E-03690

Introduction

Microbial fuel cells (MFCs) have recently gained great importance in the field of emerging technologies for bioenergy production. The metabolism of the microbial community present as organic and waste substrates is used by these devices to produce electricity. This operating mechanism allows MFCs to simultaneously treat wastewater while generating electricity. Other advantages include good efficiency at room temperature and self-powering (Table 13.1) (Rinaldi et al. 2008, Logan et al. 2006, Santoro et al. 2017).

Table 13.1. Benefits of MFCs as both power supplier and wastewater treatment system (Hernández-Fernández et al. 2015)

MFC advantages	
Bioenergy production	**Wastewater treatment**
Direct conversion of the substrate into electricity. They are self-sufficient systems	MFCs generate lower amounts of sludge
MFCs are efficient under ambient conditions	High values of chemical oxygen demand removal
Do not require gas treatment	Capacity to remove a broad range of pollutants, including heavy metals

*Corresponding author: victor.ortiz@upct.es, mariajose.salar@upct.es

Notwithstanding the advantages of MFCs, this technology still shows some limitations related to its up-scaling (Salar-García et al. 2016). Hence, big efforts have been made in order to optimize MFCs from an experimental point of view by developing new materials and designs, and also by using alternative substrates (Oh and Logan 2005, Oliveira et al. 2013a, Ortiz-Martínez et al. 2016, Scott et al. 2008, Touach et al. 2016). Modeling has been widely employed for the study of fuel cell technology. By way of example, Fig. 13.1 summarizes the classification of several models reported in the literature for chemical fuel cells. In the specific case of MFCs, the interest in modeling has grown exponentially in only the last few years (Gao et al. 2012, Oliviera et al. 2013b, Ortiz-Martínez et al. 2015).

MFCs are complex systems that involve a large number of bioelectrochemical reactions and physical-chemical processes, whose behavior is difficult to predict. These systems have been widely investigated through experimental work over the last two decades. The number of MFC modeling works is considerably lower. However, modeling is a useful tool that enables the in-depth study of MFC technology. The use of modeling techniques is very helpful to analyze the behavior of MFC devices in order to optimize their performance as well as finding the bottlenecks of the system. Among the most important benefits of mathematical modeling are time and cost savings. The complexity and accuracy of a model will be determined by the dimension/order and assumptions and simplifications used when explaining the existing processes of the system, among other factors. Mass transport through the system, microbial growth, and the anodic and cathodic reactions are some of the key phenomena that take place in a microbial fuel cell (see Fig. 13.2) (Min and Logan 2004, Lu et al. 2009). The predictive

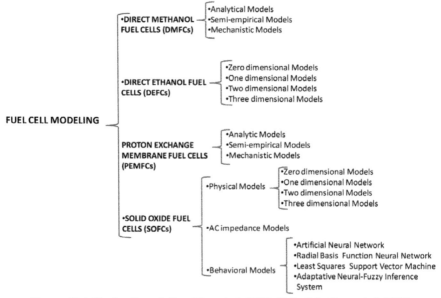

Figure 13.1. Fuel cell modeling (Gao et al. 2012, Ortiz-Martínez et al. 2015).

capacity as well as the precision of the results and the time required to obtain them will determine the robustness of a given model.

In this chapter, the models reported in the literature are grouped into (a) conventional modeling approach, which studies the overall MFC performance by modeling the main phenomena occurring in the systems by using well-established laws and mathematical equations, and (b) special models, which include models to study specific factors of the system and based on advanced modeling techniques, such as artificial intelligence. The works belonging to the group that use a conventional modeling approach can be in turn categorized depending on the focus of the model in i) bulk liquid and electrochemical models, and ii) biofilm-based models. Table 13.2 shows the main laws and equations that are usually used in conventional modeling approach and in models that analyze specific factors.

Advanced techniques, such as artificial intelligence are complex modeling tools for the optimization of MFC performance, including non-linear processes. For instance, fuzzy logic and artificial neural networks have been successfully employed for the study of these systems (Sewsynker-Sukai et al. 2017). In this chapter, particular attention will be paid to the works dealing with these types of techniques.

MFC Models

Conventional Modeling Approach

This section includes those works that model MFC performance using well-established laws and mathematical equations that describe the main

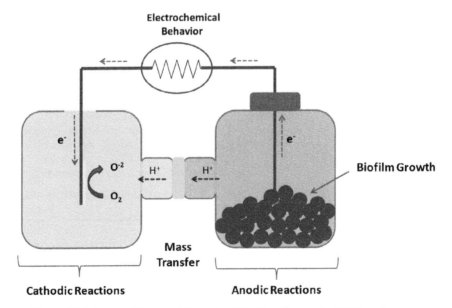

Figure 13.2. Some of the most important limiting factors in MFC performance.

Table 13.2. MFC modeling (modified after Hernández-Fernández et al. 2015)

Monod equation $$\mu = {}_m\frac{[S]}{K_s + [S]}$$	μ: specific microbial growth rate; μ_m: maximum specific growth; $[S]$: substrate concentration; K_s: Monod constant.
Tafel equation $$E - E_e = \frac{RT}{(1-\alpha)zF}\ln\frac{I}{I_0}$$	E: potential; E_e: equilibrium potential; R: universal gas constant; T: temperature; I: electrical current density; I_0: exchange current density; z: n° electrons; F: Faraday constant; α: charge transfer coefficient.
Butler-Volmer equation $$i = i_0\left(e^{\alpha_a\left(\frac{zF}{RT}\right)(E-Eeq)} - e^{-\alpha_c\left(\frac{zF}{RT}\right)(E-Eeq)}\right)$$	i: electrical current density; i_0: exchange current density; E: potential; E_{eq}: equilibrium potential; $\alpha_{a,c}$: anodic and cathodic charge transfer coefficients; z: n° electrons; F: Faraday constant; R: universal gas constant; T: temperature.
Nernst equation $$E = E^\circ - \frac{RT}{zF}\ln Q$$	E: non-standard potential; E°: standard potential; R: universal gas constant; T: temperature; z: number of electrons; F: Faraday constant; Q: reaction quotient.
Nernst-Planck equation $$\frac{\partial C}{\partial t} = \nabla\left[D\nabla C - vC + \frac{Dzq}{kT}\left(\nabla\phi + \frac{\partial A}{\partial t}\right)\right]$$	C: concentration; t: time; D: diffusivity; v: fluid velocity; z: valence; q: elementary charge; k: Boltzmann constant; T: temperature; ϕ: potential; A: magnetic vector potential.
Fick's first law $$J = -D\Delta C$$	J: diffusion flux; D: diffusivity; C: concentration.
Maxwell-Stefan diffusion equation (multicomponent case) $$\frac{\nabla_i}{RT} = \sum_{\substack{j=1 \\ j\neq i}}^{n}\frac{\chi_i\chi_j}{D_{ij}}\left(\overline{v_j} - \overline{v_i}\right) = \sum_{\substack{j=1 \\ j\neq i}}^{n}\frac{c_i c_j}{c^2 D_{ij}}\left(\frac{\overline{J_j}}{c_j} - \frac{\overline{J_i}}{c_i}\right)$$	i, j: sub-indexes for components; μ: chemical potential; T: temperature; R: universal gas constant; χ: mole fraction; D_{ij}: Maxwell-Stefan diffusion coefficient; $v_{j,i}$: velocity diffusion; $c_{i,j}$: molar concentration; c: total molar concentration; J: flux of component $i;j$.
Ohm's law $$V = IR$$	V: voltage; I: electrical current; R: ohm resistance.

phenomena in the system, including anodic/cathodic kinetics, microbial growth at the anode chamber, mass transfer, and electrochemical principles. Among them, biofilm-focused models can be considered as a separate group in which the conduction through the biofilm region is thoroughly analyzed and becomes particularly crucial.

Bulk liquid and electrochemical models

This group of models uses the description of the bulk region of MFCs and basic electrochemical theory (Xia et al. 2018). The first attempts in the literature consisted of a one-dimensional model for the description of an MFC system using a redox chemical mediator, which is responsible for electron transfer from the bulk region to the anodic electrode (Zhang and Halme 1995). This work mainly uses mass balance for the oxidation of carbon source (marine sediment), Monod equation for microbial growth, and Faraday's law and Nernst's equation to calculate current output. However, the model neglects cathode reaction and thus describes the performance of the systems by focusing exclusively on the anode compartment. This approach implies that the anode reaction is the main constraining factor and will be followed by later works in the literature.

Other more advanced MFC modeling works were reported in the late 2000s. Picioreanu et al. (2007) designed a 3-D model, which is built on mass balances along the system with microbial and biomass growth, in the last case represented by hard spherical particles (Recio-Garrido et al. 2016). In fact, the work integrates two respective well-differentiated approaches, electrochemical and biofilm growth approaches. The first one is based on the Bultler-Volmer equation (see Table 13.2) in order to obtain the total current density collected from the anode electrode. The second approach considers the biofilm growth on the basis of the Monod equation. In this respect, it is worth noting that the model distinguishes between suspended cells at the anode chamber (bulk solution) and the biofilm itself that is fixed to the anode electrode. Local charge balance is also addressed in other to take into account the influence of pH over the development of the biofilm, since abrupt pH changes greatly affect microorganism activity.

Well-established anaerobic digestion models have been integrated in MFC modeling works. This is the case of the ADM1 (Anaerobic Digestion Model N1) model by the International Association Water (IWA). ADM1 set a benchmark in the field of modeling of wastewater treatment. This is a generalized model that includes biochemical processes (homogenous particulate disintegration, extracellular hydrolysis, acidogenesis, acetonesis, methanogesis, etc.), as well as physico-chemical phenomena (ion dissociation/association, phase transfer, etc). These processes are described as differential and algebraic equations (Batstone et al. 2002). This model can be adapted to describe the anaerobic digestion of organic matter in the anode chamber of MFCs. For instance, in a later work, Picioreanu et al. (2008) employed the ADM1 to study anodophilic/methanogenic bacteria competition in MFC devices. In this sense, MFC performance devices greatly depends upon the presence of anodophilic bacteria in the anode chamber, since they are responsible for the direct transference of the electrons to the anode. Competition from microbial populations (e.g., growth of methanogenic microorganism) can significantly reduce the overall power output obtained (Islam et al. 2017).

The competition from these types of bacteria has also been studied in other modeling works. Pinto et al. (2010) presented a two population-based model in which Monod equations are used to describe the above mentioned competing metabolism processes. In the first case, microorganisms act as redox mediators themselves, with no need of any type of additional chemical mediator, and therefore the model comprises the oxidized/reduced species of the microorganism (intracellular mass balance). The electrochemical approach of the work is developed using common equations, such as Nernst equation, including the quantification of potential drops. The optimal values of substrate concentration with which the MFC systems are fed (acetate) is carefully analyzed, since the amount of substrate in the feed flow can influence power performance. Within the specific substrate range of 10-800 mg/L, it is concluded that high concentration values contribute to boosting methane production, while low substrate concentrations lead to insufficient growth rates of the biofilm. In sum, these types of parameters need to be carefully balanced to optimize the efficiency of MFC performance and, accordingly, modeling tools can greatly help to set the optimal values in each case.

Extended versions of the works presented until now have been proposed by the earlier mentioned authors for the study of the impact of pH and electrode arrangement on MFC efficiency. For example, these parameters were analyzed in the work of Picioreanu et al. (2010), whose approach involves mass balances for the anodic bulk region, and respective mass/charge/momentum balances in the domain of the biofilm. The model is simulated for simple and complex substrates (acetate and wastewater, respectively), obtaining several conclusions in line with some of the ideas expressed before, and giving special attention to the effect of the pH on the system. Wastewater-fed MFC systems, which are of special interest for the practical implementation of this technology, include heterogeneous communities and the study of competing microbial processes in favor and to the detriment of electricity generation is analyzed. Other geometry parameters include the surface properties of the anode electrode, e.g., porosity of the material employed and total surface area.

The models discussed so far are centered on the anode, while the phenomena taking place at the cathode are virtually ignored. There are other approaches that describe the systems from a comprehensive perspective by modeling the reactions and processes occurring in the cathode compartment. The work presented by Zeng et al. (2010) belongs to this group. The implemented model is based on a dual-chamber system, considering both anodic and cathodic reactions. Specifically, both compartments consist of backmix reactors or CSTRs, the electrochemical behavior being described using the Bulter-Volmer equation. According to a sensitivity analysis performed in this work for several operational variables, the authors suggest that the cathodic reaction acts as bottleneck for the system efficiency in terms of power generation.

Sirinutsomboon (2014) also studied both anodic and cathodic reactions in one-chamber system that lacks a selective proton exchange separator. In this type of set-up, several advantages, such as lowered costs are offered. In this configuration, the cathode includes a polytetrafluoroethylene (PTFE) layer permeable to oxygen but not to the water in the anode. Oxygen reduction (usually abbreviated as ORR) takes place within the liquid-electrode (cathode) interface. The diffusion of oxygen is modeled by Fick's second law and the rate of reduced oxygen at the cathode (r_{O2}) by the Butler–Volmer equation, while Nernst's equation is employed to calculate cathode potential. The anode compartment is addressed by Nernst-Monod expression to describe electrodonor oxidation rate, taking into consideration both exogenous and endogenous respiration of bacteria.

More recently, Ou et al. (2016) reported a one-dimensional and multi-species modeling work for an air-cathode MFC device. One interesting feature of this work is that the model comprises the formation of biofilm over the cathode. To this regard, the work includes multi-substance transport analysis and the description of electrochemical behavior over the cathode considering a multilayered biofilm formed by multiple biomass over its surface. As expected, the results obtained with the simulation of the system clearly show that the content of dissolved oxygen and its diffusion over the cathode are key parameters that greatly affect the power density achieved.

Another interesting modeling work is that of Oliveira et al. (2013b), which introduces the study of heat transfer in a double-chamber MFC system, making it possible to obtain temperature profiles across the system. Heat transfer analyses are set for both anodic and cathodic compartments, considering several energy parameters involved (activation energies, electrochemical reactions, heat streams, heat losses, etc.). As regards the rest of electrical variables, they are approached in a similar way as commented in other modeling works. For example, the anode is addressed as a continuous stirring tank reactor, the density of electrical current generated from anode (oxidation) is calculated with the Monod/Tafel expressions, and cathode compartment descriptions lie on the ORR.

Biofilm-based models

These type of works are generally half-cell models that primarily focus on the study of the formation and properties of the biofilm in the anode chamber to describe the whole MFC system (Xia et al. 2018). One of the most relevant MFC biofilm-based (1-D) model was presented by Marcus et al. (2007), and since then other extended version have been based on this approach when addressing the modeling of the biofilm region in depth. In this work, the biofilm is defined as a biological conductor with an associated conductivity parameter-matrix called k_{bio}. The work considers two regions or types of biofilm, active and inert biomass. The model covers two electron transfer mechanisms, endogenous respiration, in addition to exogenous electron-donor oxidation. Among those factors that can be analyzed with simulated results are biofilm thickness, coulombic yield, and current density. The first parameter has an important impact on the system efficiency. An excessive biofilm thickness and a high accumulation of inactive biomass lead to the limitation of MFC performance. As regards the main mathematical equations used, Monod and Nernst equations are employed for the description of the electron production rate on the basis of substrate oxidation.

Other models have been reported extending or considering that of Marcus et al. (2007) as a starting point. For example, a 2-D model based on this work was built to delve into the performance of the anode (Merkey and Chopp 2012). In this case, in addition to the biofilm region characterized by the conductivity matrix-parameter k_{bio} commented above, the model comprises the bulk liquid region and the solid electrode as well as the corresponding interfaces. The mathematical complexity of the model makes it possible to analyze the development of spatial patterns of the biofilm and its growth depending on several anode geometry arrangements.

Genome-scale metabolic models (GEMs) have become a popular tool in biological systems, being used in many fields, such as industrial biotechnology. To complete the model of Marcus et al. (2007), other authors have employed a version of it along with a GEM approach considering flux balance analysis (FBA) (Jayasinghe et al. 2014). In addition to the two regions of biofilm comprised by the original work (inactive and active), in this case, Jayasinghe et al. (2014) also consider a so-called 'respiring zone', which is responsible for energy conservation at the anode surface. The model enables a thorough analysis of the metabolic pathways of the *G. sulfurreducens* bacteria.

An integrated genetic algorithm was also combined by Sedaqatvand et al. (2013) into the model developed by Marcus et al. (2007) and subsequently applied to analyze the performance of a single-chamber system. A genetic algorithm is used to determine optimal parameters of the model. Among other analyses, the model was employed by the authors to obtain profiles of electrical potential and of wastewater concentration across the biofilm.

An adapted version of such work to describe the performance of microfluidic MFCs was also presented by Mardanpour et al. (2017). Unlike the above-mentioned models, the formation of the biofilm on the basis of chemotactic motility quantification is comprised, which enables the investigation of the spatial distribution of suspended cells throughout the anodic bulk liquid. Furthermore, the attachment of such cells to the electrode surface (biofilm formation) can be better explained. Among other analyses, the work also covers substrate variation, bioenergy generation rate, and impact of the external load values on biofilm development.

Special models

Along with critical factors, such as reaction kinetics or biofilm growth, other less influencing factors have also been used to optimize the system. This specific type of model is still scarce in literature, although they provide a valuable point of view of the process. The models addressed in this section can be grouped in models that involve the study of special factors, such as the polarization around a specific area (separator), and models designed by using special modeling tools.

Model based on special factors

Hardly ever, models focus on special factors for analyzing the processes involved in MFC devices. One of them is that of Harnisch et al. (2009), who designed a model focusing on the polarization process around a commercial proton exchange membrane. This model provided valuable information on the mechanism of the ionic transport through the membrane and its limitations. The MFC assembly studied consists of a double-chambered system with both anode and cathode electrodes made of platinum mesh. The evolution of the pH in the anodic and cathodic chambers was controlled as well as the polarization around the membrane by using two electrodes and two Luggin capillaries, respectively.

This work investigates the transference of the chemical species present in the system through the separator by employing the Nernst-Planck expression. This equation depicts flux density in terms of diffusion and migration transport mechanisms, and considers the transfer through the convection process as negligible. Figure 13.3 shows the three different regions considering the 1-D model in the whole MFC, i.e., cathodic and anodic volumes, zone around the membrane, membrane itself, a piece of Nafion-117, respectively. An accumulation of opposite charges takes place

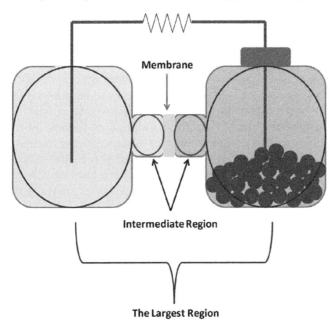

Figure 13.3. MFC regions considered by Harnisch et al. (2009) to design their one-dimensional model.

around both faces of the membrane. The measurement of this polarization phenomenon allows the determination of the internal resistance of the separator, which directly affects the electric performance of the system. According to the analysis performed with the model, it can be concluded that high pH differences between the cathode and the anode compartments caused by the presence of concentrated electrolytes directly increase the resistance of the separator, and therefore the performance of the MFC is reduced.

Other models focus on the polarization/power curves of the system, such as that of Wen et al. (2009), in which a single-chamber MFC set up with a cathode exposed to the air and made of carbon cloth coated with platinum is studied. The anode chamber was fed with a continuous flow of brewery wastewater. In addition to the approach of the model, the main novelty is the use of real wastewater as substrate in MFC modeling. The power and the polarization curves of an MFC can be obtained by applying a wide range of external load resistances. The graphical representation of the voltage *vs* the current intensity provides a polarization curve, while the power curve is obtained by plotting power output *vs* current intensity (Logan et al. 2006). Three types of regions can be distinguished in polarization curves, being related to different processes: i) activation, ii) ohmic losses, and iii) concentration. The main purpose of the model is to reproduce the performance of the MFC in terms of voltage, which is calculated with the following equation:

$$V = E_{thermo} - \eta_{act} - \eta_{ohm} - \eta_{conc} \qquad (1)$$

where E_{thermo} is the thermo-dynamically predicted voltage, while η_{act} includes activation losses, η_{ohm} represents ohmic losses, and η_{conc} concentration losses. Moreover, both the experimental and simulated data are fitted taking into account the parasitic losses caused by the leakage of electrical current. Besides, maximum power output of the system can be inferred from power curves. Both polarization and power curves provide crucial information about the performance of MFCs. The validation of the model consisted of comparing the simulated values of ohmic resistance and experimental values obtained by impedance spectroscopy techniques (i.e. EIS). The individual contribution of several factors on voltage cell are also analyzed, according to which concentration and kinetics (activation) losses are those that display the highest impact on voltage output.

Model Based on Special Modeling Tools

Alternative methods have been used to predict the behavior of MFCs. The relation between several characteristic biological parameters in the anodic chamber and MFC performance in terms of energy output was established by Stratford et al. (2014). Three different indexes were employed, i.e., Shannon index, related to the abundance of microbial communities, Simpson index, related to their uniformity, and richness, which determines bacterial diversity. The tests were performed in single-chamber devices with the cathode exposed to the air and continuously fed with sucrose solution. An initial maturing stage was running in order to obtain reproducible voltages. After that, the DNA of both biofilm and microorganisms in suspension was analyzed in order to determine the indexes described above. The power output and the bacterial indexes were related by using bivariate analysis. The highest impact on MFC performance (power density) was displayed by Shannon index. Three regression models were designed by using lineal regression analysis. Each model contains two variables as prediction tools: i) the diversity of the biofilm, determined by the three indexes, and ii) the amount of DNA per unit of area in the biofilm. Once again, Shannon index-based model showed the highest impact on power performance. According to this, Shannon index is a suitable tool to predict the power output of a given MFC. Regarding the diversity of bacterial communities, this factor has a positive effect on the energy produced by the system while the bacteria in suspension have virtually no impact on this parameter.

Another group of models are those designed by using artificial intelligence algorithms. This approach allows models to detect hidden interactions between input and output variables, which is very advantageous in data-saturated domains, and improves its accuracy (Lesnik and Liu 2017). Artificial intelligence (AI) focuses on developing intelligent software in order to solve specific problems. This technique has potential application in broad research domains, such as health, business, environment pollutions, etc. In recent years, AI has been successfully used in the field of renewable energy technologies for their optimization (Russel and Norvig 1995, Jha et al. 2017). AI encompasses a wide variety of mathematical methods, such as fuzzy control, artificial neural networks (ANNs), supported vector machines (SVMs), etc.

Since the internal functioning of MFCs is complex due the broad range of processes and mechanisms involved, alternative modeling methods based on an intelligent approach show potential applications to predict and optimize the behavior of these systems. A part of the data to create a model (learning process) and another part to confirm the reliability of the model are used in this method. They are commonly known as black box-based models. In contrast to those model based on classical approaches, these methods are capable of establishing a correlation between input and output factors without in-depth understanding of the MFC behavior (Xia et al. 2018).

A classical approach with mathematical algorithms were combined by Yan and Fan (2013) to design an MFC model, which in turn is based on the work previously reported by Zeng et al. in 2010. The new model is based on a combination of fuzzy control with classical proportional and derivative control (PID), which is applied to a double-chambered MFC. Fuzzy control is used as tool to approach the limitation posed by conventional PID control in high non-linear systems, such as MFCs fed with very complex substrates, e.g., real wastewater. Fuzzy logic enables it to build a controller even when the system is only partially understood and some of the benefits are robustness and model-free and universal approximation theorem (Chan 2010). The model is programmed by using Matlab and Simulink software tools in order to keep the voltage generated by the system at a constant level. The results show that the combination of both techniques delivers more accuracy than traditional PID strategies, because fuzzy-PID control acts faster to correct any disturbance in the system response (e.g., voltage).

To overcome the limitation of classical approach models related to long computational times, a data-driven Gaussian Process Regression (GPR)-based model was designed by He and Ma (2016) to study a double-chambered MFC. This method plots the nonlinear parameters of a complex system based on data-driven models that need less information of the process, but a huge amount of experimental data. This model aims at optimizing the voltage of MFCs by using three input variables, substrate concentration, current density, and feed flow. The results show that an appropriate selection of experimental data allows the model to reach an accurate prediction of the MFC behavior under the off-line mode.

AI has been recently employed by Garg et al. (2014) to model the voltage generated by an MFC according to two input variables: temperature and concentration of ferrous sulfate in the anode. Three mathematical tools are used in order to design the model: 1) Multigene Genetic Programming (MGGP), 2) Artificial Neural Networks (ANNs), and 3) Support Vector Machines (SVMs). Voltage data were collected from double-chambered MFCs operating in batch mode in two series of tests. The first series were carried out at different temperatures and the second one at different substrate concentrations. Although the three mathematical methods used to model the system provide good determination coefficients ($R^2 > 0.95$), MGGP exhibits the strongest correlation (versus experimental data) at describing voltage response as function of temperature of the system ($R^2 = 0.98$). When the influence of other parameters, such as the concentration of ferrous sulfate ($FeSO_4$) in the anodic chamber (which can greatly influences biofilm formation) are comprised, MGGP also offers higher performance ($R^2 = 0.98$) in comparison to the other two techniques (ANN and SVR), whose correlation factor is significantly lower ($R^2 = 0.83$). Although these techniques do not offer valuable information on the internal processes of MFC as classical approach, they are suitable to predict the behavior of this system according to different input variables. ANNs are inspired by biological neuron processing and their main advantage is that this computational tool does not need a phenomenological model of the system in order to establish a relationship between different input and output factors. They are capable of predicting the behavior of a complex system with a suitable selection of experimental data in a low processing time (Khataee and Kasiri 2010).

Artificial neural network was also used by Esfandyari et al. (2016) to model the performance of MFC technology. Important factors, such as pH, concentration of nitrogen, ionic strength, and temperature were chosen as input variables in a five-level model to predict power output and coulombic efficiency. The adaptive network-based fuzzy inference system (ANFIS) method is also used by the authors to describe the system and determine the correlation between the input and output factors. Although both methods provide an accurate and predictable model, ANN offers a simple structure and training procedure.

The effect of the anode position on both power generation and wastewater capacity of a mediator-less MFC has been recently analyzed by Jaeel et al. (2016). The inclination of the anode (with respect to flow direction) was carried out in a system continuously fed with two different feed flow of dairy wastewater. Experimental tests show that the maximum power output (486 mW.m^{-2}) and COD removal (92%) were obtained using low feed rate (in the order of 1 mL/min) when the anode is placed perpendicular to the flow direction. In this case, the Levenberg-Marquardt back-propagation algorithm (LMA) was employed in order to design a three-layer ANN-based model to predict MFC performance. Besides, the correlation between the experimental data and model results was very high.

More recently, ANN was also used by Lesnik and Liu (2017) to determine the correlation between the characteristics of the substrate (conductivity, nitrogen and phosphorus concentration, type and concentration of substrate) and the microbial communities and the power output. ANN-based models including both abiotic and biotic factors are useful to predict the bacterial communities and power performance of MFCs. However, those models that take into account biotic interactions offer a better prediction of MFC performance, obtaining an average error percentage of the power density predicted at 16%.

In order to acquire a large number of experimental data and reduce the acquisition time, experiment design (ED) plays an important role. A few years ago, Fang et al. (2013) optimized the performance of MFC in terms of power output by using a multivariable modeling method. The design of the experiments was based on a uniform approach including four input variables (ionic strength, temperature, nitrogen concentration, and initial pH) in five levels and two output variables (power output and coulombic efficiency). A mathematical model to describe the relationship between input and output variables, obtained from sixteen tests, was established by using relevance vector machine (RVM). An accelerating genetic algorithm was employed with the purpose of accelerating the model optimization rate, optimizing the reaction conditions, and obtaining the maximum values of coulombic efficiency and power output. The model was compared to another work based on support vector machine (SVM). The results show that RVM offers a more accurate model than SVM for the prediction of the maximum MFC performance.

More recently, the response surface design methodology (RSM) was employed by Martínez-Conesa et al. (2017) to design the experiments of their work. The authors studied the power output of double-chambered MFC by the optimization of four input variables (temperature, substrate concentration, external resistance load, and pH of the anolite) and using a three-level Box-Behnken design. The results show that a quadratic mathematical model of the system is suitable to maximize the power output.

Final Thoughts

The present chapter attempts to provide an insight into the optimization and study of MFC technology by modeling tools. MFC models have been classified into several groups in accordance with the approach followed and are widely discussed. MFC technology has gained much attention in recent years because of the need to produce clean energy to address the effects of the climate change. In terms of experimental work, big efforts have been exerted in order to enhance MFC performance by designing novel assemblies and tailoring more efficient materials that form the components of the system. Due to the multiple benefits of modeling in terms of optimization in addition to cost and time saving, the number of MFC models in literature is expected to increase in coming years. Furthermore, the application of information and communication technologies (ICTs), such as Artificial Intelligence has opened up a broad range of possibilities for the study of MFC systems.

Acknowledgement

This work is partially supported by the Spanish Ministry of Science and Innovation (MICINN) (Ref: TIN2014-53067-C3-1-R). V.M. Ortiz-Martínez, M.J. Salar-García and A. de Ramón-Fernández thank the Ministry of Education and the Ministry of Economy and Competitiveness for supporting their doctoral theses (Refs. FPU12/05444, BES-2012-055350 and BES-2015-073611, respectively).

REFERENCES

Batstone, D.J., J. Keller, I. Angelidaki, S.V. Kalyuzhnyi, S.G. Pavlostathis, A. Rozzi, W.T. Sanders, H. Siegrist and V.A. Vavilin. 2002. The IWA Anaerobic Digestion Model No 1 (ADM1). Water Sci. Technol. 45: 65-73.

Chan, A.H.S. 2010. Advances in industrial engineering and operations research. Springer-Verlag New York Inc., New York.

Esfandyari, M., M.A. Fanaei, R. Gheshlaghi and M.M. Akhavan. 2016. Neural network and neuro-fuzzy modeling to investigate the power density and coulombic efficiency of microbial fuel cell. J. Taiwan Inst. Chem. Eng. 58: 84-91.

Fang, F., G.-L. Zang, M. Sun and H.-Q. Yu. 2013. Optimizing multi-variable of microbial fuel cells for electricity generation with an integrated modeling and experimental approach. Appl. Ener. 110: 98-103.

Gao, F., B. Blunier and A. Miraoui. 2012. Proton exchange membrane fuel cell modeling. Wiley, London.

Garg, A., V. Vijayaraghavan, S.S. Mahpatra, K. Tai and C.H. Wong. 2014. Performance evaluation of microbial fuel cell by artificial intelligence methods. Expert Syst. Appl. 41: 1389-1399.

Harnisch, F., R. Warmbier, R. Scheneider and U. Schröder. 2009. Modeling the ion transfer and polarization of ion exchange membranes in bioelectrochemical systems. Bioelectrochem. 75: 136-141.

He, Y.J. and Z.F. Ma. 2016. A data-driven Gaussian Process Regression model for two-chamber microbial fuel cells. Fuel Cells 16: 365-376.

Hernández-Fernández, F.J., A. Pérez de los Ríos, M.J. Salar-García, V.M. Ortiz-Martínez, L.J. Lozano-Blanco, C. Godínez, F. Tomás-Alonso and J. Quesada-Medina. 2015. Recent progress and perspectives in microbial fuel cells for bioenergy generation and wastewater treatment. Fuel Process. Technol. 138: 284-297.

Islam, M.A., B. Ethiraj, C.K. Cheng, A. Yousuf and M.M.R. Khan. 2017. Electrogenic and antimethanogenic properties of bacillus cereus for enhanced power generation in anaerobic sludge-driven microbial fuel cells. Energy Fuels 31: 6132-6139.

Jaeel, A.J., A.I. Al-wared and Z.Z. Ismail. 2016. Prediction of sustainable electricity generation in microbial fuel cell by neural network: effect of anode angle with respect to flow direction. J. Electroanal. Chem. 767: 56-62.

Jayasinghe, N., A. Franks, K.P. Nevin and R. Mahadevan. 2014. Metabolic modeling of spatial heterogeneity of biofilms in microbial fuel cells reveals substrate limitations in electrical current generation. Biotechnol. J. 10: 1350-1361.

Jha, S.K., J. Bilalovic, A. Jha, N. Patel and H. Zhang. 2017. Renewable energy: present research and future scope of artificial intelligence. Renew. Sust. Energ. Rev. 77: 297-317.

Khataee, A.R. and M.B. Kasiri. 2010. Artificial neural networks modeling of contaminated water treatment processes by homogeneous and heterogeneous nanocatalysis. J. Mol. Catal. A Chem. 331: 86-100.

Lesnik, K.L. and H. Liu. 2017. Predicting microbial fuel cell biofilm communities and bioreactor performance using artificial neural networks. Environ. Sci. Technol. Lett. 51: 10881-10892.

Logan, B.E., B. Hamelers, R. Rozendal, U. Schröder, J. Keller, S. Freguia, P. Aelterman, W. Verstraete and K. Rabaey 2006. Microbial fuel cells: methodology and technology, Environ. Sci. Tecnol. 40: 5181-5192.

Lu, N., S. Zhou, L. Zhuang, J. Zhang and J. Ni. 2009. Electricity generation from starch processing wastewater using microbial fuel cell technology. Biochem. Eng. J. 43: 246-251.

Marcus, A.K., C.I. Torres and B.E. Rittmann. 2007. Conduction-based modeling of the biofilm anode of a microbial fuel cell. Biotechnol. Bioeng. 98: 1171-1182.

Mardanpour, M.M., S. Yaghmaei and M. Kalantar. 2017. Modeling of micro-fluidic microbial fuel cells using quantitative bacterial transport parameters. J. Power Sources 342: 1017-1031.

Martínez-Conesa, E.J., V.M. Ortiz-Martínez, M.J. Salar-García, A.P. de los Ríos, H.F. Hernández-Fernández, L.J. Lozano and C. Godínez. 2017. A Box-Behnken design-based model for predicting power performance in microbial fuel cells using wastewater. Chem. Eng. Commun. 204: 97-104.

Merkey, B.V. and D.L. Chopp. 2012. The performance of a microbial fuel cell depends strongly on anode geometry: a multidimensional modeling study. B Math. Biol. 74: 834-857.

Min, B. and B.E. Logan. 2004. Continuous electricity generation from domestic wastewater and organic substrates in a flat plate microbial fuel cell. Environ. Sci. Technol. 38: 5809-5814.

Oh, S.E. and B.E. Logan. 2005. Hydrogen and electricity production from a food processing wastewater using fermentation and microbial fuel cell technologies. Water Res. 39: 4673-4682.

Oliveira, V.B., M. Simões, L.F. Melo and A.M.F.R. Pinto. 2013a. Overview on the developments of microbial fuel cells. Biochem. Engin. J. 73: 53-64.

Oliveira, V.B., M. Simões, L.F. Melo and A.M.F.R. Pinto. 2013b. A 1D mathematical model for a microbial fuel cell. Energy 61: 463-471.

Ortiz-Martínez, V.M., M.J. Salar-García, A.P. de los Ríos, F.J. Hernández-Fernández, J.A. Egea and L.J. Lozano. 2015. Developments in microbial fuel cell modeling. Chem. Eng. J. 271: 50-60.

Ortiz-Martínez, V.M., I. Gajda, M.J. Salar-García, J. Greenman, F.J. Hernández-Fernández and I. Ieropoulos. 2016. Study of the effects of ionic liquid-modified cathodes and ceramic separators on MFC performance. Chem. Eng. J. 291: 317-324.

Ou, S., Y. Zhao, D.S. Aaron, J.M. Regan and M.M. Mench. 2016. Modeling and validation of single-chamber microbial fuel cell cathode biofilm growth and response to oxidant gas composition. J. Power Sources 328: 385-396.

Picioreanu, C., I.M. Head, K.P. Katuri, M.C.M. Loosdrecht and K. Scott. 2007. A computational model for biofilm-based microbial fuel cells. Water Res. 41: 2921-2940.

Picioreanu, C., K.P. Katuri, I.M. Head, M.C.M. Loosdrecht and K. Scott. 2008. Mathematical model for microbial fuel cells with anode biofilms and anaerobic digestion. Water Sci. Technol. 57: 965-971.

Picioreanu, C., M. van Loosdrecht, T.P. Curtis and K. Scott. 2010. Model based evaluation of the effect of pH and electrode geometry on microbial fuel cell performance. Bioelectrochem. 78: 8-24.

Pinto, R.P., B. Srinivasan, M.-F. Manuel and B. Tartakovsky. 2010. A two-population bio-electrochemical model of a microbial fuel cell. Bioresour. Technol. 101: 5256-5265.

Recio-Garrido, D., M. Perrier and B. Tartakovsky. 2016. Modeling, optimization and control of bioelectrochemical systems. Chem. Eng. J. 289: 180-190.

Rinaldi, A., B. Mecheri, V. Garavaglia, S. Licoccia, P. Di Nardo and E. Traversa. 2008. Energy Environ. Sci. 1: 417-429.

Russel, S.J. and P. Norvig. 1995. Artificial intelligence: a modern approach. New Jersey: Prentice Hall.

Salar-García, M.J., V.M. Ortiz-Martínez, Z. Baicha, A.P. de los Ríos and F.J. Hernández-Fernández. 2016. Scaled-up continuous up-flow microbial fuel cell based on novel embedded ionic liquid-type membrane-cathode assembly. Energy 101: 113-120.

Santoro, C., C. Arbizzani, B. Erable and I. Ieropoulos. 2017. Microbial fuel cells: from fundamentals to applications: a review. J. Power Sources 356: 225-244.

Scott, K., I. Cotlarciuc, I. Head, K.P. Katuri, D. Hall, J.B. Lakeman and D. Browning. 2008. Fuel cell power generation from marine sediments investigation of cathode materials. J. Chem. Technol. Biotechnol. 83: 1244-1254.

Sedaqatvand, R., M.N. Esfahany, T. Behzad, M. Mohseni and M.M. Mardanpoura. 2013. Parameter estimation and characterization of a single-chamber microbial fuel cell for dairy wastewater treatment. Bioresour. Technol. 146: 247-253.

Sewsynker-Sukai, Y., F. Faloye and E.B.G. Kana. 2017. Artificial neural networks: an efficient tool for modeling and optimization of biofuel production (a mini review). Biotechnol. Biotechnol. Equip. 31: 221-235.

Sirinutsomboon, B. 2014. Modeling of a membraneless single-chamber microbial fuel cell with molasses as an energy source. Int. J. Energy Environ. Eng. 5: 1-9.

Stratford, J.P., N.J. Beecroft, R.T.C. Slade, A. Grüning and C. Avignone-Rossa. 2014. Anode microbial community diversity as a predictor of the power output of microbial fuel cells. Bioresour. Technol. 156: 84-91.

Touach, N., V.M. Ortiz-Martínez, M.J. Salar-García, A. Benzaouak, F. Hernández-Fernández, A.P. de los Ríos, N. Labjar, S. Louki, M.E. Mahi and E.M. Lotfi. 2016.

Influence of the preparation method of MnO_2-based cathodes on the performance of single-chamber MFCs using wastewater. Sep. Pur. Technol. 171: 174-181.

Wen, Q., Y. Wu, D. Cao, L. Zhao and Q. Sun. 2009. Electricity generation and modeling of microbial fuel cell from continuous beer brewery wastewater. Bioresour. Technol. 100: 4171-4175.

Xia, C., D. Zhang, W. Pedrycz, Y. Zhu and Y. Guo. 2018. Models for microbial fuel cells: a critical review. J. Power Sources 373: 11-131.

Yan, M. and L. Fan. 2013. Constant voltage output in two-chamber microbial fuel cells under fuzzy PID control. Int. J. Electrochem. Sci. 8: 3321-3332.

Zeng, Y., Y.F. Choo, B.H. Kim and P. Wu. 2010. Modeling and simulation of two-chamber microbial fuel cell. J. Power Sources 195: 79-89.

Zhang, X.-C. and A. Halme. 1995. Modeling of a microbial fuel cell process. Biotechnol. Lett. 17: 809-814.

Development of Alternative Proton Exchange Membranes Based on Biopolymers for Microbial Fuel Cell Applications

Srinivasa R. Popuri*[1], Alex J.T. Harewood[1], Ching.-Hwa Lee[2] and Shima L. Holder[2]

[1] The University of the West Indies, Faculty of Science and Technology, Cave Hill Campus, St. Michael, Bridgetown 11000, Barbados

[2] Da-Yeh University, Department of Environmental Engineering, 168 University Rd., Dacun 515, Changhua, Taiwan, ROC

Introduction

Biopolymers

Biopolymer materials-based applications in science and industry have been increasing significantly recently due to stringent environmental regulations on the use of artificial polymers. Biopolymers are derived from living organisms and have been present on earth for billions of years. For example, biopolymers, such as proteins, carbohydrates, DNA, RNA, lipids, nucleic acids, peptides, and polysaccharides constitute a major part of the human body and the ecosphere (http://www.chemistrylearner.com/biopolymer.html). These polymers are classified as four types namely, starch, sugar, cellulose, and synthetic materials, based on their source from agricultural plant or animal products (Aravamudhan et al. 2014). Biopolymers offer numerous advantages, as they are degradable, renewable, sustainable, compostable, and maintain carbon neutrality in the atmosphere (Chen 2014).

Biopolymer Membranes

Biopolymer materials possess several intrinsic properties, such as hydrophilic nature, easy film formation, biocompatibility, bio-degradability,

*Corresponding author: popurishrinu@gmail.com

anti-bacterial properties, and easy chemical modification of the functional groups. These properties attract the use of biopolymers in membrane-based separation applications, such as wastewater treatment (reverse osmosis), separation of liquid-liquid mixtures (pervaporation) and gaseous mixtures (gas separation), purification of pharmaceutical compounds, capturing of toxic metals and biocatalysts (Houghton and Quarmby 1999, Kim and Hoek 2007, Rao et al. 2007, Catherine 2009 , Lee et al. 2011, Lee et al. 2013, Cadogan et al. 2016, Su et al. 2017). There has been continuous research in the synthesis of innovative and effective biopolymers in the field of sustainable membrane applications, but current needs of chemical, biological, and ecological industries demand more emphasis on efficient biopolymer membranes with well-defined structures for various applications (Kumar et al. 2007).

Proton Exchange Membranes in Microbial Fuel Cell Technology

The microbial fuel cell (MFC) is revealed as one of the promising and alternative renewable energy technologies, which produces electrical energy from a wide range of waste soluble substrates, by simultaneous treatment of wastewater (Pant et al. 2010, Hou et al. 2014, Logan and Rabaey 2012, Modin and Gustavsson 2014). Research continues on several components and operating conditions of MFCs, such as electrode materials, electrode catalysts, biocatalysts, separators (membrane), substrates, temperature, pH, cell configuration, and architecture (Yang et al. 2010, Belleville et al. 2011, Zhou et al. 2011, Kim et al. 2014). Proton conductive materials used in MFCs allow proton transfer from the anode to the cathode effectively, while inhibiting fuel (substrate) and electron acceptor (oxygen) crossover. A proton exchange membrane (PEM) is the most commonly used proton conductive material in MFC due to its low internal resistance (Kim et al. 2007, Zhang et al. 2009, Li et al. 2011, Ghasemi et al. 2013). Additionally the ideal PEM must have characteristics of consistent ionic transport, high Coulombic efficiency (CE), and efficient and sustainable stability during long operation periods. However, PEMs are expensive and approximately cost 40% of overall MFC system cost (Rozendal et al. 2008). They also exhibit high internal resistance that lowers the bioelectricity generation due to chemical and biological fouling of MFCs (Choi et al. 2011, Xu et al. 2016) and decrease cathode cell performance due to high transfer ratio of cations to protons (Kim et al. 2007, Rozendal et al. 2008). Optimizing the PEM or obtaining more alternative separators is one of the issues that needs to be addressed for the benefit of the MFC technology. Albeit, several types of membranes have been synthesized in the literature. These membranes are classified as i) Nafion, ii) non-Nafion fluorinated, iii) composite fluorinated, and iv) non-fluorinated membranes (Neburchilov et al. 2007).

Nafion is the widely used proton exchange membrane in fuel cell technology due to its advantages of proton conducting ability and high physical (mechanical and structural) and chemical stability (Devanathan 2008). In fact, most of the commercial PEMs are developed using perfluorosulfonic acid (PFSA) polymer membranes, such as Acipex®, Flemion®, and Nafion®. Walther Grot of DuPont discovered Nafion® membrane in 1970 and it consists of a hydrophobic $-CF_2-CF_2-$ fluorocarbon backbone with hydrophilic sulfonate groups (SO_3H^-) (Oh and Logan 2006, Jana et al. 2010, Liu et al. 2016). However, Nafion membranes have two significant issues yet to be resolved, i.e., they are expensive to produce (Hernandez-Flores 2014, Shahgaldi et al. 2014) and permit methanol crossover. In addition, Nafion could not be ideal for MFC at neutral or higher pH, because other cations are dominant in moving from anode to cathode compared to protons (Rozendal et al. 2006).

Alternative PEMs, such as salt bridges, composite membranes, and porous materials have been developed in order to address the issues associated with Nafion. Salt-bridge PEM exhibits low oxygen diffusion and a higher cation exchange than Nafion membranes, but it produces a low power production due to high internal resistance (Min et al. 2005). Composite membranes like sulfonated polyether ether ketone (SPEEK) PEM displays superior removal of chemical oxygen demand (COD) compared to Nafion membranes; however, the overall power production is low. Porous material PEMs could be the alternatives for Nafion membranes in MFCs due to low cost, durability, decreased ohmic loss, and greater power production than Nafion, but it allows non-selective charge transfer (Choi et al. 2011, Kim et al. 2012). As Nafion possesses corrosive behavior, it can harm the MFC system, and its preparation involves the use of environmentally hazardous chemicals, so the use of biopolymeric materials is considered as suitable replacement for Nafion as the PEM in MFC technology due to their eco-friendliness and water sorption ability (Shaari and Kamarudin 2015).

Most of the PEMs developed for fuel cell applications are modified partially fluorinated sulfonated aromatic (PFSA) polymers (Ghassemi and McGrath 2004, Asano et al. 2006, Hickner et al. 2006, Dai et al. 2007, Liu et al. 2007, Miyatake et al. 2007, Zhong et al. 2007, Matsumoto et al. 2009, Zhang et al. 2009, Sun et al. 2010, Bae et al. 2011, Choi et al. 2016) and acid-based blends (Herranen et al. 1995, Park et al. 2008). However, these membranes did not provide solutions to the unresolved issues, such as low proton conductivity under low humidity conditions, membrane stability in the oxidative environment, and poor mechanical stability either in long-term operation or at higher temperatures. Polymer membranes with hydrophilic property are also attempted as PEMs in fuel cells (Bai et al. 2015). Hydrophilic polymers are a diverse class of polymers in which polynucleotides, polypeptides, and polysaccharides being biological in origin are recognized as (Shalaby et al. 1991) alternative eco-friendly membrane material for the PEMs (Shaari and Kamarudin 2015). Considering the fact that water uptake behavior in PEMs is central to the membrane's ability to transfer protons and cation exchange capacity (Peighambardoust et al. 2010, Rudra et al. 2015), polysaccharides (often referred as biopolymers) are a plausible consideration as PEM materials.

Biopolymer membranes have been used as PEMs in fuel cell technology for decades in energy production and this lead researchers to focus on the utilization of biopolymer, especially polysaccharide-based membranes as PEMs in microbial fuel cell technology (Smitha et al. 2004, Smitha et al. 2005). Among several polysaccharides, the most abundant, naturally produced polymers are: i) cellulose, ii) chitin (as chitosan), and iii) starch. These materials possess hydrolyzable molecules which contain large amounts of energy and can enhance hydrophilicity via the addition of hydroxyl, carboxyl, phosphate, and sulphate groups (Lynd et al. 2005, Ye et al. 2012).

Cellulose

Among several renewable biopolymers, cellulose is the most plenteous material available in the natural environment. Cellulose is produced in billions of tons per year through photosynthesis of plants and is a condensation polymer with reiterating units of D-anhydroglucose (Zhang et al. 2006, French et al. 2007). Cellulose is naturally crystalline having a regular structure and is able to form strong intra- and inter-molecular hydrogen bonding with hydroxyl groups (Ruan et al. 2004, Jie et al. 2005). This renders it recalcitrant to decomposition and insoluble in common solvents, such as acetone, N-methyl pyrrolidone, and dimethyl sulfoxide. Cellulose membranes are widely used in membrane technology due to their high chemical, structural and mechanical properties, biological compatibility, and high permeability of the material (Sokolnicki et al. 2006, Sukma and Çulfaz-Emecen 2018).

Starch

Starch is an important, abundant, and renewable polysaccharide used to prepare biodegradable films (Ghanbarzadeh et al. 2010). Starch is a polymer that acts as an energy bank present in plant biomass and it is analogous with structural polymers cellulose and chitin, but differs in monomers. Starch comprises of a combination of amylose (30%) and amylopectin (70%). Starch has an alpha glucose monomer, whereas cellulose has a beta glucose monomer (Knill and Kennedy 2005). Starch is highly hydrophilic towards water and has poor mechanical properties. This can be compensated by blending starch with other hydrophilic polymers to enhance its physical and functional properties (Vasconez et al. 2009).

Chitosan

Chitosan is a naturally occurring, biodegradable, environmentally friendly biomaterial primarily produced from the alkaline deacetylation of chitin. Chitin is the most abundant (second to cellulose) natural biopolymer present in shells of shrimps and crabs, cartilage of squid, and outer cover of insects (Wan et al. 2003, Chandur et al. 2011, Al-Manhel et al. 2016). Chitosan is a linear, semi-crystalline polysaccharide comprised of N-acetyl D-glucosamine and D-glucosamine units (Rinaudo 2006). Among the naturally occurring polysaccharides, chitosan is a highly basic polysaccharide compared to cellulose, alginic acid, agar, agarose, and carrageenan, which are neutral or acidic in nature. Native chitosan is soluble in water or organic solvents in acidic conditions with a high charge density. Due to protonation of amino groups, chitosan's physical state can be modified easily (Nechita 2017), to fibers, membranes, beads, etc., further enhancing its usefulness (Wan et al. 2003). The existence of very reactive amino ($-NH_2$) and hydroxyl ($-OH$) groups in the chitosan molecule backbone allows particular modifications to build polymers for specific applications. The free amino groups provide the positive charge to chitosan, which enables it to make a great number of electrostatic interactions with negatively charged molecules. This is an additional advantage of using chitosan as a PEM in MFCs, which not only transport the protons, but can also treat wastewater by metal ion removal.

Several studies have been conducted on utilization of biopolymers as substrate, electrode material, and PEM in MFCs. Ahmad et al. (2013) conducted a review on the performance of cellulose, starch, and chitin substrates in power production and reported that MFCs with starch substrate have stronger electrical performance than cellulose and chitin due to its high rate of hydrolysis. The average coulombic efficiency (%) of starch, cellulose, and chitin are 38, 37, and 13, respectively.

Chitosan is also utilized as bioelectrodes to immobilize enzymes in MFCs. The key functional side groups of chitosan, such as hydroxyl and amine, serve as protein-binding ligands for enzyme immobilization (Wei et al. 2002, Falk et al. 2004, Liu et al. 2005). Chitosan arranged as a three-dimensional scaffold in enzymatic electrodes possesses a large pore size which allows the growth of bacterial colonization without increasing flow resistance or glucose oxidation, yet generates effective bio electricity (Cooney et al. 2008, Higgins et al. 2011a). Researchers stated that chitosan- nanotube composite materials in electrodes could increase the power production of MFCs (Higgins et al. 2011b, Katuri et al. 2011, Liu et al. 2011a, Liu et al. 2005).

Blending of chitosan with various water soluble polymers, composites, and inorganic fillers enhances the thermal stability, water/methanol sorption, and proton conductivity of the membranes in fuel cells (Yamada and Honma 2005, Soontarapa and Intra 2006, Wu et al. 2007, Kariduraganavar et al. 2009, Mohanapriya et al. 2009, Lau et al. 2010). Crosslinked chitosan membranes offer a substantial reduction in methanol permeability and competitive proton conductivity as compared to Nafion (Stambouli and Traversa 2002, Xiang et al. 2009). The ionic crosslinking of chitosan during blending with anionic polymers to form a polyelectrolyte can avoid the usage of other crosslinking agents. Although Chitosan PEMs are studied extensively in fuel cell technology (Smitha et al. 2004, Smitha et al. 2005), limited research has been conducted on their usage in microbial fuel cells. Utilization of chitosan as a PEM in MFC technology is novel and challenging because the sources of these biological products are usually considered as waste. However, the low production cost of chitosan compared to Nafion is an attractive feature to use in MFC technology. Hernandez-Flores et al. (2015a) prepared low-cost agar PEM for the treatment of sanitary landfill leachate in MFC, and reported an average power density of 550 mW/m^2 with the removal of 30-40% organic matter. Further, the membrane compared favorably with Nafion in both performance and cost aspect. The volumetric power output of Agar PEM was two times more than the output of Nafion PEM and achieved high COD removal. These are clear advantages of using biopolymer PEMs in MFC compared to synthetic materials. For example, Agar PEM obtained an outstanding power to cost ratio of 39 mW/US$ compared to a low value of 0.11 for Nafion MFCs, and agar membranes do not need any pretreatment, whereas Nafion membranes require a pretreatment with H_2O_2 and H_2SO_4, which produces hazardous waste. Additionally, the fabrication cost of Agar membranes ($14/m^2) is low compared to Nafion membranes ($1733/m^2). Bioelectricity performance of biopolymer-based PEMs in terms of power production and power to cost ratio in MFCs are presented in Table 14.1.

Livinus et al. (2013) demonstrated the feasibility of using cassava starch and modified cassava starch with sodium alginate PEMs in dual-chambered MFCs using swine house effluent inoculated with mangrove forest consortium of soil bacteria. These biopolymers are used in place of a salt bridge and showed a marked improvement in performance in proton conductivity, besides the reduction of overall cell cost. Addition of sodium alginate not only improves the performance of the PEM, but also promotes cell durability due to its hydrophilic properties and immobilizes enzymes by inclusion and encapsulation, thereby stopping bacteria from degrading the starch. The maximum power densities from the cells are 945.69 mW/m^2 and 570.83 mW/m^2 for cassava starch and cassava starch/sodium alginate blend PEMs, respectively, along with 78.2% COD removal efficiency. Recently Srinophakun et al. (2017) synthesized crosslinked chitosan membrane with glutaraldehyde and sulfosuccinic acid as a PEM in a dual-chambered MFC containing tryptone-yeast-NaCl cultural medium as anolyte. As the content of sulfosuccinic acid increased, the proton conductivity of the membrane increased; however, the maximum voltage attained was 790 mV for both plain chitosan and chitosan-sulfosuccinic acid PEMs.

Chemical Modification of Chitosan

Grafting

The grafting technique involves chemical modification of monomers on the membrane surface to form a branched copolymer with side-chain components that are structurally different from that of the main chain (Bhattacharya and Misra 2004). In the process, the grafted material is immobilized through the bonding interaction. Graft co-polymerization is a complicated synthesis procedure, which can be initiated by photo-irradiation, plasma, chemical treatment, high-energy radiation technique (e.g., UV photo grafting and

Table 14.1. Performance of MFCs equipped with various Proton Exchange Media (Hernandez-Flores et al. 2016)

Separator/membrane	Substrate	Average power density (mW/m²)	Internal resistance (Ω)	Power-to-cost (mW/ US$)	Cell operation time (days)	Reference
Agar (salt bridge)	Sodium acetate-G. metallirreducens	2.2	19,920	0.16	17	Min et al. 2005
Agar (salt bridge)	Chocolate industry	145	NR	10.35	15	Patil et al. 2009
Salt-agar slab	Synthetic wastewater with molasses	2.9	NR	0.21	4	Kargi and Eker 2007.
Cation exchange membrane	Sodium acetate	286	55	1.43	1	Zuo et al. 2008
Cellulose acetate microfiltration membrane	Sodium acetate	831	263	2.77	1	Tang et al. 2010
Cellulose acetate microfiltration membrane	Poultry wastewater	746	250	2.50	3	Tang et al. 2010
Agar (2%)	Leachate	547	47	39	15	Hernandez-Flores et al. 2015a

(Contd.)

Table 14.1. (*Contd.*)

Separator/membrane	Substrate	Average power density (mW/m²)	Internal resistance (Ω)	Power-to-cost (mW/US$)	Cell operation time (days)	Reference
Agar (2%)	Semisynthetic leachate	67	90	1.63	1	Hernandez-Flores et al. 2016
Hybrid membrane (Agar 2% - Nafion 1%)	Semisynthetic leachate	96	88	0.89	1	Hernandez-Flores et al. 2016
Nafion 117	Leachate	195	82	0.11	15	Hernandez-Flores et al. 2015a
Sulfonated polyether ether ketone (SPEEK) membrane	Palm oil mill effluent	77.3	811	0.09	1	Shahgaldi et al. 2004.

*NR: Not reported

electron beam), and enzymatic reactions (Liu et al. 2011a, Nady et al. 2011). Compared to membrane coating, graft co-polymerization is a more stable modification method (Kato et al. 2003); however, it incurs a high cost.

Blending

Blending is one of the modification processes that involves the mixing of at least two polymers in various ratios to form a membrane in order to develop preferred functional properties. This method allows for both surface and intra-structural membrane modification. While blending can be cheap and simple, the resulting chitosan membrane can still be limited by low mechanical strength (Nady et al. 2011). Plasticization is also a part of the blending modification method that reduces rigidity and increases the flexibility of the polymer chains. As the plasticity of the chitosan increases through the use of plasticizing agents, such as glycerol, sorbitol, sucrose, and polyethylene glycol, interfacial formation is promoted and the crystallinity of the plasticized chitosan is expected to decrease (Yuan et al. 2007).

Nanocomposite Membrane

The development of high-performance chitosan biopolymers can be achieved through the incorporation of filler materials to obtain significant mechanical reinforcement (Li et al. 2009). Polymer nanocomposites have been reinforced by blending with inorganic nano-sized particles, such as carbon nanotubes and graphene oxide (GO) to improve its physical and mechanical properties (Pan et al. 2011, Venkatesan and Dharmalingam 2013, Bai et al. 2014, Azarniya et al. 2016, Bakangura et al. 2016, Thakur and Voicu 2016, Shirdast et al. 2016). Addition of nano materials into the chitosan matrix not only improves the mechanical or structural properties, but also enhances the permeability properties due to the formation of interfacial gaps between fillers and the polymer matrix.

Crosslinking

Crosslinking is one of the chemical modifications of chitosan in order to increase its mechanical, thermal, and chemical stability, and ionic exchange capacity due to the formation of a three-dimensional (3-D) structure (Smitha et al. 2004, Ma and Sahai 2013). Common classes of crosslinking agents used in membrane modifications are inorganic acids (sulfuric acid, phosphoric acid), aldehydes (Glutaraldehyde), and organic acids (1,4 butanedial diglycol ether, epichlorohydrin, citric acid).

Crosslinking is best suited for large-scale applications compared to the grafting process in terms of costs and efficiency (Liu et al. 2011b). Most of the crosslinking modification studies focused on improving specific properties of chitosan for specific applications; however, there have been no substantial investigations into the potential of crosslinked chitosan-based proton exchange membranes in an MFC system. Due to low-cost associated with chitosan and easy modification through blending, sulfonation, phosphorylation, and increased ion conductivity of the membrane with crosslinking, the authors conducted investigations into the development of chitosan-based proton exchange membranes for MFC technology.

Chitosan-based PEMs

As chitosan can be modified by several ways to improve its performance and structural properties, the authors approached various strategies, such as plasticization, blending, crosslinking, and inorganic fillers on a chitosan matrix (mixed matrix) to synthesize: i) plasticized chitosan membrane with sorbitol (CS-S), ii) copolymer-blended chitosan membrane with poly(malic acid-citric acid) (CS-PMC), iii) crosslinked chitosan membrane using phosphoric acid (P-CS) and sulfuric acid (S-CS), and iv) graphene oxide filled chitosan (CS-G) membranes followed by crosslinking with phosphoric acid and sulfuric acid as PEMs in MFC technology (Holder et al. 2016, Harewood et al. 2017, Holder et al. 2017).

These membranes are prepared by the casting-evaporation method using 2% CHT solution. The detailed procedure for membrane preparation is reported elsewhere (Holder et al. 2016, Harewood et al. 2017, Holder et al. 2017). Briefly, sufficient quantity of CS (2-4 g) is dissolved in 2% acetified water and the obtained homogenous solution is converted into membranes by casting onto a clean glass plate, followed by evaporation of the solvent and drying at 45°C. Plasticized CS membranes (CS-S) are prepared by adding 29.4 g of sorbitol to a homogenous 2% CS solution and prepared in a similar way as above. CS-G composite membranes are prepared by mixing synthesized graphene oxide nano-material (5 wt.%) with the CS (2%) solution in 1:1 ratio and the resulting bubble-free CS-G solution is transformed into membranes. The above membranes are crosslinked using phosphoric acid and sulfuric acid in an isopropyl alcohol (IPA) bath (Holder et al. 2016, Holder et al. 2017). CS-PMC blend membranes are also prepared in a similar fashion by mixing both CS polymer and PMC copolymer in 1:1 ratio (Harewood et al. 2017). The PMC copolymer is prepared via the melt-condensation process by combining malic acid and citric acid in a 3:1 ratio (Popuri et al. 2014).

Experiments are conducted in laboratory-constructed, dual-chambered MFCs with two different wastewaters as anolyte: primary clarifier wastewater (for CS-S, P-CS, S-CS and CS-G membranes) and brewery wastewater (for CS-PMC membrane). L-cysteine hydrochloride (0.5 g/L) is added to the anode chamber immediately after wastewater addition in order to scavenge dissolved oxygen present in the medium, ensuring anaerobic growth conditions. Potassium ferricyanide (50 mM, pH 7) is used as the catholyte under aerobic conditions and carbon rods with projected surface area: 0.019 m^2 (for primary clarifier wastewater) and carbon brushes with projected surface area of 0.0024 m^2 (for brewery wastewater) as electrodes for both anode and cathode. Cell polarizations are obtained by varying the external resistance from 3 kΩ to 10 Ω at time intervals of 3-5 minutes after the MFCs is allowed to reach their steady state open circuit voltage. The respective voltage and current outputs are recorded at each resistance. The organic content of the anolyte throughout the experiment (Chemical oxygen demand) is measured using a standard closed reflux titrimetric method.

Structural Interactions during Blending and Crosslinking

Blending modification using a plasticizer preceding membrane development relaxes the rigid CS structure and increases its elastic behavior, which promotes interfacial formation (Yuan et al. 2007). Chitosan has characteristic hydroxyl (-OH) and amino (-NH$_2$) functional groups. Upon plasticization with sorbitol, surface -OH groups of sorbitol react with the -NH$_2$ groups of CS to form hydrogen bonds. Strong interactions of hydrogen bonding are also observed with this tertiary amino group in CS and water. During crosslinking with phosphoric acid, the ionic interactions between amino and phosphoric acid groups create an assembly exhibiting a polymeric cross-linked structure (Rao et al. 2007, Shanmugam et al. 2016). The CS-S membrane also exhibits this behavior when it is crosslinked with phosphoric acid, but the existing ionic sorbitol interactions are weakened because of the introduced phosphoric acid with water. The addition of sorbitol further assists the extent of crosslinking by ionic interactions of sorbitol and of the phosphate ion, as shown in Fig. 14.1, which postulates the probable ionic interactions taking place between CS and modifying agents, such as sorbitol, phosphoric acid, GO, and PMC. Membranes (CS, CS-S) that are crosslinked with phosphoric acid show an increase in mechanical strength due to electrostatic interactions that take place during crosslinking.

The degree of deacetylation (DDA) plays an important role on chitosan's performance for its applications. A high DDA strongly suggests the degree of bonding within CS and with other substances will be increased due to the high density of CS amino groups (Ziani et al. 2008, Feng et al. 2012). The increased DDA also amplifies the flexible behavior of CS, as there are greater

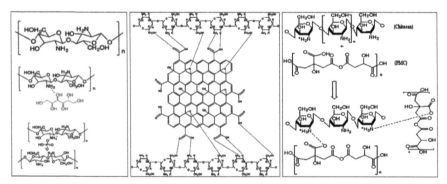

Figure 14.1. Interactions of the chitosan with sorbitol (A), graphene oxide (B) and PMC (C) in the synthesized membranes (Holder et al. 2016, Harewood et al. 2017, Holder et al. 2017).

amounts of –OH and –NH$_2$ sites available (Clasen et al. 2006). The DDA of CS used in our work is determined as 79.80 ± 0.16%.

Several researchers recently investigated the use of CS nanocomposite membranes by filling the CS matrix with inorganic materials (such as graphene oxide and carbon nanotubes) in order to improve their physico-chemical properties (Ramanathan et al. 2008, Zhao et al. 2016, Venkatesan and Dharmalingam 2013, Bai et al. 2014, Bakangura et al. 2016, Thakur and Voicu 2016). The main interaction between inorganic nanomaterials and polymer is electrostatic attraction. However, the interactive sites are low due to the location of these reactive groups in the inorganic nanomaterials. In CS-G membranes, the structures GO is amphipathic in nature compared to the highly hydrophilic chitosan due to hydrophobic carbon and hydrophilic polar groups. During blending of the two, there is a possibility of formation of ionic crosslinking between the polymers if they possess an opposite ionic charge. For example, in CS-PMC membranes, the amino groups of CS may interact ionically with the carboxyl groups of PMC, resulting in the formation of a poly-ion complex presented in Fig. 14.1, as CS and PMC hydrophilic polymers preferentially absorb water through dipole–dipole interactions and intra and inter-molecular hydrogen bonding.

Sorption Behavior of Biopolymer PEMs

Sorption behavior is one of the parameters that indicate the membrane's ability to transport proton from one chamber to the other chamber in an MFC. The ideal PEM should be able to limit oxygen and wastewater crossover without compromising proton transport. This can be achieved through allowing the membranes sufficient sorption under controlled conditions. Prototrophic movement across a PEM occurs either by proton hopping/ Grotthuss transport or diffusion mechanisms. In the Grotthus mechanism, protons are delivered from one carrier site (ex: –NH$_2$, or –SO$_3$H), to a neighboring one through hydrogen bonding. In proton hopping (vehicle) mechanism, protons form hydronium ions after combining with water molecules (ex: H$_3$O$^+$). More specifically, proton transport occurs by selective sorption and diffusion of protons through the membrane due to differences in pH, solubility, and diffusivity of the permeating constituents (Bai et al. 2015). As the proton transportation occurs at the carrier sites of the polymer, the mobility of protons depends on the level of hydration of the membrane.

Sorption trends shown in Table 14.2 indicate that the chitosan-based PEMs used in our study have a greater affinity towards wastewater. Sorption of wastewater by the CS-PMC membrane is higher than all other pristine and crosslinked CS membranes due to availability of free surface external carboxylic, hydroxyl, and amino groups, which interact with water molecules. Among the membranes used for municipal wastewater, CS and CS-S PEMs exhibit higher (>90%) swelling percentage than that of P-CS (59.5%) and P-CS-S (18%) membranes due to the hydrophilic nature of chitosan and sorbitol. Furthermore, the crosslinking process makes the polymer chains in the membrane hold together tightly and compactly, and so reduce the accommodation for feed solution. As crosslinking time increases, the reduction of free volume further decreases for feed solution molecules. This affects proton transfer and power production, which is noticed for P-CS membranes with increased crosslinking time. The sorption percentage of the 24 hour crosslinked P-CS, S-CS, S-CS-G, and P-CS-G membranes are: 60.0, 59.3, 49.3, and 36.7, respectively. However, the un-crosslinked CS-G membrane sorption percentage was 57.4, which is less than the crosslinked CS, S-CS and higher than the S-CS-G and P-CS-G membranes. This was due to the restriction of polymer chain movement with the addition of inorganic nanomaterials (Bai et al. 2015). The low sorption values of the membranes with sulfuric acid crosslinking is possibly due to the formation of strong ionic bonding that may reduce the proton transportation.

Tensile Strength and Cation Exchange Capacity (CEC)

The membrane utilized in MFCs compared to the standard fuel cell is in contact with the solution, therefore the PEM must have a reasonable tensile strength in order to sustain the MFC for long-term operation. Tensile strength values presented in Table 14.2 for all chitosan-based PEMs and P-CS demonstrated the highest mechanical strength. Crosslinking with phosphoric acid creates ionic bonds through electrostatic interactions between amide and phosphate groups in the membrane, reducing movement along the polymer chain, and henceforth increase the mechanical strength of the membrane. In the case of CS-G membrane, the development of interfacial interactions and hydrogen bonding between graphene oxide and chitosan causes an improvement in mechanical strength. In contrast, CS-S membranes exhibit low mechanical strength due to the addition of a plasticizer.

Cationic Exchange Capacity of a membrane indicates its ability for proton transfer from the anodic chamber to the cathodic chamber. CEC values of synthesized chitosan-based PEMs are presented in Table 14.2 and the results indicate that all modified CS membranes display higher CEC values than the pure CS membrane due to the release of protons during hydration and incorporation of phosphate groups. As the crosslinking time increases, the CEC of the membrane also increases, as there are a greater number of ionic clusters and proton hopping sites within the membrane (Seo et al. 2009, Pandey and Shahi 2015, Shaari and Kamarudin 2015).

The synthesized CS and CS-based PEMs maximum voltages are shown in Table 14.2. The pure CS membrane produced a maximum voltage of 152.60 mV, and subsequently failed due to excessive swelling. Crosslinking CS with phosphoric acid improves the performance of the P-CS and P-CS-G membranes, but this effect is inversely proportional to crosslinking time as P-CS (24) produced a maximum voltage of 147 mV. When CS-S and CS-G membranes are crosslinked (with phosphoric acid and sulfuric acid, respectively) they showed decrease in maximum voltage.

Table 14.2. MFC performance and structural properties of chitosan-based PEMs

Membrane	Maximum voltage (mV)	Maximum power density (mW/m²)	Resistance (Ω)	Tensile strength (N/mm²)	Swelling %	Cation exchange capacity (meq/g)	COD removal	Reference
CS	152.6	7.42	1000	39.3	97.4	0.24	NM	Holder et al. 2016
CS-S	447	94.59	1000	0.3	96.6	0.30	NM	Holder et al. 2016
P-CS	504	130.03	1000	49.9	59.5	1.43	49.07	Holder et al. 2016
P-CS-S	425	20.76	1000	0.4	18	0.52	NM	Holder et al. 2016
P-CS(24)	147	11.73	1000	39	60.0	1.35	83.47	Holder et al. 2016
CS-S(24)	87	8.70	1000	27.1	59.3	1.02	NM	Holder et al. 2016
P-CS-G(24)	420	181.56	OCV	25.13	49.3	1.18	89.52	Holder et al. 2016
S-CS-G(24)	89	13.0	OCV	13.07	36.7	1.08	31.99	Holder et al. 2016
CS-PMC	30.6	3822	100	24.67	275	1.87	70	Harewood et al. 2017
Nafion 117	32	4245	100	NM	14.29	NM	75	Harewood et al. 2017
Agar (Salt Bridge)	12.4	637	100	NM	NM	NM	55	Harewood et al. 2017

NM: Not Measured; OCV: Open Circuit Voltage

The P-CS membrane is best suited for MFC operation and the open cell voltage of this MFC unit can be observed over several operational cycles, as shown in Fig. 14.2. A good parameter to evaluate performance without power analysis is the measurement of open circuit voltage (OCV), which is a thermodynamic measurement. The maximum voltages obtained in three operational cycles are close (745, 763, and 735 mV respectively), and it demonstrates the sustainability and reliability of the P-CS MFC system.

Higher operating voltages and power densities are observed in MFCs with CS crosslinked with phosphoric acid PEMs compared to those crosslinked with sulfuric acid, CS-PEMs. Power production decreased with crosslinking time and this trend is in full accordance with those observed throughout the physico-chemical analyses performed, whereby phosphorylated CS membranes exhibited superior CEC and sorption properties. The P-CS (24)-operated MFC showed maximum voltages of 147 mV (day 7), 375 mV (day 15), and 409 mV (day 22) during feed-batch cycles 1, 2, and 3, respectively. Results indicate that the system was able to maintain its performance soon after the addition of fresh anolyte, suggesting the active nature of the bacteria in the anolyte and on the anode electrode. The OCV, maximum current density, and maximum power density obtained were 654 mV, 689.02 mA/m³, and 257 mW/m³, respectively. It is noted that the P-CS membrane (two hour crosslinking time) exhibited higher power density (577.92 mW/m³) than the P-CS (24) membrane. The two systems are not operated over the same time, hence the conditions of influent wastewater COD strengths appeared as 134.57 and 2227.64 mg/L for P-CS (2) and P-CS (24), respectively. The reduction in power density in the P-CS (24) may therefore be due to increased concentration of sludge in the anode chamber. As the electron-transfer bacteria present are unable to completely convert the organic matter into bioelectricity, this creates the favorable environment for the growth and increased activity of methanogens (observed in near neutral pH systems), and an overall reduction in current produced. This process subsequently affected water treatment, as the COD removal rate of P-CS (24) membrane MFC is 83.47% (Mohan et al. 2008).

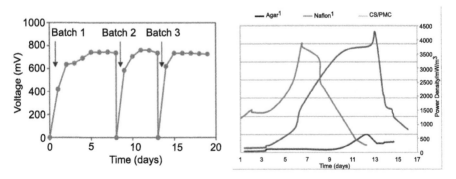

Figure 14.2. MFC performance of CS-P membrane and CS/PMC membrane (Holder et al. 2016, Harewood et al. 2017).

The effects of different crosslinking agents in inorganic mixed matrix CS membranes are seen in Table 14.2 with the phosphorylated CS-G membrane outperforming the sulfonated CS-G membrane. This is due largely on the variances in physicochemical properties of the PEMs, with enhanced sorption and CEC properties by the addition of PO_4^{3-} groups in the CS-G membrane, since SO_4^{2-} inhibits prototrophic transfer. The differences in maximum power densities are large 181.56 mW/m³ for P-CS-G, compared to 13.00 mW/m³ with the S-CS-G (24) membrane. For brewery wastewater, three MFC systems are operated with synthesized CS-PMC, Nafion, and agar (salt bridge) proton exchange media at room temperature. The MFCs are operated for 18 days with continuous monitoring and the power densities of the MFCs with the aforementioned proton exchange media are presented in Fig. 14.2 B. The CS-PMC membrane produced an initial power density of 1226 mW/m², which is higher than the Nafion membrane (159 mW/ m²), but the Nafion membrane produced the maximum amount of power (4245 mW/m²), while CS-PMC generated less (3882 mW/m²). Compared to the CS-PMC and Nafion proton exchange media, the Agar salt bridge generated a maximum power density of 637 mW/m², the relatively low bioelectricity generated due to the protons having to travel farther in Agar, along with the high resistance. The exponential increase in voltage in MFC operation gives evidence to the microbes in the substrate, which directly contributed to electrical generation, suggesting that there is a greater amount of exoelectrogenic bacteria present in the substrate (Wang et al. 2009). The CS-based synthesized membrane CS-PMC converts wastewater chemical energy into electricity in half the time of a Nafion MFC unit and generates higher energy in the initial days, as CS-PMC is more hydrophilic than Nafion. This facilitates the prompt movement of protons through the membrane at an earlier timeframe. In addition, carboxyl and amino groups of the CS-PMC membrane would be able to attract and release protons at a quicker rate than Nafion, which possesses highly electronegative functional groups that will retain protons for longer. The Nafion membrane takes a significant amount of time to be hydrated and this would inhibit prototrophic transfer across the membrane in the MFC unit in initial operating stages (Moilanen et al. 2008).

Effect of pH on Power Generation and COD Removal

Anolyte pH plays a vital role in power production of the MFC system. Generally neutral pH (7) is more suitable for MFC operation, as the microbial activity of the substrate is ideal at neutral, and sluggish just below or above seven (Rozendal et al. 2008). The effect of pH on MFC performance and COD removal is studied for P-CS PEM, and the results indicate that the primary clarifier wastewater produces a maximum current at pH 7 at 1000 Ω and COD removal is 49.07%. At acidic and basic pH values, the bioelectricity generation is reduced due to stunted metabolic activity, and a lesser amount of H⁺, respectively (Oliveira 2013). This can be evidenced by COD removal rates, as the COD removal values are higher at pH 7 than acidic and basic pHs during MFC operation.

COD removal rates are also affected by external resistances applied in MFC operation. As the external resistance varies, there will be a change in catalytic activity of the electrode in the anodic chamber and a change in the microbial metabolic mechanism within the MFC. For example, COD removal by P-CS membrane MFCs is higher (49.07%) at 1000 Ω resistance than in open circuit MFCs (9.07%).

CS-based inorganic mixed matrix PEMs are crosslinked with phosphoric and sulfuric acid to investigate the effect of MFC operation on anolyte properties. The anolyte pH, solution conductivity and COD would provide an insight into MFC operational mechanisms within the anodic chamber. It is expected that the generation of H^+ in the anodic chamber will lower the pH of the anolyte (He et al. 2008, Mohan et al. 2014, Rozendal et al. 2008), which is seen in the case of P-CS-G (24) and S-CS-G (24) membranes, as they both showed a decrease from neutral pH to 5.29 and 3.01, respectively. There can be losses due to the pH imbalance and can ultimately affect power output of the MFC. It is postulated that the equimolar consumption in the cathodic chamber was unable to regulate the pH imbalance due to the presence of graphene oxide, phosphate, and sulfate ions in the CS-based membranes, which inhibit proton transport. Solution conductivity of both CS-based P-CS-G (24) and S-CS-G (24) MFCs rose from 0.510 mS to 1.571 and 1.692 mS respectively. The preeminent COD performance is 89.52% from P-CS-G (24) membrane, while S-CS-G (24) membrane showed 31.99% COD removal. The dissimilarities in COD removal rates between these membranes are due to the differences in physicochemical properties of each membrane, with the sulfonated CS-based membrane retaining protons more than its phosphorylated counterpart. For the CS-PMC blend biopolymer PEM-MFC system, neutral pH is maintained for brewery wastewater throughout the study in order to provide ideal conditions for bacterial growth to achieve maximum bioelectricity generation. In COD removal, both Nafion and CS/PMC PEMs achieved similar results of 75% and 70%, respectively (Table 14.2).

Membrane Fouling

Membrane fouling is a critical issue in MFCs with the need to be addressed as it affects the MFC performance significantly during long-term operation. When MFCs are operated for long durations, microbes present in wastewater attach and grow to the PEM's surface and reduce the effective area of the membrane, causing a decrease in proton transfer (Choi et al. 2012, Chae et al. 2008). Extracellular polymeric substances (foulants) present in microbial clusters and sulfonate groups with multivalent cations in the membrane foster the formation of biofilms, which causes membrane fouling (Beveridge and Doyle 1989, Wingender et al. 1999). Metal elements present in wastewater are absorbed onto the membrane throughout the early stages of fouling, and act as a binding agent between microbes and biopolymers to bolster the biofouling layer. Metal elements are also absorbed on biopolymers and Nafion PEMs due to anionic groups present on hydrophilic surface of the former and negatively charged sulfonate groups of the latter, enhancing the fouling layer on the membrane surface (Hernandez-Fernandez et al. 2015b, Wang et al. 2008a, Wang et al. 2008b). Membrane fouling deteriorates the performance of MFCs due to an actual blockage of proton transport by the biofilm. Growth of biofilm on membrane surfaces indicates the increased level of carbon present in organic matter, which eventually covers the original properties of the membrane. This can be seen using Scanning Electron Microscopy (SEM) in Fig. 14.3, which shows CS-PMC before and after MFC operation. The surface of the membranes appeared clean and clear with no major protrusions or cavities (Fig. 14.3A and B) before MFC operation. However, after MFC operation the surface morphology had clearly changed (Fig. 14.3 C and D), with many pollutants accumulated on the surface of the membrane, causing a significant reduction of effective membrane area. As the time of MFC operation increases, growth of the biofouling layer and complexity of the bacterial structure on the membrane surface also increases (Jorge and Livingston 2000, Chae et al. 2008, Rabiller-Baudry et al. 2012). There are developments to combat membrane fouling by employing the use of anti-biofouling membranes using ionic liquids or biofouling cleaners.

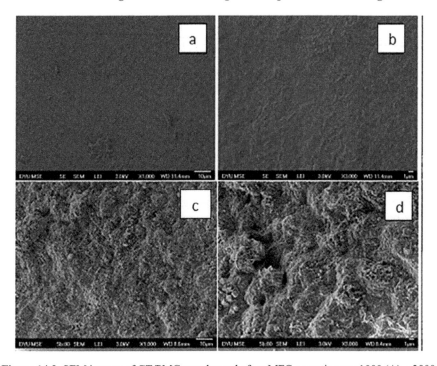

Figure 14.3. SEM images of CT/PMC membrane before MFC operation at x1000 (A), x3000 (B) and after MFC operation at x1000 (C), x3000 (D) (Harewood et al. 2017).

These are not toxic to the microorganisms present inside the wastewater and present a possible solution to maintaining MFC performance for long term operation (Hernandez-Fernandez et al. 2015b, Miskan et al. 2016)

Conclusions

Chitosan (CS) biopolymer can be modified in varied ways using different methods to develop PEMs for use in MFCs in this study. The PEMs under investigation displayed competitive structural and performance attributes for electrical generation and organics removal. The highest power densities in generating electricity from wastewater were by using these PEMs: CS-PMC (3882 mW/m^2) and Nafion (4245 mW/m^2). The anolyte in these MFCs is brewery wastewater, which possesses a high degree of organic content, and an elevated microbial load compared to the other PEMs under investigation which used Primary Clarifier as the anolyte and possessed different microbial attributes. The degree of sorption and the CEC for the CS-PMC membrane is remarkably high in comparison to the other PEMs, and this is due to the type of crosslinking in this membrane, which facilitates high hydration levels and prototrophic transfer. CS-PMC has an interesting property as it can decrease the standard time for maximum hydration of the membrane and can improve start-up times for MFC electrical generation. This high level of hydration did not significantly affect its stability due to the ionic bonding between the PMC copolymer and the CS membrane, as its tensile strength is one of the highest of the PEMs under investigation.

The least stable PEMs under investigation during operation are the membranes plasticized by sorbitol. Sorbitol is advantageous because it increases CEC, as we can see in Table 14.2; however, plasticization decreases the durability of the membrane and allows oxygen crossover, which can decrease the bioelectricity generation of the MFC. Even when the membrane is crosslinked with phosphoric acid for two hours, the addition of phosphate ions does not improve the membrane significantly. In fact, the crosslinked S-CS-P membrane loses a large degree of sorption and this contributes to low bioelectricity generation.

The incorporation of PO_4^{3-} groups using crosslinking proved to be one of the best ways to enhance membrane sorption behavior and bioelectricity production. Phosphorylation did improve the maximum power density (130.03 mW/m^2) of P-CS, which is significantly higher to that of the CS membrane (7.42 mW/m^2). The MFC unit demonstrated that a COD removal of nearly 50% could be achieved with the P-CS as PEM. The PEM S-CS is crosslinked with sulphuric acid to obtain high tensile strength, but exhibited one of the lowest MFC performances due to a low CEC. These outcomes point to the appeal in using environment-friendly biopolymers, especially CS to replace with more expensive PEMs that are used in MFC technology. Adaptation and alteration of low-cost biopolymers to exhibit a combination of improved MFC performance properties is an effective and feasible strategy for bioelectricity generation and wastewater treatment.

Acknowledgements

Authors are thankful to Prof. Sean Carrington, Department of Biological and Chemical Sciences, The University of the West Indies, Cave Hill Campus, Barbados for editing and technical review of this chapter.

REFERENCES

Ahmad, F., M.N. Atiyeh, B. Pereira and G.N. Stephanopoulos. 2013. A review of cellulosic microbial fuel cells: performance and challenges. Biomass Bioener. 56: 179-188.

Al-Manhel, A.J., A.R. Al-Hilphy and A.K. Niamah. 2016. Extraction of chitosan, characterisation and its use for water purification. J. Saudi Soc. Agri. Sci. (In Press).

Aravamudhan, A., D.M. Ramos, A.A. Nada and S.G. Kumbar. 2014. Natural polymers: polysaccharides and their derivatives for biomedical applications. pp. 67-89. *In*: Kumbar, S., C. Laurencin and M. Deng (eds.). Natural and Synthetic Biomedical Polymers. Elsevier, Brazil.

Asano, N., M. Aoki, S. Suzuki, K. Miyatake, H. Uchida and M. Watanabe. 2006. Aliphatic/aromatic polyimide ionomers as a proton conductive membrane for fuel cell applications. J. Am. Chem. Soc. 128: 1762-1769.

Azarniya, A., N. Eslahi, N. Mahmoudi and A. Simchi. 2016. Effect of graphene oxide nanosheets on the physico-mechanical properties of chitosan/bacterial cellulose nanofibrous composites. Compos. Part A. Appl. Sci. 85: 113-122.

Bae, B., T. Hoshi, K. Miyatake and M. Watanabe. 2011. Sulfonated block poly(arylene ether sulfone) membranes for fuel cell applications via oligomeric sulfonation. Macromole. 44: 3884-3892.

Bai, H., H. Zhang, Y. He, J. Liu, B. Zhang and J. Wang. 2014. Enhanced proton conduction of chitosan membrane enabled by halloysite nanotubes bearing sulfonate polyelectrolyte brushes. J. Membr. Sci. 454: 220-232.

Bai, H., Y. Li, H. Zhang, H. Chen, W. Wu, J. Wang and J. Liu. 2015. Anhydrous proton exchange membranes comprising of chitosan and phosphorylated graphene oxide for elevated temperature fuel cells. J. Membr. Sci. 495: 48-60.

Bakangura, E., L. Wu, L. Ge, Z. Yang and T. Xu. 2016. Mixed matrix proton exchange membranes for fuel cells: state of the art and perspectives. Prog. Polym. Sci. 57: 103-152.

Belleville, P., P.J. Strong, P.H. Dare and D.J. Gapes. 2011. Influence of nitrogen limitation on performance of a microbial fuel cell. Water Sci. Technol. 63: 1752-1757.

Beveridge, T.J. and R.J. Doyle. 1989. Metal ions and bacteria. U.S.A: John Wiley & Sons, Inc.

Bhattacharya, A. and B.N. Misra. 2004. Grafting: a versatile means to modify polymers: techniques, factors and applications. Prog. Polym. Sci. 29(8): 767-814.

Cadogan, E.I., C.H. Lee, S.R. Popuri and H.Y. Lin. 2016. Characterization, fouling and performance in bacterial removal of synthesized chitosan-glycerol membranes. Desali. Water Treat. 57: 17670-17682.

Catherine, C. 2009. A review of membrane processes and renewable energies for desalination. Desali. 245: 214-231.

Chae, K.J., M. Choi, F.F. Ajayi, W. Park, I.S. Chang and I.S. Kim. 2008. Mass transport through a proton exchange membrane (Nafion) in microbial fuel cells. Energy Fuels. 22: 169-176.

Chandur, V.K., A.M. Badiger and R.K. Shambashiva. 2011. Characterizing formulations containing derivatized chitosan with polymer blending. Int. J. Res. Pharm. Chem. 4(1): 950-967.

Chen, Y.J. 2014. Bioplastics and their role in achieving global sustainability. J. Chem. Pharma. Res. 6(1): 226-231.

Choi, J., R. Patel, J. Han and B. Min. 2016. Proton conducting composite membranes comprising sulfonated poly(1,4-phenylene sulfide) and zeolite for fuel cell. Ionics 16: 403-408.

Choi, M.J., K.J. Chae, F.F. Ajayi, K.Y. Kim, H.W. Yu, C.W. Kim and I.S. Kim. 2011. Effects of biofouling on ion transport through cation exchange membranes and microbial fuel cell performance. Bioresour. Technol. 102: 298-303.

Choi, T.H., Y.R. Won, J.W. Lee, D.W. Shin, Y.M. Lee, M. Kim and H.B. Park. 2012. Electrochemical performance of microbial fuel cells based on disulfonated poly(arylene ether sulfone) membranes. J. Power Sour. 220: 269-279.

Clasen, C., T. Wilhelms and W.M. Kulicke. 2006. Formation and characterization of chitosan membranes. Biomacromole. 7: 3210-3222.

Cooney, M.J., C. Lau, M. Windmeisser, B.Y. Liaw, T. Klotzbach and S.D. Minteer. 2008. Design of chitosan gel pore structure: towards enzyme catalyzed flow through electrodes. J. Mater. Chem. 18: 667-674.

Dai, H., R. Guan, C. Li and J. Liu. 2007. Development and characterization of sulfonated poly(ether sulfone) for proton exchange membrane materials. Solid State Ionics. 178: 339-345.

Devanathan, R. 2008. Recent developments in proton exchange membranes for fuel cells. Energy Environ. Sci. 1: 101-119.

Falk, B., S. Garramone and S. Shivkumar. 2004. Diffusion coefficient of paracetamol in a chitosan hydrogel. Mater. Lett. 58: 3261-3265.

Feng, F., Y. Liu, B. Zhao and K. Hu. 2012. Characterization of half N-acetylated chitosan powders and films. Procedia Engg. 27: 718-732.

French, A.D., N.R. Bertoniere, R.M. Brown, H. Chanzy, D. Gray, K. Hattori and W. Glasser. 2007. Cellulose. p. 429. In: Seidel, A. (ed.). Kirk-othmer Concise Encyclopedia of Chemical Technology. New York: Wiley-Interscience.

Ghanbarzadeh, B., H. Almasi and A.A. Entezami. 2010. Physical properties of edible modified starch/carboxymethyl cellulose films. Inno. Food Sci. Emer. Technol. 11: 697-702.

Ghasemi, M., W.R.W. Daud, S.H. Hassan, S.E. Oh, M. Ismail, M. Rahimnejad and J.M. Jahim. 2013. Nano-structured carbon as electrode material in microbial fuel cells: a comprehensive review. J. Alloy. Compo. 580: 245-255.

Ghassemi, H. and J.E. McGrath. 2004. Synthesis and properties of new sulfonated poly(p-phenylene) derivatives for proton exchange membranes. I. Polym. 45: 5847-5854.

Harewood, A.J.T., S.R. Popuri, E.I. Cadogan, C.H. Lee and C.C. Wang. 2017. Bioelectricity generation from brewery wastewater in a microbial fuel cell using chitosan/biodegradable copolymer membrane. Int. J. Environ. Sci. Technol. 14: 1535-1550.

He, Z., Y. Huang, A.K. Manohar and F. Mansfeld. 2008. Effect of electrolyte pH on the rate of the anodic and cathodic reactions in an air-cathode microbial fuel cell. Bioelectrochem. 74(1): 78-82.

Hernandez-Flores, G. 2014. Interim report: bioelectricity generation in microbial fuel cells that use enriched inocula and graphite flakes as anodic material. ScD Thesis. Mexico, D.F: CINVESTAV-IPN.

Hernandez-Flores, G., H.M. Poggi-Varaldo, O. Solorza-Feria, T. Romero-Castanon, E. Ríos-Leal, J. Galíndez-Mayer and F. Esparza-García. 2015a. Batch operation of a microbial fuel cell equipped with alternative proton exchange membrane. Int. J. Hydrogen Ener. 40: 17323-17331.

Hernandez-Flores, G., A. Perezdelos-Ríos, F. Mateo-Ramírez, C. Godínez, L.J. Lozano-Blanco, J.I. Moreno and F. Tomas-Alonso. 2015b. New application of supported ionic liquids membranes as proton exchange membranes in microbial fuel cell for waste water treatment. Chem. Eng. J. 279: 115-119.

Hernandez-Flores, G., H.M. Poggi-Varaldo and O. Solorza-Feria. 2016. Comparison of alternative membranes to replace high cost Nafion ones in microbial fuel cells. Int. J. Hydrogen Ener. 41: 23354-23362.

Herranen, J., J. Kinnunen, B. Mattsson, H. Rinne, F. Sundholm and L. Torell. 1995. Characterisation of poly(ethylene oxide) sulfonic acids. Solid State Ionics. 80: 201-212.

Hickner, M.A., C.H. Fujimoto and C.J. Cornelius. 2006. Transport in sulfonated poly(phenylene)s: proton conductivity, permeability, and the state of water. Polymer 47: 4238-4244.

Higgins, S.R., D. Foerster, A. Cheung, C. Lau, O. Bretschger and S.D. Minteer. 2011a. Fabrication of macroporous chitosan scaffolds doped with carbon nanotubes and their characterization in microbial fuel cell operation. Enzyme Microb. Tech. 48: 458-465.

Higgins, S.R., C. Lau, P. Atanassov, S.D. Minteer and M.J. Cooney. 2011b. Hybrid biofuel cell: microbial fuel cell with an enzymatic air-breathing cathode. ACS Catal. 1: 994-997.

Holder, S.L., C.H. Lee and S.R. Popuri. 2017. Simultaneous wastewater treatment and bioelectricity production in microbial fuel cells using cross-linked chitosan-graphene oxide mixed-matrix membranes. Environ. Sci. Pollut. Res. 24(15): 13782-13796.

Holder, S.L., C.H. Lee, S.R. Popuri and M.X. Zhuang. 2016. Enhanced surface functionality and microbial fuel cell performance of chitosan membranes through phosphorylation. Carbohyd. Polym. 149: 251-262.

Hou, J., Z. Liu, S. Yang and Y. Zhou. 2014. Three-dimensional macroporous anodes based on stainless steel fiber felt for high-performance microbial fuel cells. J. Power Sour. 258: 204-209.

Houghton, J.I. and J. Quarmby. 1999. Biopolymers in wastewater treatment. Curr. Opin. Biotechnol. 10(3): 259-262.

Jana, P.S., M. Behera and M.M. Ghangrekar. 2010. Performance comparison of up-flow microbial fuel cells fabricated using proton exchange membrane and earthen cylinder. Int. Hydrogen Ener. 35(11): 5681-5686.

Jie, X., Y. Cao, J.J. Qin, J. Liu and Q. Yuan. 2005. Influence of drying method on morphology and properties of asymmetric cellulose hollow fiber membrane. J. Membr. Sci. 246: 157-165.

Jorge, R.M.F. and A.G. Livingston. 2000. Microbial dynamics in an extractive membrane bioreactor exposed to an alternating sequence of organic compounds. Biotechnol. Bioeng. 70: 313-322.

Kargi, F. and S. Eker. 2007. Electricity generation with simultaneous wastewater treatment by a microbial fuel cell (MFC) with Cu and Cu-Au electrodes. J. Chem. Technol. Biotechnol. 82: 658-662.

Kariduraganavar, M.Y., J.G. Varghese, S.K. Choudhari and R.H. Olley. 2009. Organic-inorganic hybrid membranes: solving the trade-off phenomenon between permeability and selectivity in pervaporation. Ind. Eng. Chem. Res. 48: 4002-4010.

Kato, K., E. Uchida, E.T. Kang, Y. Uyama and Y. Ikada. 2003. Polymer surface with graft chains. Prog. Polym. Sci. 28(2): 209-259.

Katuri, K., M. Luisa Ferrer, M.C. Gutierrez, R. Jimenez, F. Monte and D. Leech. 2011. Three-dimensional microchanelled electrodes in flow-through configuration for bioanode formation and current generation. Energy Environ. Sci. 4: 4201-4210.

Kim, D., J. An, B. Kim, J.K. Jang, B.H. Kim and I.S. Chang. 2012. Scaling-up microbial fuel cells: configuration and potential drop phenomenon at series connection of unit cells in shared anolyte. Chem. Sus. Chem. 5(6): 1086-1091.

Kim, J.R., S. Cheng, S.E. Oh and B.E. Logan. 2007. Power generation using different cation, anion, and ultrafiltration membranes in microbial fuel cells. Environ. Sci. Technol. 41(3): 1004-1009.

Kim, S. and E.M.V. Hoek. 2007. Interactions controlling biopolymer fouling of reverse osmosis membranes. Desalination 202(1-3): 333-342.

Kim, Y., S.H. Shin, I.S. Chang and S.H. Moon. 2014. Characterization of uncharged and sulfonated porous poly(vinylidene fluoride) membranes and their performance in microbial fuel cells. J. Membr. Sci. 463: 205-214.

Knill, C.J. and J.F. Kennedy. 2005. Starch: commercial sources and derived products. pp. 605-624. In: Dumitriu, S. (ed.). Polysaccharides: Structural Diversity and Functional Versatility. Second edn. New York: Marcel Dekker.

Kumar, A., A. Srivastava, I.Y. Galaev and B. Mattiasson. 2007. Smart polymer: physical forms and bioengineering applications. Prog. Polym. Sci. 32: 1205-1237.

Lau, C., G. Martin, S.D. Minteer and M.J. Cooney. 2010. Development of a chitosan scaffold electrode for fuel cell applications. Electroanal. 22: 793-798.

Lee, C.H., H.Y. Lin, E.I. Cadogan, S.R. Popuri and C.Y. Chang. 2013. Biosorption performance of biodegradable polymer powders for the removal of gallium(III) ions from aqueous solution. Polish J. Chem. Technol. 17: 124-132.

Lee, K.P., T.C. Arnot and D. Mattia. 2011. A review of reverse osmosis membrane materials for desalination-development to date and future potential. J. Membr. Sci. 370: 1-22.

Li, Q., J. Zhou and L. Zhang. 2009. Structure and properties of the nanocomposite films of chitosan reinforced with cellulose whiskers. J. Polym. Sci. Part B: Polym. Phy. 47(11): 1069-1077.

Li, W.W., G.P. Sheng, X.W. Liu and H.Q. Yu. 2011. Recent advances in the separators for microbial fuel cells. Bioresour. Technol. 102: 244-252.

Liu, B., G.P. Robertson, D.-S. Kim, M.D. Guiver, W. Hu and Z. Jiang. 2007. Aromatic poly(ether ketone)s with pendant sulfonic acid phenyl groups prepared by a mild sulfonation method for proton exchange membranes. Macromole 40: 1934-1944.

Liu, F., N.A. Hashim, Y. Liu, M.R.M. Abed and K. Li. 2011b. Progress in the production and modification of PVDF membranes. J. Membr. Sci. 375(1-2): 1-27.

Liu, L., W. Chen and Y. Li. 2016. An overview of the proton conductivity of Nafion membranes through a statistical analysis. J Membr. Sci. 504: 1-9.

Liu, X., X. Sun, Y. Huang, G. Sheng, S. Wang and H. Yu. 2011a. Carbon nanotube/chitosan nanocomposite as a biocompatible biocathode material to enhance the electricity generation of a microbial fuel cell. Energy Environ. Sci. 4: 1422-1427.

Liu, Y., M. Wang, F. Zhao, Z. Xu and S. Dong. 2005. The direct electron transfer of glucose oxidase and glucose biosensor based on carbon nanotubes/chitosan matrix. Biosens. Bioelectron. 21: 984-988.

Livinus, A.O., C.O. Charles, O. Ken and O. Akuma. 2013. Performance of cassava starch as a proton exchange membrane in a dual chambered microbial fuel cell. Greener J. Biolog. Sci. 3(2): 74-83.

Logan, B.E. and K. Rabaey. 2012. Conversion of wastes into bioelectricity and chemicals by using microbial electrochemical technologies. Science 337: 686-690.

Lynd, L.R., W.H. van Zyl, J.E. McBride and M. Laser. 2005. Consolidated bioprocessing of cellulosic biomass: an update. Curr. Opin. Biotechnol. 16(5): 577-583.

Ma, J. and Y. Sahai. 2013. Chitosan biopolymer for fuel cell applications. Carbohyd. Polym. 92: 955-975.

Matsumoto, K., T. Higashihara and M. Ueda. 2009. Locally sulfonated poly(ether sulfone)s with highly sulfonated units as proton exchange membrane. J. Polym. Sci. Part A Polym. Chem. 47: 3444-3453.

Min, B., S. Cheng and B.E. Logan. 2005. Electricity generation using membrane and salt bridge microbial fuel cells. Water Res. 39(9): 1675-1686.

Miskan, M., M. Ismail, M. Ghasemi, J. Md. Jahim, D. Nordin and M.H.A. Bakar. 2016. Characterization of membrane biofouling and its effect on the performance of microbial fuel cell. Int. J. Hydrogen Ener. 41: 543-552.

Miyatake, K., Y. Chikashige, E. Higuchi and M. Watanabe. 2007. Tuned polymer electrolyte membranes based on aromatic polyethers for fuel cell applications. J. Am. Chem. Soc. 129: 3879-3887.

Modin, O. and D.J. Gustavsson. 2014. Opportunities for microbial electrochemistry in municipal wastewater treatment: an overview. Water Sci. Technol. 69(7): 1359-1372.

Mohan, S.V., G. Mohanakrishna and P.N. Sarma. 2008. Effect of anodic metabolic function on bioelectricity generation and substrate degradation in single chambered microbial fuel cell. Environ. Sci. Technol. 42: 8088-8094.

Mohan, S.V., G. Velvizhi, J.A. Modestra and S. Srikanth. 2014. Microbial fuel cell: critical factors regulating bio-catalyzed electrochemical process and recent advancements. Renew. Sus. Ener. Rev. 40: 779-797.

Mohanapriya, S., S.D. Bhat, A.K. Sahu, S. Pitchumani, P. Sridhar and A.K. Shukla. 2009. A new mixed-matrix membrane for DMFCs. Energy Environ. Sci. 2: 1210-1216.

Moilanen, D.E., D.B. Spry and M.D. Fayer. 2008. Water dynamics and proton transfer in Nafion fuel cell membranes. Langmuir 24: 3690-3698.

Nady, N., M.C. Franssen, H. Zuilhof, M.S.M. Eldin, R. Boom and K. Schroën. 2011. Modification methods for poly(arylsulfone) membranes: a mini-review focusing on surface modification. Desalination 275(1): 1-9.

Neburchilov, V., J. Martin, H. Wang and J. Zhang. 2007. A review of polymer electrolyte membranes for direct methanol fuel cells. J. Power Sour. 169: 221-238.

Nechita, P. 2017. Applications of chitosan in wastewater treatment. Pp. 209-228. *In*: Shalaby, E.A. (ed.). Biological Activities and Application of Marine Polysaccharides. InTech Open Ltd. London SE19SG UK.

Oh, S.E. and B.E. Logan. 2006. Proton exchange membrane and electrode surface areas as factors that affect power generation in microbial fuel cells. Appl. Microbial. Biotechnol. 70(2): 162-169.

Oliveira, V.B., M. Simões, L.F. Melo and A.M.F.R. Pinto, 2013. Overview on the developments of microbial fuel cells. Biochem. Engg. J. 73: 53-64.

Pan, Y., T. Wu, H. Bao and L. Li. 2011. Green fabrication of chitosan films reinforced with parallel aligned graphene oxide. Carbohyd. Polym. 83: 1908-1915.

Pandey, R.P. and V.K. Shahi. 2015. Phosphonic acid grafted poly(ethyleneimine)-silica composite polymer electrolyte membranes by epoxide ring opening:

improved conductivity and water retention at high temperature. Int. J. Hydro. Ener. 40: 14235-14245.

Pant, D., G.V. Bogaert, L. Diels and K. Vanbroekhoven. 2010. A review of the substrates used in microbial fuel cells (MFCs) for sustainable energy production. Bioresour. Technol. 101: 1533-1543.

Park, J.K., Y.S. Kang and J. Won. 2008. Characterization of deoxyribonucleic acid/ poly(ethylene oxide) proton-conducting membranes. J. Membr. Sci. 313: 217-223.

Patil, S.A., V.P. Surakasi, S. Koul, S. Ijmulwar, A. Vivek, Y.S. Shouche and B.P. Kapadnis. 2009. Electricity generation using chocolate industry wastewater and its treatment in activated sludge based microbial fuel cell and analysis of developed microbial community in the anode chamber. Bioresour. Technol. 100: 5132-5139.

Peighambardoust, S.J., S. Rowshanzamir and M. Amjadi. 2010. Review of the proton exchange membranes for fuel cell applications. Int. J. Hydrogen Energ. 35: 9349-9384.

Popuri, S.R., C. Hall, C.C. Wang and C.Y. Chang. 2014. Development of green/ biodegradable polymers for water scaling applications. Int. Biodeterior. Biodegrad. 95: 225-231.

Rabiller-Baudry, M., F. Gouttefangeas, J.L. Lannic and P. Rabiller. 2012. Coupling of SEM-EDX and FTIR-ATR to (quantitatively) investigate organic fouling on porous organic composite membranes. Curr. Microscopy Contri. Adv. Sci. Technol. 1066-1076.

Ramanathan, T., A.A. Abdala, S. Stankovich, D.A. Dikin, M. Herrera-Alonso, R.D. Piner D.H. Adamson, H.C. Schniepp, X. Chen, R.S. Ruoff, S.T. Nguyen, I.A. Aksay, R.K. Prud'Homme and L.C. Brinson. 2008. Functionalized graphene sheets for polymer nanocomposites. Nature Nanotechnol. 3(6): 327-331.

Rao, P.S., S. Sridhar, A. Krishnaiah and M.Y. Wey. 2007. Pervaporation separation of ethylene glycol/water mixtures by using cross-linked chitosan membranes. Ind. Eng. Chem. Res. 46: 2155-2163.

Rinaudo, M. 2006. Chitin and chitosan: properties and applications. Prog. Polymer Sci. 31(7): 603-632.

Rozendal, R.A., H.V.M. Hamelers, K. Rabaey, J. Keller and C.J.N. Buisman. 2008. Towards practical implementation of bioelectrochemical wastewater treatment. Trends Biotechnol. 26(8): 450-459.

Rozendal, R.A., H.V.M. Hamelers and C.J.N. Buisman. 2006. Effects of membrane cation transport on pH and microbial fuel cell performance. Environ. Sci. Technol. 40: 5206-5211.

Ruan, D., L. Zhang, Y. Mao, M. Zeng and X. Li. 2004. Microporous membranes prepared from cellulose in NaOH/Thiourea aqueous solution. J. Membr. Sci. 241: 265-274.

Rudra, R., V. Kumar and P.P. Kundu. 2015. Acid catalysed cross-linking of poly vinyl alcohol (PVA) by glutaraldehyde: effect of crosslink density on the characteristics of PVA membranes used in single chambered microbial fuel cells. RSC Advances 5: 83436-83447.

Seo, J.A., J.H. Koh, D.K. Roh and J.H. Kim. 2009. Preparation and characterization of crosslinked proton conducting membranes based on chitosan and PSSA-MA copolymer. Solid State Ionics 180: 998-1002.

Shaari, N. and S.K. Kamarudin. 2015. Chitosan and alginate types of bio-membrane in fuel cell application: an overview. J. Power Sour. 289: 71-80.

Shahgaldi, S., M. Ghasemi, W.R. Wan Daud, Z. Yaakob, M. Sedighi, J. Alam and A.F. Ismail. 2014. Performance enhancement of microbial fuel cell by PVDF/Nafion nanofibre composite proton exchange membrane. Fuel. Process. Technol. 124: 290-295.

Shalaby, S.W., C.L. McCormick and G.B. Buttler. 1991. Water-soluble polymers, synthesis, solution properties and applications. ACS Symposium Series, Vol. 467. American Chemical Society, Washington, DC, USA.

Shanmugam, A., K. Kathiresan and L. Nayak. 2016. Preparation, characterization and antibacterial activity of chitosan and phosphorylated chitosan from cuttlebone of Sepia kobiensis (Hoyle, 1885). Biotechnol. Rep. 9: 25-30.

Shirdast, A., A. Sharif and M. Abdollahi. 2016. Effect of the incorporation of sulfonated chitosan/sulfonated graphene oxide on the proton conductivity of chitosan membranes. J. Power Sour. 306: 541-551.

Smitha, B., S. Sridhar and A.A. Khan. 2004. Polyelectrolyte complexes of chitosan and poly (acrylic acid) as proton exchange membranes for fuel cells. Macromole. 37: 2233-2239.

Smitha, B., S. Sridhar and A.A. Khan. 2005. Solid polymer electrolyte membranes for fuel cell applications—a review. J. Membr. Sci. 259(1-2): 10-26.

Sokolnicki, A.M., R.L. Fisher, T.P. Harrah and D.L. Kaplan. 2006. Permeability of bacterial cellulose membranes. J. Membr. Sci. 272(1-2): 15-27.

Soontarapa, K. and U. Intra. 2006. Chitosan based fuel cell membranes. Chem. Eng. Comm. 193: 855-868.

Srinophakun, P., A. Thanapimmetha, S. Plangsri, S. Vetchayakunchai and M. Saisriyoot. 2017. Application of modified chitosan membrane for microbial fuel cell: roles of proton carrier site and positive charge. J. Cleaner Prod. 142: 1274-1282.

Stambouli, A.B. and E. Traversa. 2002. Solid oxide fuel cells (SOFCs): a review of an environmentally clean and efficient source of energy. Renew. Sustain. Energy Rev. 6: 433-455.

Su, Z., T. Liu, W.Z. Yu, X. Li and N.J.D. Graham. 2017. Coagulation of surface water: observations on the significance of biopolymers. Water Res. 126(1): 144-152.

Sukma, F.M. and P.Z. Çulfaz-Emecen. 2018. Cellulose membranes for organic solvent nanofiltration. J. Membr. Sci. 545: 329-336.

Sun, F., T. Wang, S. Yang and L. Fan. 2010. Synthesis and characterization of sulfonated polyimides bearing sulfonated aromatic pendant group for DMFC applications. Polym. 51: 3887-3898.

Tang, X., K. Guo, H. Li, Z. Du and J. Tian. 2010. Microfiltration membrane performance in two-chamber microbial fuel cells. Biochem. Engg. J. 52: 194-198.

Thakur, V.K. and S.I. Voicu. 2016. Recent advances in cellulose and chitosan based membranes for water purification: a concise review. Carbohyd. Polym. 146: 148-165.

Vasconez, M.B., S.K. Flores, C.A. Campos, J. Alvarado and L.N. Gerschenson. 2009. Antimicrobial activity and physical properties of chitosan-tapioca starch based edible films and coatings. Food Res. Inter. 42: 762-769.

Venkatesan, P.N. and S. Dharmalingam. 2013. Characterization and performance study on chitosan-functionalized multi walled carbon nano tube as separator in microbial fuel cell. J. Membr. Sci. 435: 92-98.

Wan, Y., K.A. Creber, B. Peppley and V.T. Bui. 2003. Synthesis, characterization and ionic conductive properties of phosphorylated chitosan membranes. Macromole. Chem. Phy., 204(5-6): 850-858.

Wang, X., Y.J. Feng and H. Lee. 2008a. Electricity production from beer brewery wastewater using single chamber microbial fuel cell. Water Sci. Technol. 57: 1117-1121.

Wang, Z., Z. Wu, X. Yin and L. Tian. 2008b. Membrane fouling in a submerged membrane bioreactor (MBR) under sub-critical flux operation: membrane foulant and gel layer characterization. J. Membr. Sci. 325: 238-244.

Wang, Z., Z. Wu and S. Tang. 2009. Extracellular polymeric substances (EPS) properties and their effects on membrane fouling in a submerged membrane bioreactor. Water Res. 43: 2504-2512.

Wei, X., J. Cruz and W. Gorski. 2002. Integration of enzymes and electrodes: spectroscopic and electrochemical studies of chitosan enzyme films. Anal. Chem. 74: 5039-5046.

Wingender, J., T.R. Neu and H.C. Flemming. 1999. Microbial extracellular polymeric substances. Springer-Verlag Heidelberg.

Wu, H., B. Zheng, X. Zheng, J. Wang, W. Yuan and Z. Jiang. 2007. Surface-modified Y zeolite-filled chitosan membrane for direct methanol fuel cell. J. Power Sour. 173: 842-852.

Xiang, Y., M. Yang, Z. Guo and Z. Cui. 2009. Alternatively chitosan sulfate blending. membrane as methanol-blocking polymer electrolyte membrane for direct methanol fuel cell. J. Membr. Sci. 337: 318-323.

Xu, L., Y. Zhao, L. Doherty, Y. Hu and X. Hao. 2016. The integrated processes for wastewater treatment based on the principle of microbial fuel cells: a review. J. Critical Rev. Environ. Sci. Technol. 46(1): 60-91.

Yamada, M. and I. Honma. 2005. Anhydrous proton conductive membrane consisting of chitosan. Electrochim. Acta. 50: 2837-2841.

Yang, Y., G. Sun and M. Xu. 2010. Microbial fuel cells come of age. J. Chem. Technol. Biotechnol. 86: 625-632.

Ye, Y.S., R. John and B.J. Huang. 2012. Water soluble polymers as proton exchange membranes for fuel cells. Polym. 4: 913-963.

Yuan, W., H. Wu, B. Zheng, X. Zheng, Z. Jiang, X. Hao and B. Wang. 2007. Sorbitol-plasticized chitosan/zeolite hybrid membrane for direct methanol fuel cell. J. Power Sour. 172(2): 604-612.

Zhang, F., S. Cheng, D. Pant, G. Van Bogaert and B.E. Logan. 2009. Power generation using an activated carbon and metal mesh cathode in a microbial fuel cell. Electrochem. Commun. 11(11): 2177-2179.

Zhang, Y., Y. Wan, C. Zhao, K. Shao, G. Zhang, H. Li, H. Lin and H. Na. 2009. Novel side-chain-type sulfonated poly(arylene ether ketone) with pendant sulfoalkyl groups for direct methanol fuel cells. Polymer 50: 4471-4478.

Zhang, Y.H.P., M.E. Himmel and J.R. Mielenz. 2006. Outlook for cellulose improvement: screening and selection strategies. Biotechnol. Adv. 24(5): 452-481.

Zhao, Y., Y. Fu, B. Hu and C. Lü. 2016. Quaternized graphene oxide modified ionic cross-linked sulfonated polymer electrolyte composite proton exchange membranes with enhanced properties. Solid State Ion. 294: 43-53.

Zhong, S., X. Cui, H. Cai, T. Fu, C. Zhao and H. Na. 2007. Crosslinked sulfonated poly(ether ether ketone) proton exchange membranes for direct methanol fuel cell applications. J. Power Sour. 164: 65-72.

Zhou, M., M. Chi, J. Luo, H. He and T. Jin. 2011. An overview of electrode materials in microbial fuel cells. J. Power Sour. 196: 4427-4435.

Ziani, K., J. Oses, V. Coma and J.I. Maté. 2008. Effect of the presence of glycerol and Tween 20 on the chemical and physical properties of films based on chitosan with different degree of deacetylation. LWT - Food Sci. Technol. 41: 2159-2165.

Zuo, Y., S. Cheng and B.E. Logan. 2008. Ion exchange membrane cathodes for scalable microbial fuel cells. Environ. Sci. Technol. 42: 6967-6972.

Using Microbial Fuel Cell System as Biosensors

Tuoyu Zhou, Shuting Zhang, Huawen Han and Xiangkai Li*

Lanzhou University, Ministry of Education,
Key Laboratory of Cell Activities and Stress Adaptations,
Tianshui South Road #222, Lanzhou 730000, China

Introduction

With the improvement of modern agriculture and industry, as well as booming of population, serious environmental pollution has become a global concern. Through the food chain's biological accumulation, the pollutants cause serious threat to human health and wildlife (Schwarzenbach et al. 2010). For example, millions of tons of agricultural chemicals are used every year (Tilman et al. 2001), which includes nitrogen, phosphorus, and natural organic compounds (Howarth 2009, Filippelli 2008, Jr 1991). These pesticides cause high nutrient loads in water, resulting in oxygen depletion, toxic algal blooms, and the death of aquatic organism (Ajani et al. 2013, Paerl and Hans 1997). Besides, the human-driven natural geological toxic chemicals, such as heavy metals and metalloid, cause a series of human diseases (Yuan et al. 2016, Michael Berg et al. 2001, Kudo et al. 1998). Considering that there are more than 100,000 kinds of chemicals registered and most of them are used everyday (Schwarzman and Wilson 2009), people can easily imagine that a mass of chemicals can enter the environment through various routes. In addition, the planet ocean circulation could promote classic persistent organic pollutants (pops) through different oceans. Canada scholars' survey for the Arctic Circle pollution shows that even so far away from areas of human activity, we can still trace the pollutants' accumulation (Schwarzman and Wilson 2009).

Therefore, the management of environmental pollution is very important. Under the process of environmental pollution treatment, monitoring pollutants is an important aspect, which includes early warning

*Corresponding author: xkli@lzu.edu.cn

of environmental pollution, control of pollutant treatment, and monitoring of pollutant emission. The warning of pollutants should be effective, timely, and quickly distinguishing different contaminants. In the process of pollutant treatment, different operation parameters have different effects. For example, in the process of wastewater treatment, the dissolved oxygen content of aeration tank is the key factor that determines the activity of aerobic microorganisms (Abrevaya et al. 2015). So real-time monitoring of dissolved oxygen can effectively control the wastewater treatment process. In order to improve the capability of pollutant treatment, it is necessary to monitor the treatment process of pollutants. In addition, when pollutants are released into the natural environment, they should also be tested to avoid the harmful effects of toxic substances on the natural environment during the discharge process.

Traditional pollutants' detection methods are usually performed by physical and chemical methods, including ultraviolet spectroscopy, gas chromatography (GC), high performance liquid chromatography (HPLC), etc. (Driscoll and Berger 1971, Berthou and Friocourt 1982, Wu et al. 2011). These kinds of methods have high accuracy and stability; however, these analytical methods have some significant limitations, including the high cost, management of inconvenience, and the samples requiring relatively cumbersome treatment. The most notable is that it's unsuitable for real-time monitoring, so it can't timely reflect the current environmental pollution situation. To solve this problem, biosensor has been proposed, and it is widely applied in different fields, from public health and environment monitoring to the food security (Zhou et al. 2017). In concept, a biosensor includes a recognition element and a physical transducer, which could provide a corresponding signal to the analyte concentration (Jiang et al. 2018, Su et al. 2011). The biometic elements could be enzymes, fluorescent molecules, cells, or microorganisms. Among these elements, enzyme is the most widely used material; nevertheless, like most of the proteins in the nature, the enzyme signal is easily disturbed by external environment factors. Besides, its high price and time-consuming purification is the main hindrance for the commercial applications (Byfield and Abuknesha 1994).

Due to the short reproductive cycle, rapid metabolism, and easy genetic manipulation, microorganisms are therefore providing an alternative to construct biosensors. Microbes are similar to factories made up of enzymes and cofactors/coenzymes that give them the ability to react to a variety of chemicals. Even if its metabolism is non-specific, by induced desired or by blocking undesired metabolic pathways through selective culture conditions, we may achieve highly selective microbial biosensors (Su et al. 2011). In addition, the latest development of molecular biology could provide a new approach to construct genetically engineered microorganisms (Urgun-Demirtas et al. 2006). In the past decade, microbial fuel cell (MFC) and its derivative technology has gradually aroused people's interest. The MFC can not only recycle energy, but also provide value-added products from wastewater. As the research progresses, microbial fuel cells also develop an aspect of water quality monitoring. This type of sensor does not require additional sensors or power to achieve self-sustaining monitoring. Till now, MFC-based biosensors have been used to monitor biological oxygen demand (BOD), chemical oxygen demand (COD), dissolved oxygen (DO), heavy metals, volatile fatty acid (VFA), and organic compounds. This chapter aims to summarize the current research, main obstacles, and future research directions of MFC sensors. In addition, the manufacturing, operation and optimization of MFC sensor will also be discussed in this chapter.

The Mechanisms Governing MFC Used as Sensor

MFC sensor is a promising new electrochemical microbial sensor, unlike other kinds of electrochemical sensors, which applies electroactive microbes as probes. The target analyte could affect the electron transfer process of microbes, resulting in different degrees of electrical signals. The MFC sensor is not concerned with high power density output, but it is concerned with the change of electricity output under different environmental conditions, so it has a very broad market application prospect (Jiang et al. 2018). MFC sensors have been used in different fields, including water quality testing, anaerobic digestion monitoring, and pathogen detection (Wu et al. 2009, Kim and Han 2013, Kim et al. 2013).

An MFC sensor usually comprises two parts: cathode chamber and anode chamber, which are usually separated by proton exchange membrane (PEM) (Fig. 15.1 A). The role of PEM is to allow migration of protons and to prevent diffusion of oxygen. The electrochemical active microbes (EAM) in the anode oxidized the organic substrate from the water sample, and then the released electrons. The leaked electrons travel through the external circuit to the cathode (Logan et al. 2006). The electron transfer process occurring in the anode biofilm is variable and complex, involving many biochemical electrochemical reactions. The total set of these processes is named extracellular electron transfer (EET) (Hernandez and Newman 2001). Based on existing research, two mechanisms of EET have been presented. One is the mediated electron transfer (MET) and the other is direct electron transfer (DET) (Fig. 15.1 B).

MET pathway relies on the secreted extracellular molecules. In *Pseudomonas*, Phenazines has been demonstrated as intrinsic electron shuttle of EET pathway (Marsili et al. 2008). Flavin and riboflavin were also proved as the main electron shuttles in *S. oneidensis* (Pham et al. 2008, Wang et al. 2010). Although some exogenous redox agents have been introduced to MFC to facilitate electron transfer, these exogenous redox mediators required continuous addition and only achieve relatively low power density (Zheng et al. 2015, Doohyun and Zeikus 2000). Furthermore, these additional electron shuttles need to consider the possible environmental pollution problems. Physical contact between MFC anode surface and bacterial cell membrane is prerequisite for DET pathway. The electron transport proteins on the EAM membrane, including OmcZ, c-type cytochrome, and heme protein, can

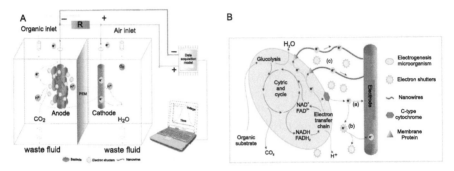

Figure 15.1. Schematic diagram of (A) MFC structure and (B) extracellular electron transfer mechanisms (Zhou et al. 2017).

transfer electrons from bacterial cells to outer membrane (OM), where the electron was transferred to electrode (Inoue et al. 2010, Lovley et al. 2009). Despite the lack of C-cell pigment in some dissimilated bacteria, they use the electrically conductive filamentous extracellular appendage, which is also known as bacterial nanowires, to transfer electron into anode instead (Reguera et al. 2005).

For BOD monitoring, due to the organic substrate used as the raw fuel for power generation, the organic matter concentration will not affect the power generation of the MFC sensor. In the case of toxicity monitoring, the presence of toxic components could inhibit or increase the activity of microbes (Zhou et al. 2017). In addition, highly selective MFC-based sensor can be achieved by applying synthetic biology or screening special strain that is sensitive to specific substrates (West et al. 2017, Chen et al. 2016). Compared to traditional biosensors, MFC sensor does not require additional sensors or power, because monitoring samples could provide the organic substrate for the anode microbe's reproduction and power generation. During operation of MFC sensors, it also does not require additional maintenance and management, which therefore reduces the associated costs.

Electrical Signal for BOD Quantification

In the ideal case, the organic substrate consumption of the anode biofilm is directly transferred to the electrons and was captured by the MFC cathode, which results in the standard linear relationship between the voltage and substrate concentration. However, since anodic matrix oxidation involves varying amount of biochemical and electrochemical reactions and may warp this relationship, such perfect proportion could not exist in actual situation. Considering the different influencing factors, Monod equation (Ortiz-Martínez et al. 2015) was used to describe the function between the organic substrate concentration and the BOD sensor electrical signal (1):

$$R_{ut} = q_{max} . X_f L_f . \frac{S}{K_{s.app} + S} \qquad (1)$$

In this equation, R_{ut} represents the substrate utilization rate, S is the substrate concentration, $K_{s,app}$ represents the apparent half saturation substrate concentration, q_{max} is the maximum rate of substrate utilization, X_f respresents the active biomass concentration, and L_f is the thickness of biofilm. According to the Monod equation, the linear correlation between the MFC coulomb yield and the BOD concentration is maintained only at low BOD concentration. When the BOD concentration reaches a high level, the current yield of MFC sensor will be saturated.

Except the current density, coulomb yield (CY) is also used to describe the MFC sensor reaction process. CY could provide a wider monitoring concentration range. Even if the reaction rate and corresponding current reached saturation state, the different concentrations of BOD would reflect as the length of the running time. However, it is hard to maintain a constant coulomb efficient when the BOD concentration.

Electric Signal Processing for MFC Toxicants Biosensor

In MFC, when the organic substrate is saturated, the toxicants could inhibit substrate consumption and microbial metabolic activity. Therefore, inhibition rates (IR) have been proposed to account for the effects of toxic substances entering MFC biosensor, which can be described by using the following equation (2) (Di et al. 2014):

$$IR(\%) = 100 \times (I_{nor} - I_{tox}) / I_{nor} \tag{2}$$

Among them, I_{nor} represents the current provided by the normal wastewater to the MFC, while I_{tox} is the current provided by the water sample containing toxic substances. The D-value between the two (ΔI) represents the degree of change in the current, and the sensitivity is calculated by the change in current per unit by toxicants. Equation (3) provides more detailed parameters to accurately characterize sensor sensitivity

$$\text{Sensetivity} = \frac{\Delta I}{A \cdot \Delta C} \tag{3}$$

A is the electrode surface area (cm^2), ΔC (mM) represents the unit change in the substrate concentration, and ΔI (μA) is the unit change in the current output. Some studies calculated the IR based on CY changes rather than current variation. In general, the former provides a change in the overall toxic effect of time, and the latter represents a variation in the rate of toxic reaction. In addition, three samples are commonly used as standard toxic substrates: iron (non-toxic metals), acetate (organic substrates), and chromium (acute toxins) (Jiang et al. 2018).

The Application and Advance of MFC-based Biosensor

At present, electrochemical-based water quality detection methods can only provide off-line detection, so it cannot immediately reflect the actual situation of water samples. To solve this problem, people have been trying to find more economical and reliable methods. Although some traditional biological sensors use electrical or optical signal to estimate the contamination degree of water body, these sensors require additional power and transducer support. By comparison, MFC sensors could provide a quick online measure by instantaneously directly converting the change of organics concentration into electrical signal. Besides, the construction of MFC sensor is relatively simple, even when considering signal acquisition and electronic demand, the MFC signal can be extended to the desired level via cheap electronics. Up till now, MFC biosensor is mainly used in three aspects. One is BOD or COD monitoring in wastewater, one is the detection of VFA content in anaerobic fermentation, and the other is monitoring of toxic substances, including heavy metals and organics. Some special MFC biosensors are also described in this section.

MFC Sensor for Water Quality Monitoring

Due to a significant increase in population and continuous strengthening of industrialization and civilization, a lot of industrial or domestic wastewater was discharged into surface water. These wastewaters usually contain a very high organic content, leading to the eutrophication of water body, and trigger a series of harmful consequences, most notably the outbreak of the algal bloom. As large numbers of breeding algae excessively consume the dissolved oxygen (DO), it usually leads to the death of fish or other aquatic organism. Chemical oxygen demand (COD) and biochemical oxygen demand (BOD) are two key parameters for water quality detection, which is widely used to estimate the water quality. Table 15.1 summarizes the detailed information about MFC-based BOD/COD sensor reviewed in this article.

MFC BOD sensor structure can be double chamber, or it can be just using an air cathode (wet ion exchange membrane against the surface of the anode chamber) and for constructing a single chamber. Its purpose is to make the system more compact and simple, and reduce operating costs. Compared to the two-compartment system, the air cathode MFC has a potential advantage because the cathode electrolyte does not need to be aerated, recycled, or chemically regenerated (Fan et al. 2007). Some single-chamber MFC sensors (SCMFC) adopt membrane-less design (Fan et al. 2007), so that it can help reduce the cost of sensors, and show a relatively shorter detection time and wider range under the low concentration of COD values. Despite the lack of cation exchange membrane, membrane-less design may be restricted to a relatively low current output and cationic gradient, but in terms of MFC sensor, this fault is not needed to be considered first. Figure 15.2 demonstrates the various MFC-based BOD biosensors.

In 1997, Karube and his colleagues first used MFC as the BOD sensor (Karube et al. 1977). In this device, clostridium butyrate bacteria were fixed on the Pt electrode as anode. Cathode (carbon electrode) was immersed in a buffer solution filled with nitrogen. As the strength of BOD increased, the

Table 15.1. Construction, function, and analytical performance of MFC BOD sensor

Items	Source inoculum	MFC type	Electrode material	Detection limit (mg L^{-1})	Response time (min)	Reference
BOD	Sludge suspension	DC	Graphite felt	150	ND	Kim et al. 2003
	MFC effluent	DC	ND	235	ND	Hsieh et al. 2015
	Activated sludge	DC	Graphite felt	100	60	Chang et al. 2004
	Mixture sludge	SC	Carbon fiber paper	1280	1200	Modin and Wilén 2012
	Livestock wastewater	SC	ND	ND.	ND	Hsieh et al. 2015
	MFC effluent	SC	Carbon fiber veil	ND	69-1008	Pasternak et al. 2017
	Domestic wastewater	Submersible	Carbon paper	ND	30-600	Peixoto et al. 2011
	Primary clarifier	Submersible	Carbon paper	250	10-40	Zhang and Angelidaki 2011
COD	Anaerobic sludge	MEA	Carbon cloth	650	79-420	Ayyaru and Dharmalingam 2014
	Marine sediment	DC	Graphite granules	4	ND	Quek, Liang Cheng, and Cord-Ruwisch 2015
	Activated sludge	MEA	ND	200	ND	Feng et al. 2013
	Anaerobic sludge	MEA	Carbon cloth	500	31-825	Lorenzo et al. 2009
Acetate	Marine sediment	DC	ND	27	ND	Quek, Liang Cheng, and Cord-Ruwisch 2015
	MFC effluent	DC	Carbon fiber brush	605	5-25	Jiang et al. 2017
	Marine sediment	DC	Graphite granules	5	<60	Cheng et al. 2014
	Secondary biofilms	SC	Gaphite rod	300	ND	Kretzschmar et al. 2016
	Anaerobic sludge	MEA	ND	ND	37-57	Chouler et al. 2017a

*ND: No data available in original work.
DC: Double chamber
SC: Single chamber
MEA: Membrane electrode assembly

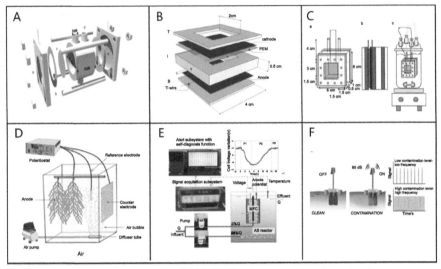

Figure 15.2. Schematic diagram of (A) single chamber MFC base BOD sensor (Fan et al. 2007), (B) 3-D printing MFC sensor (Di et al. 2014), (C) submerged MFC sensor (Peixoto et al. 2011), (D) open-type MFC sensor (Yamashita et al. 2016), (E) MFC as activated sludge process sensor (Xu et al. 2014), and (F) floating autonomous MFC sensor (Schievano et al. 2016).

hydrogen and formic acid produced by immobilized bacteria also increased, causing the current of the electrode to increase significantly with the concentration.

Subsequently, the authors found that the different feed rate of artificial wastewater (AW) will influence the BOD linear relationship (Karube et al. 2009). When MFC sensor is starving, the time of restoring the original current value varies according to the duration of starvation. During starvation period, MFC sensor still produces a background level of voltage, possibly due to endogenous metabolism.

Most BOD biosensors are used to monitor high BOD values in industrial wastewater; however, since secondary effluent and non-polluted rivers or lakes usually contain low concentrations of organic compounds, it still has great practical significance to detect low BOD values in water bodies. A method for measuring low BOD concentration is enriching low-concentration anode microorganisms. In a study, through the operation of the low BOD concentration wastewater, the enrichment of oligonucleotides nutritional microbial community was applied for the MFC sensor (Kang et al. 2004). The results showed the BOD detection limit is low at 1mg L^{-1}. This oligotrophic MFC sensor is described as highly stable and repeatable, but it may lose efficiency when the sensor suddenly suffers a high organic load or the high redox potential electron acceptor shock.

One study has shown that SCMFC sensors could provide higher output signals than double chamber MFC sensors (Fan et al. 2007). The smaller reactor (12.6 mL) could provide higher coulomb efficiency (56% efficiency versus 6% coulomb efficiency) than the larger reactors (50 mL) (Fig. 15.2 A). Moreover, the system only takes 40 minutes to achieve a stable current, in contrast to the four hours required for a large SCMFC biosensor. In terms of measuring range, SCMFC also showed superior performance. Compared to the double-compartment MFC sensor, its dynamic range increased about 133%.

The progress of science and technology obviously provides new vitality for the construction and assembly of sensor. In 2014, a SCMFC biosensor was constructed by using rapid prototyping and 3-D printing (Di et al. 2014). With the overall design, this 3-D printing MFC greatly improves the response rate of the sensor (Fig. 15.2 B). By using acetate as substrate, the MFC sensor response time is quickly reduced to 2.8 minutes. Furthermore, in the saturated concentration of acetic acid (COD > 164 ppm), this SCMFC sensor can quickly monitor the presence of cadmium with low detection limit (1 µg L^{-1}) and high sensitivity (0.2 µg L^{-1} cm^{-2}). When SCMFC supplies fresh water, the initial steady state current could recover only requiring 12 minutes.

Considering the demand for *in situ* detection, submerged MFC (SMFC)-based BOD sensor has been developed in recent years (Peixoto et al. 2011). As shown in Fig. 15.2 C, it is usually designed to be closed to create anaerobic conditions. This simple and compact SMFC sensor demonstrates great potential for direct and inexpensive BOD monitoring. The experimental results show in the condition of temperature for 22±2°C, hydraulic retention time in 1.53 mS cm^{-1}, and pH in 6.9, the current density is linear with BOD value ranging from 17 mg O$_2$ L^{-1} to 78 mg O$_2$ L^{-1}.

Besides, a study was also performed to explore whether the opening SMFC sensor could be used in aeration tank. The open anode was directly inserted in an intermittent aeration tank (Fig. 15.2 D) (Yamashita et al. 2016). Interestingly, this biosensor produced similar current levels both under aerated and non-aerated conditions. Anode community sequencing showed the anaerobic bacteria (*Geobacter* spp. et al.) could be detected in the covered anode. This may be because the suspended solids adhere and cover the entire anode, which leads to the generation of anaerobic conditions inside the coated anode.

Many cities lack a continuous supply of fresh water; therefore, reverse osmosis desalination has become a common technique. However, one of the problems of this technique is the pollution of the permeable membrane. The biggest factor of membrane fouling is the biofilm formed by marine bacteria on the permeable membrane. To solve this problem, MFC sensor is used to detect organic carbon concentration (AOC) to monitor the accumulation of biofilms. In seawater, the organic matter content is much lower than others, which makes the interference of DO more obvious when using MFC sensor to monitor AOC. Quek et al. therefore designed an AOC monitoring device integrated with a electrochemical cell and MFC (EC-MFC) (Quek, Cheng, and Cord-Ruwisch 2015). Electrochemical batteries were used to restore DO from seawater, and MFC was applied for AOC monitoring of anoxic seawater. At the 12.5 mL min^{-1} velocity, the sensor can detect 0 to 75 mg L^{-1} concentration of AOC. Although the long-term operation may cause the biomass accumulation in deoxygenation electrochemical cell and affect the AOC value of seawater, this simple device provides an economical and effective solution for continuous seawater AOC monitoring.

Another attractive idea is to use hexacyanoferrate-adapted microbes as anode probes (Liang et al. 2014). By adding potassium ferricyanide as a mediator in the anode to competitively acquire electrons, it can avoid the suspension of the MFC sensor due to the presence of DO, so the MFC sensor can detect AOC in seawater under the oxygen containing condition. As the potassium ferricyanide is oxidized back to trivalent potassium ferricyanide at the electrode, this sensor can continuously monitor the seawater AOC. However, given the toxicity of potassium ferricyanide and its price, this balance of extra expense and benefits will need to be considered in the practical application.**

SMFC was also applied in sludge (AS) process monitoring (Xu et al. 2014). The whole system included an early warning subsystem, a single chamber MFC, and signal acquisition subsystem (Fig. 15.2 E). Since there was no CEM (cation exchange membrane), this system avoided the potential harm of anode chamber pH drop to the anode microbial activity. The MFC voltage change is relatively stable under normal conditions and fluctuates within only 5% (0.41 V) in six months without any maintenance. Not only that, this sensor shows a high degree of sensitivity to different types of impacts.

In 2016, Italian researchers used two floating SCMFC sensors as self-powered water quality online sensors (Schievano et al. 2016). The sensor system incorporates new rectifier and data acquisition system and the actual testing was performed at the wastewater treatment plant (Milano-Nosedo, Italy) and the water ecosystem area (Bosco-in-citta). The study results show that the SCMFC performance and COD detection range in these two areas is different. These results provide practical operating experience for the future development of the MFC water quality sensor platform. Another study also developed a floating SMFC COD sensor (Schievano et al. 2016). It comprises an energy management system and four MFCs. When the urine concentration exceeds the appropriate limit (i.e., 57 mg O$_2$ L^{-1}), the sensor system will emit audible and visual alarms, and continue for two days till the organic load drops to 15.3 ± 1.9 mg O$_2$ L^{-1} (Fig. 15.2 F). This study also manifested the frequency of the sensor signal depends on the pollutant concentration, which can provide grading early warning for different levels of pollution events.

MFC as VFA Biosensor

Fossil fuels are non-renewable resources, and large amounts of fossil fuels cause severe environmental pollution. Energy prices are also becoming more expensive. Hence, we urgently need a new type of clean energy to replace the former energy. Biogas is considered an ideal alternative energy and has good application prospect (Maie et al. 2017). Degradation of biomass and fermentation can produce biosynthetic gas (Deublein and Steinhauser 2011). With the large-scale application of the fermentation industry and anaerobic digestion, there is a need for an efficient and convenient method to evaluate and control the process. Volatile fatty acids (VFAs) are intermediately produced during anaerobic digestion. When the fermentation process becomes unstable, the concentration of VFAs will change significantly. Therefore, VFAs can be used as a parameter to evaluate the stability of anaerobic digestion process (Falk et al. 2015). The traditional method for measuring VFAs concentration are HPLC and GS (Raposo et al. 2013). However, these methods need to consume more labor and cannot be monitored online.

For continuous, on-line monitoring, Kaur et al. (2013) used an MFC to monitor the concentration of VFAs. They explored the correlation between VFA concentration and voltage response, and evaluated the electrochemical performance of MFC. The results demonstrate that there is a clear linear relationship between the specific VFA concentration and the voltage within a certain range. Subsequently, based on the microbial desalination cell (MDC), Zhang et al. improved MFC-based VFA sensor by introducing the third chamber (Jin et al. 2015). An anion exchange membrane (AEM) segregates the middle chamber from the anode chamber (Fig. 15.3 A). The AEM passes through the ionized VFA and sequesters other complex organic materials, such as proteins and lipids in the intermediate chamber. As a result, the current is only related to the VFA concentration in the intermediate chamber. This innovation improves the detection range and accuracy.

To simplify the sensor architecture and reduce the sensor response time, microbial electrolysis cell (MEC) was also used to monitor VFA concentrations (Jin et al. 2017). The microbial electrolytic cell consists of two chambers and an external power source (Fig. 15.3 B). External power was supplied to accelerate the VFA from the cathode chamber through the AEM

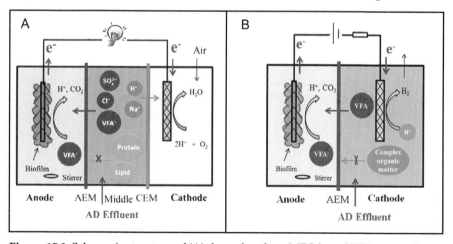

Figure 15.3. Schematic structure of (A) three chambers MFC-based VFA sensor (Jin et al. 2015) and (B) MEC-based VFA biosensor (Jin et al. 2017).

to the anode chamber. Besides, Schievano et al. (2017) created a SCMFC-based VFA biosensor. In the air cathode filmless MFC, the electrical signal moves instantaneously from the stability value with the accumulation of the VFAs on anode. The MFC can serve as an early warning sensor for anaerobic digestion system changes. However, only a few MFC-based VFA biosensors have been reported. There is still a great need to explore the performance of this type of biosensor, especially in long term operation and online monitoring.

MFC as Toxicants Biosensor

Monitoring different poisons in domestic or industrial wastewater online is a necessary condition for water resource recycling and public health safety. MFC sensor provides long-term stable and low maintenance solution, which is because toxic substances may influence the activity of electrogenesis microbes in the biofilm, resulting in sudden changes in voltage (decrease or increase). Figure 15.4 shows the principle of MFC-based toxicant biosensor.

MFC sensors have been applied to detect various substances, including toxic organic compounds, heavy metals, and other components that affect biofilm, such as microbe's activity, wastewater acidity, or ammonia concentration.

MFC sensor for heavy metal monitoring

Heavy metal is a widely distributed source of pollution, which could lead to a series of tissues and organs damage. Aim to solve this problem, MFC has been used to explore the detection of heavy metals in actual wastewater. Table 15.2 provides a detailed content of MFC-based heavy metals biosensor.

In 2007, Kim et al. used a two-compartment MFC to detect diazinon, lead, mercury, and polychlorinated biphenyls (Mia et al. 2007). The results demonstrate that the inhibition rate of the MFC power output was 61%, 46%, 28%, and 38%, respectively. In addition, when actual wastewater was applied to the MFC as the substrate, the author observed a higher inhibition rate than the lab test after importing the toxicants. This may be due to the interaction between complex composition of the actual wastewater and the toxic substances, which finally contributes to a more serious toxic effect. Figure 15.5 shows different MFC-based heavy metal biosensors.

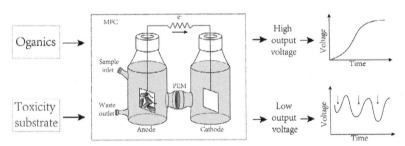

Figure 15.4. MFC used as a toxicity biosensor (Zhou et al. 2017).

Table 15.2. Characteristics summary of MFC-based heavy metals sensor

Items	Source inoculum	MFC type	Electrode material	Detection limit (mg L^{-1})	Reference
Ni(I)	MFC effluent	Membrane	Carbon ink	20	Xu et al. 2015
	MFC effluent	DC	ND	13.2	Stein et al. 2012a
Pb(II)	ND	DC	Carbon cloth	0.1	Kim et al. 2006
Cd(II)	MFC effluent	MEA	Carbon cloth	0.001	Di et al. 2014
Hg(II)	Activated sludge	DC	ND	1	Mia et al. 2007
Cu(II)	MFC effluent	DC	Flat graphite	85	Stein et al. 2010
	Primary wastewater (biofilm)	SC	ND	6	Patil et al. 2010
	MFC effluent	DC	Graphite felt	2	Jiang et al. 2015
	Domestic wastewater	MEA	Carbon cloth	5	Shen et al. 2013
Fe(II)	Natural mud	DC	Graphite granules	167.5	Tran et al. 2015
Fe(III)	*Geobacter sulfurreducens*	SC	Graphite rod	2.8	Li et al. 2016
	Domestic wastewater	SC	Carbon cloth	8	Liu et al. 2014
As(III)	*Enterobacter cloacae*	DC	Carbon felt	0.33	Michelle and Minteer 2015
	Mutant *S. oneidensis*	SC	Graphite rod	3	Webster et al. 2014
As(V)	*E. cloacae*	DC	Carbon felt	3.45	Michelle and Minteer 2015
Cr(VI)	MFC effluent	Membrane	Carbon ink	10	Xu et al. 2015
	Domestic wastewater	SC	Carbon cloth	1	Liu et al. 2014

* ND: No data available in original work.
DC: Double chamber
SC: Single chamber
MEA: Membrane electrode assembly
Membrane: membrane-based electrodes

Subsequently, an MFC was studied for detecting the soil Cu(II) concentration (Deng et al. 2015) In this experiment, 5% glucose was added in soil samples and then was inoculated to the anode chamber (Fig. 15.5 A). The results show that with the increase in Cu(II) concentration, the output voltage of the MFC sensor is restrained and delayed. 16S DNA sequencing of the post-operative anode biofilm revealed that the Firmicutes and bacillus were increased with the Cu(II) concentration, but the total bacterial abundance gradually decreased with the Cu(II) concentration.

Environmental enrichment method can be used to obtain anodic biofilms adapted to specific pollutants (Tran et al. 2015). Tran et al. enriched iron oxide bacteria by using ferrous iron as the only carbon source. The iron oxide bacteria group was used to monitor ferrous and manganese. As the concentration of iron and manganese gradually increased, these bacteria used iron or manganese as the sole electron donor to enhance MFC production, and their detection ranges were 3-20 mM and 0-3 mM, respectively. It can be noticed that the detection limit of manganese is much lower than iron's, which may be due to the fact that manganese has a greater inhibitory effect on bacteria than ferrous ions. The presence of organic compounds in the anode (if not excessive) will not affect Fe(II) oxidation. Although the presence of other metals (Ni(II), Pb(II), etc.) influences the current, MFC still has a significant specificity for the iron response.

Another experiment uses sediment MFC to monitor heavy metal ions (Fig. 15.5 B). It provides a relatively simple device and just needs a glass beaker as MFC container (Zhao et al. 2018). The anode is embedded in sediments and the cathode is placed above the anode, and part of it is full of water. The sludge from the sewage treatment plant is directly cultivated as an anode inoculation. It not only simplifies the anode biofilm inoculate steps, but also can enhance the sensor system stability due to the diversity of sludge flora. When Cr(VI) concentration is between 0.2-0.7 mg L^{-1}, the MFC sensor can obtain a stable linear voltage relationship. In addition, the authors sequenced the sludge community after and before inoculation Cr(VI). The results showed that in the sediment MFC added by Cr(VI), the *Proteobacteria* was about 14% higher than control, which may be due to the fact that most electrogenesis microorganisms, such as *Geobacter* and *Pseudomonas*, belong to this Phylum. At genus level, *Flauobaterium* (32%) increased specially, possibly because other species are unable to tolerate the toxicity of Cr(VI), which made the *Flauobaterium* gradually become the main bacteria.

In the MFC-based heavy metals biosensor, Liu et al. developed a modular SCMFC sensor (Fig. 15.5 C). The cathode and anode are placed on both sides of a cube, allowing the MFC to be assembled freely and easily as a building block. Four toxic substances (chromium, iron, nitrate, and sodium acetate) were tested in this SCMFC sensor (Liu et al. 2014). These four toxic substances represent acute, chronic metal toxicity, common nutrients, and organic pollutants, respectively. The results demonstrated 8 mg L^{-1} Cr(VI) inhibited electrochemical activity of electrogenesis microorganism, while until 48 mg L^{-1}, Fe(III) could inhibit the bacteria activity. In addition, NO_3^-

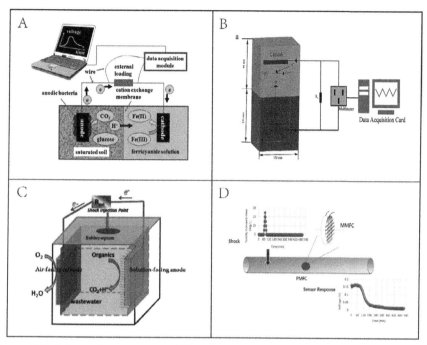

Figure 15.5. Schematic structure of (A) soil heavy metal MFC biosensor (Deng et al. 2015), (B) sediment MFC (Zhao et al. 2018), (C) SCMFC as the shock sensor (Liu et al. 2014), and (D) PMFC as heavy metal biosensor (Xu et al. 2015).

shock resulted in a slow decline in MFC voltage, while the shock of acetate caused a rapid increase in the voltage of CMFCs. These experimental results show the stability and practicability of this MFC sensor structure.

In term of structure improvement, it is particularly notable that Xu et al. (2015) developed the flat microsoft-based MFC. This MFC sensor is very compact and consists of only two sticker filters (Fig. 15.5 D). To be specific, the authors use a brush to coat the membrane with carbon ink lines to effectively transfer the biofilm generated electrons. The cathode film is added with the platinum layer above carbon ink, while the anode film is coated with ordinary carbon ink. Next, stack the anode and the cathode layer together and the side coated with the carbon ink is outwards. Copper wires are glued to the top of the membrane for connecting external resistance. The high microporosity and hydrophilicity of the membrane provide the MFC sensor with only a few hours of adjustment period. In addition, this device not only has the simple and compact configuration of the microliter size, but also has high sensitivity and stability. During the month-long running time, the signal is very stable before the heavy metal shock, and the electric signal generated by the toxicant shock is obviously higher than the noise signal.

Although most MFC sensors have achieved success in the laboratory and actual water samples testing, but in general, the MFC sensor is vulnerable to changes in the bacterial community, and has great flexibility in the substrate specificity. Most of the time, it is difficult to distinguish between different substrates. It brought difficulty for the actual water monitoring, because the wastewater always contains a variety of pollutants. The content below does not strictly belong to the MFC sensor, but it provides a practical case for the future development of MFC-based specific heavy metals sensor. By using genetically engineered *shewanella*, Webster et al. (2014) has developed a biological electrochemical system for arsenic detection. They constructed *shewanella* electrogenesis relative gene *mtrB* in front of the arsenic response promoter *aesR*, making engineering *shewanella* specific to monitor arsenic ion concentration in the water samples.

MFC-based organic toxin biosensor

Organic toxicants are common contaminants that often pose a threat to public safety and cause eutrophication. Table 15.3 lists the performances and characteristics of MFC-based organic toxin biosensors.

In 2010, Dávila et al. (2011) used silicon plates to design a small device for formaldehyde testing. It consists of a PEM located between two miniature silicon plates (Fig. 15.6 A). The volume of each chamber is only 144 µL. The results show the maximum power density of this sensor can reach 6.5 µW cm^{-2}, which is much higher than macro size MFC. This miniature MFC sensor can be further used in the manufacture of toxicity monitoring equipment.

Another particularly intriguing experiment was using MFC sensors to detect the β-lactamide antibiotics (Fig. 15.6 B). In most cases, antibiotic therapy is experiential because antibioticogram test could take a long time. Schneider et al. used printed circuit board (PCB) technology to construct a fairly compact MFC system and test the effects of β-lactam antibiotics on two common pathogens (*Staphylococcus aureus* and *Escherichia coli*) (Schneider et al. 2015). This method is more visualized and faster than traditional paper diffusion method. The sensitivity of the tested bacterial strains can be revealed within four hours. Besides, the experimental results show that the antibacterial effect of this MFC sensor is consistent with the paper diffusion method.

Artificial neural network (ANN) is a network of computational models similar to natural neurons as it has the ability to learn like the human brain (Yao 2002). ANN maps the correlation between output and input data, and can adapt to changing behaviors. These models associate output and input data with a series of hidden layers, training with real data, and then testing against individual subsets of data. In 2014, King et al. (2014) detected three organic contaminants by using ANN and the MFC coupling system, including aldicarb, dimethyl methyl phosphonate (DMMP), and bisphenol A (BPA). Overall, ANN system has been proved more reliable than directly observing the types and concentrations of chemical substances. Through multiple learning MFC signal changes caused by different pollutant inputs, the ANN system can accurately capture the signal characteristics of these three different contaminants. In addition, although acetate can mask the

Table 15.3. Function and performance of MFC-based organic toxin sensors

Items	Source inoculum	MFC type	Electrode material	Detection limit (mg L^{-1})	Reference
Bentazon	MFC effluent	DC	Graphite plate	1	Stein et al. 2012b
Nitrate	Domestic wastewater	SC	Carbon cloth	48	Liu et al. 2014
Cyanide	Primary wastewater	SC	Carbon felt	10.4	Ahn and Schröder 2015
Azide	Primary wastewater	SC	Carbon felt	1.3	Ahn and Schröder 2015
Imidazole	Primary wastewater	SC	Carbon felt	1.4	Ahn and Schröder 2015
p-nitrophenol	*Pseudomonas monteilii LZU-3*	DC	Carbon felt	9	Chen et al. 2016
Ticarcillin	*E. coli / Staphylococcus aureus*	DC	Graphite	75	Schneider et al. 2015
Tobramycin	MFC effluent	SC	Carbon cloth	100	Wu et al. 2014
Levofloxacin	ND	SC	Carbon cloth	0.0001	Zeng et al. 2017
Acidic	MFC effluent	DC	Graphite felt	pH 6	Jiang et al. 2017
SDS	MFC effluent	DC	Graphite plate	10	Stein et al. 2012a
	MFC effluent	DC	Graphite plate	50	Stein et al. 2012c
Formaldehyde	*G. sulfurreducens*	DC	Graphite felt	0.10%	Dávila et al. 2011
	G. sulfurreducens	DC	Graphite rod	0.01%	Li et al. 2016
	Shewanella oneidensis	SC	Graphite rod	0.01%	Wang et al. 2013

$_{\circ}$ND: No data available in original work.
DC: Double chamber
SC: Single chamber

influence of the contaminants on the current, the chemical substances were also accurately determined by ANN system.

For specific substrate detection, Chen et al. (2016) took the lead in applying MFC sensor to specific substrate detection. They screened strains (*P. mongolicus* LZU-3) that specifically degraded PNP (*p*-nitrophenol). LZU-3 can use PNP as the sole carbon source. As the PNP concentration (15 mg L^{-1} to 44 mg L^{-1}) gradually increases, the MFC output voltage also becomes relatively enhanced (120 mV to 160 mV). In addition, when sensor system is used to measure PNP in wastewater containing different metal ions and other aromatic compounds, the sensor capability is not influenced. They also developed a portable sensor system that consists of a suitcase-sized box, including a modular MFC and sensing circuitry for future commercial use (Fig. 15.6 C).

In order to make sure everyone on the planet has access to safe drinking water, testing on-site water supply devices should be low cost and simple. MFC sensor offers this opportunity, but its relatively expensive materials and inconvenient assembly methods prevent it from becoming popular. Paper electronic technology provides a new way of thinking for acquiring a simple MFC sensor. Earlier research had developed paper-based multi-anode microbial fuel cells (PMMFC) (Xu et al. 2016). Nevertheless, such a

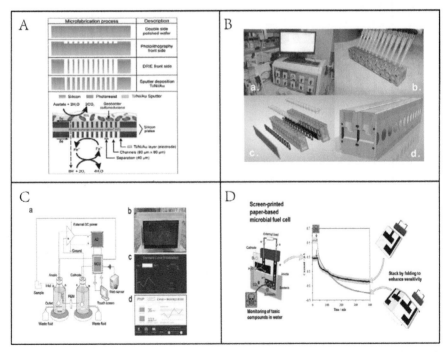

Figure 15.6. Diagram of (A) silicon plate-based MFC biosensor (Dávila et al. 2011), (B) MFC-based β-lactamide antibiotics biosensor (Schneider et al. 2015), (C) portable MFC biosensor for specific substrate detection (Chen et al. 2016) and (D) paper-based MFC sensor (Chouler et al. 2017b).

design is still not suitable for carrying; it only simplifies the construction of the anode, so it is very inconvenient in practical applications. Recently, a completely paper-based MFC sensor has been developed (Chouler et al. 2017b).

It also uses PCB technology to fabricate the sensor's cathode and anode. The device has only a 4 cm paper blade and an external resistor (Fig. 15.6 D). By connecting two back-to-back PMMFCs in parallel, the performance of the sensor can be further improved. This conceptual innovation allows the MFC sensor to perform water quality testing in most locations and is affordable for everyone. 0.1% formaldehyde was used to simulate organic contaminants for testing. The results show that this paper-based MFC is sensitive enough.

Other Applications of MFC Sensor

MFC have been used to monitor microbial activity since the metabolic activity of microorganisms could affect the power output of MFC sensors (Tront et al. 2008). Similarly, it's the same for evaluating the degree of acidification (Shen et al. 2012). Different pH will affect the electrochemical activity of the microbial community, and thus we can evaluate the pH value in water sample. Some studies have also explored the application of MFC sensor for the detection of the number of *E. coli* colonies (Kim and Han 2013). The results show that MFC can quickly estimate the number of *E.coli* cells. However, these sensors require a relatively stable single substrate, making the MFC sensor unaffected by other factors.

MFC sensors were also used to detect photosynthetic metabolism (Figueredo et al. 2015). Photosynthetic microorganisms were inoculated into the anode. As photosynthesis occurs and extinguishes, the MFC sensor signal also periodically rises and falls with alternating light and dark.

In addition, MFC is used to detect target DNA in serum samples (Asghary et al. 2016). In this MFC system, an Au nanoparticle-modified graphite electrode was used as the cathode. The ssRNA probe (classic tumor suppressor gene P53) can be incorporated on the surface of the AuNP/ graphite electrode. When there is a corresponding target cell in the serum, the voltage output of the MFC is weakened due to the DNA-interacting formation of DsRNA, which is used for rapid detection of the target DNA.

Performance Parameters and Improvement of MFC-based Biosensor

The performance parameters of MFC sensor include detection range, sensitivity, response time, etc. These parameters could be influenced by various factors, such as DO concentration, external resistance, or outflow amount of wastewater sample. To remain a stable and correct signal output of MFC sensor, it is essential to eliminate these adverse effects. In monitoring of industrial wastewater, high redox potential electron acceptors (nitrate or oxygen), could dramatically reduce the anode signal from MFC

BOD sensor (Chang et al. 2005). Chang et al. suggest eliminating these two electron acceptors by using the respiratory inhibitors azide and cyanide. The experimental results expressed that the respiratory inhibitors can improve MFC performance under nitrates and oxygen-enriched conditions.

A study has focused on the improvement of ion-exchange membranes. Since Nafion® also transports cationic substances other than protons, in MFC sensor, the concentration of other cationic substances is usually 10 times than the concentration of protons (Chang et al. 2005). In addition, because electrons are depleted in the anode reaction, the transport of cationic will lead to lower pH and reduce the sensor sensitivity (Rozendal et al. 2008). The experiment of Ayyaru and Dharmalingam (2014) showed that using sulfonated polyether ketone membrane instead of Nafion® could enhance the responsiveness of MFC-based BOD sensor. The detection limit of this MFC-based BOD biosensor can reach 650 mg L^{-1}, compared to the 400 mg L^{-1} BOD concentration using Nafion® membrane. In addition, the sulfonated polyether ether ketone also showed less oxygen permeability, thereby reducing the MFC sensor resistance (39 Ω relative to 67 Ω) and the substrate consumption rate.

Anode modification could also improve the performance of MFC sensor. Six different polymers or electropolymers were tested as the MFC sensor polymer (Kaur et al. 2014). The results demonstrate poly (pyrrole alkyl ammonium) could provide better repeatability and shorter signal response and enhance start-up speed of MFC sensor.

The composition of the anode solution obviously has different effects on the MFC sensor sensitivity. Hsieh et al. (2015) explored the influence of various substances on MFC BOD sensor. The results show that monosaccharide is excellent fuel for MFC sensor while methionine, phenylalanine, and ethanol are inferior fuel for MFC. In addition, the ions in the wastewater will not remarkably influence the capability of MFC sensor.

The response time of MFC sensor is described as the duration of the steady signal output. Some of the MFC sensors were operated under the injection impact mode. The sudden shock caused the signal of MFC exhibit as current peak instead of the steady state signal. Therefore, Jiang et al. (2018) suggested that in the impact mode, the time period between the sample injection and peak height should be used to calculate the response time.

To shorten the response time, the researchers dedicated to adjust the structure of the MFC, external circuit, hydraulic retention time (HRT), and anode initial status of biological membrane (Moon et al. 2004). They proposed MFC fuel supply should be kept at 0.53 mL min^{-1}, which would give the shortest response time. This experiment also demonstrates that when MFC anode volume decreases from 25 mL to 5 mL, response time can significantly reduce to 5 minutes from 36 minutes, which reveals miniature anode volume can help improve the response speed of the MFC sensor. External resistance is also a factor affecting the response time of MFC water quality sensor. While external resistance decreases from 100 Ω to 10 Ω, it can noticeably reduce sensor response time to 1.4 hours from 2.1 (Moon et al. 2004); however, although small external resistor can cut short the response time, it is hard to acquire sensor signals when external resistance is at a very low level.

Although in actual field applications, MFC biosensor has been proved to be a fast and practical method for on-line monitoring, it still needs to improve its detection rate and accuracy. To achieve this goal, several methods have been presented, such as reducing the internal resistance of the sensor, preventing excessive pH gradient and oxygen leakage, and using more efficient cathode receptors. Enhance the actual performance of MFC sensor need to further understand the MFC structure and material improvement. Compared with the traditional MFC, the introduction of MEA in MFC structure enable to increase sensor sensitivity and reduce the difficulty of MFC assembly (Zhang and Angelidaki 2012). MEA comprises the electrodes and the membrane. It uses high pressure and high temperature to combine catalytic electrode, membrane, and air cathode together to minimize overpotential of gas diffusion and metal load, implement effective heat and water management, and extend the life of the MFC sensor (Lefebvre et al. 2011).

Another area to improve MFC sensors is selection of sustainable and affordable cathode catalysts. So far, most of MFC sensors reported in the literatures used expensive platinum as cathode catalyst. Considering corrosion degradation and catalytic activity over time lead to high costs and low durability of platinum, Kharkwal et al. (2017) used manganese dioxide (MnO_2) as air cathode for oxygen reduction. The results reveal that the best performance of the sensor can be achieved by using the β-MnO_2 as catalyst, and the R^2 value is 0.99 in sodium acetate solution and 0.98 in real life sewage, respectively.

Although MFC sensor has great self-sustainability potential, changes in the substrate and concentration in the aqueous environment will influence the metabolic activity of electrogenic microbes. The main performance focus is on its sensitivity, repeatability, and recovery capability. The most important parameters that affect MFC sensitivity is extracellular polymeric material (EPS) levels, biofilm porosity, and density (Shen et al. 2013). One study showed that applying a nitrogen jet to a biofilm at a certain flow rate could enhance the MFC sensor's response to Cu(II). That is because at low water retention rates, the injected nitrogen reduces the biofilm's EPS content and the pores of the biofilm. Different configurations and control methods also affect the sensitivity of the MFC sensor. Compared to flow-by anodes, using flow-through anodes can increase MFC sensitivity by approximately 14 to 15 times. At the same time, the constant external resistance (ER) mode shows lower sensitivity than the controlled anode potential (CP) mode (Jiang et al. 2015).

A study has shown that using MFC biosensor as sensing elements is much more sensitive than using bioanode as sensing elements (Shen et al. 2013). The biocathode can detect formaldehyde at a low detection limit

(0.0005%), while the bioanode is more suitable for higher concentrations (> 0.0025%). As the DO and conductivity increase, the biocathode base MFC sensor could respond faster. In addition, the biocathode has higher response intensity than the bioanode in the monitoring of the combined organic toxic impact.

Using micro MFC can improve the performance of biosensors. The miniaturized structure shortens the response time and accelerates the attachment of cells to the anode electrode. However, microbubbles in a narrow room often interfere with the sensitivity of the micro MFC sensor. Yang et al. (2016) therefore designed a new type of miniature MFC that includes a bubble trap so that undesired bubbles can be trapped by this trap, thereby providing the MFC with a stable anode potential. In fact, in addition to reducing costs and increasing sensitivity, the miniaturization of the MFC also reduces the analyte concentration difference between the input and the biofilm, thereby making the output signal of the sensor more reliable.

For MFC-based toxicant sensors, the rapid and complete recovery of electrobiological microbial activity is important. Ideally, the inhibition of microbial activity should be completely reversible. Actually, this reduction could be reversible or irreversible. In order to assess the integrity of the recovery, the term degree of recovery is presented, which is described as the percentage of signal output after the recovery process. The degree of recovery is affected by anode biofilm, operation of the sensor, and the toxicant type and concentration (Jiang et al. 2018). In general, too high concentrations of toxicants will slow the recovery rate of MFC signal or directly inhibit the activity of electrogenic microorganisms. Jiang et al. showed that ER mode controlled MFC sensor has faster recovery compared to CP controlled mode, which is because the ER mode has less interference with electrogenesis microbes than the previous two modes, allowing it to reconstruct the anode potential and current (Jiang et al. 2015) .

In addition to the overall toxicity, some MFC sensors have been developed to monitor specific toxicants. This requires the MFC sensor responding specifically to certain substrates, which is generally much more difficult. Some researchers use computer and MFC sensor coupling systems to distinguish between different substrates. For example, the ANN system is used to train to distinguish three different substrates (King et al. 2014). In addition, MFC sensors monitoring specific substrates can be further realized by genetically engineered microorganisms in either specific or pure culture.

Challenges and Prospects

In the past decade, MFC has been extensively studied as the rapid water quality monitoring biosensor. As mentioned above, MFC sensor provides alternative options for detecting multiple substances. Nevertheless, there are some crucial challenges that limit their practical applications. Future improvements to MFC sensors should include reducing response times, using more economical materials, and developing uniform standards. The MFC design should focus on reducing the internal resistance. In addition, it is also important to screen for new anode microorganisms or microbial flora with efficient substrate utilization. The voltage generation of a single MFC may be limited to actual appliance. By stacking MFCs appropriately, we can obtain the required power output.

The electrical signal output of the MFC sensor depends on the water quality parameters, which include pH, BOD, conductivity, temperature, toxic agents, and electron acceptors concentration. Although a lot of studies have investigated the MFC biosensor performance in the laboratory, long-term operation may change the system parameters, so the application of MFC in a real environment is still necessary. For hybrid microbial communities, understanding its composition and dynamics under different substrates can provide theoretical support for sensor detection.

In addition, MFC biosensors should be simplified and portable for easy maintenance and manufacture. 3-D printing devices can provide an effective design method. From an economic point of view, expensive cathode catalysts and PEM should be altered by other more cheap materials. In addition, as MFC's electron transport mechanism is gradually elucidated, genetically engineered bacteria can be used as anode catalysts to effectively monitor specific substrates. This is one of the promising methods for improving the selectivity of MFC sensors.

Currently, there have been many studies aimed to improve the capability of MFC sensor. However, these efforts only focused on a part of the sensor. It must be pointed out that sensor reactor is a system; some of the performance may not be directly affected by other components. Therefore, when designing MFC biosensors, we should adopt an overall strategic approach. For the early warning system of the MFC sensor, a perfect detection algorithm has not yet been established. Therefore, in the future, we need to develop scalable data-driven estimation models and related detection algorithms to improve the applications of MFC sensor in various fields.

Conclusion

This chapter summarizes the MFC-based biosensor in contaminant monitoring systems. It has become potential alternative tools to rapid detect various substrates. In order to further develop MFC sensor, we should focus on how to improve the stability and sensitivity of MFC-based biosensor. These two aspects are major obstacles for the practical application. Laboratory and field tests show that MFC sensor has the potential to become automatic sensing devices. With an integrated remote monitoring system, MFC sensors can provide an affordable solution for remote monitoring water quality in the future. For specific detection, MFC sensors are also faced with many problems. Further research should explore effective kinetic models or use computers to analyze signals from different substrates to distinguish specific substrates.

REFERENCES

Abrevaya, X.C., N.J. Sacco, M.C. Bonetto and A. Hildingohlsson. 2015. Analytical applications of microbial fuel cells. Part II: Toxicity, microbial activity and quantification, single analyte detection and other uses. Biosensors 63: 591-601.

Ahn, Yoomin and Uwe Schröder. 2015. Microfabricated, continuous-flow, microbial three-electrode cell for potential toxicity detection. BioChip J. 9(1): 27-34.

Ajani, Penelope, Steve Brett, Martin Krogh, Peter Scanes, Grant Webster and Leanne Armand. 2013. The risk of harmful algal blooms (HABs) in the oyster-growing estuaries of New South Wales, Australia. Environ. Monit. Assess. 185(6): 5295-5316.

Asghary, M., J.B. Raoof, M. Rahimnejad and R. Ojani. 2016. A novel self-powered and sensitive label-free DNA biosensor in microbial fuel cell. Biosensors 82: 173-176.

Ayyaru, S. and S. Dharmalingam. 2014. Enhanced response of microbial fuel cell using sulfonated poly ether ether ketone membrane as a biochemical oxygen demand sensor. Anal. Chim. Acta. 818(818): 15-22.

Berthou, F. and M.P. Friocourt. 1982. Combination of high-performance chromatographic methods for the analysis of aromatic hydrocabon pollutants in marine biota. Anal. Tech. Environ. Chem. 221-230.

Byfield, M.P. and R.A. Abuknesha. 1994. Biochemical aspects of biosensors. Biosensors 9(4-5): 373.

Chang, I.S., H. Moon, J.K. Jang and B.H. Kim. 2005. Improvement of a microbial fuel cell performance as a BOD sensor using respiratory inhibitors. Biosensors 20(9): 1856-1859.

Chang, In Seop, Jae Kyung Jang, Geun Cheol Gil, Mia Kim, Hyung Joo Kim, Byung Won Cho and Byung Hong Kim. 2004. Continuous determination of biochemical oxygen demand using microbial fuel cell type biosensor. Biosensors 19(6): 607-613.

Chen, Z., Y. Niu, S. Zhao, A. Khan, Z. Ling, Y. Chen, P. Liu and X. Li. 2016. A novel biosensor for P-nitrophenol based on an aerobic anode microbial fuel cell. Biosensors 85: 860.

Cheng, Liang, Soon Bee Quek and Ralf Cord-Ruwisch. 2014. Hexacyanoferrate adapted biofilm enables the development of a microbial fuel cell biosensor to detect trace levels of assimilable organic carbon (AOC) in oxygenated seawater. Biotechnol. Bioeng. 111(12): 2412.

Chouler, Jon, Isobel Bentley, Flavia Vaz, Annabel O'Fee, Petra J. Cameron and Mirella Di Lorenzo. 2017a. Exploring the use of cost-effective membrane materials for microbial fuel cell based sensors. Electrochim. Acta. 231: 319-326.

Chouler, Jon, Aacute, Lvaro Cruz-Izquierdo, Saravanan Rengaraj, Janet L. Scott and Mirella Di Lorenzo. 2017b. A screen-printed paper microbial fuel cell biosensor for detection of toxic compounds in water. Biosensors 102: 49.

Dávila, D., J.P. Esquivel, N. Sabaté and J. Mas. 2011. Silicon-based microfabricated microbial fuel cell toxicity sensor. Biosensors 26(5): 2426-2430.

Deng, H., Y.B. Jiang, Y.W. Zhou, K. Shen and W.H. Zhong. 2015. Using electrical signals of microbial fuel cells to detect copper stress on soil microorganisms. Eur. J. Soil Sci. 66(2): 369-377.

Deublein, Dieter and Angelika Steinhauser. 2011. Biogas from waste and renewable resources: an introduction. Wiley-VCH.

Di, Lorenzo M., A.R. Thomson, K. Schneider, P.J. Cameron and I. Ieropoulos. 2014. A small-scale air-cathode microbial fuel cell for on-line monitoring of water quality. Biosensors 62(62C): 182-188.

Doohyun, Park and J.G. Zeikus. 2000. Electricity generation in microbial fuel cells using neutral red as an electronophore. Appl. Environ. Microbiol. 66(4): 1292-1297.

Driscoll, J.N. and A.W. Berger. 1971. Improved chemical methods for sampling and analysis of gaseous pollutants from the combustion of fossil fuels. Vol. I. Sulfur oxides. Final Report.

Falk, H.M., P. Reichling, C. Andersen and R. Benz. 2015. Online monitoring of concentration and dynamics of volatile fatty acids in anaerobic digestion processes with mid-infrared spectroscopy. Bioprocess Biosyst. Eng. 38(2): 237-249.

Fan, Yanzhen, Hongqiang Hu and Hong Liu. 2007. Enhanced coulombic efficiency and power density of air-cathode microbial fuel cells with an improved cell configuration. J. Energy Power Sources 171(2): 348-354.

Feng, Y.H., O. Kayode and W.F. Harper Jr. 2013. Using microbial fuel cell output metrics and nonlinear modeling techniques for smart biosensing. Sci. Total Environ. 449(2): 223-228.

Figueredo, F., E. Cortón and X.C. Abrevaya. 2015. In situ search for extraterrestrial life: a microbial fuel cell-based sensor for the detection of photosynthetic metabolism. Astrobio. 15(9): 717.

Filippelli, Gabriel M. 2008. The global phosphorus cycle: past, present, and future. Elements 4(2): 89-95.

Hernandez, M.E. and D.K. Newman. 2001. Extracellular electron transfer. Cell Mol Life Sci 58(11): 1562-1571.

Howarth, Robert W. 2009. Coastal nitrogen pollution: a review of sources and trends globally and regionally. Harmful Algae 8(1): 14-20.

Hsieh, Min Chi, Chiu Yu Cheng, Man Hai Liu and Ying Chien Chung. 2015. Effects of operating parameters on measurements of biochemical oxygen demand using a mediatorless microbial fuel cell biosensor. Sensors 16(1): 35.

Inoue, K., X.L. Qian, L. Morgado, B.C. Kim, T. Mester, M. Izallalen, C.A. Salgueiro and D.R. Lovley. 2010. Purification and characterization of OmcZ, an outer-surface, octaheme c-type cytochrome essential for optimal current production by geobacter sulfurreducens. Appl. Environ. Microbiol. 76(12): 3999-4007.

Jiang, Y., P. Liang, C. Zhang, Y. Bian, X. Yang, X. Huang and P.R. Girguis. 2015. Enhancing the response of microbial fuel cell based toxicity sensors to Cu(II) with the applying of flow-through electrodes and controlled anode potentials. Bioresour. Technol. 190(14): 367-372.

Jiang, Yong, Peng Liang, Panpan Liu, Xiaoxu Yan, Yanhong Bian and Xia Huang. 2017. A cathode-shared microbial fuel cell sensor array for water alert system. Int. J. Hydrogen Energy 42(7): 4342-4348.

Jiang, Yong, Xufei Yang, Peng Liang, Panpan Liu and Xia Huang. 2018. Microbial fuel cell sensors for water quality early warning systems: fundamentals, signal resolution, optimization and future challenges. Renewable Sustainable Energy Rev. 81: 292-305.

Jin, X., X. Li, N. Zhao, I. Angelidaki and Y. Zhang. 2017. Bio-electrolytic sensor for rapid monitoring of volatile fatty acids in anaerobic digestion process. Water Research 111: 74-80.

Jin, Xiangdan, Irini Angelidaki and Yifeng Zhang. 2015. Microbial electrochemical monitoring of volatile fatty acids during anaerobic digestion. Environ. Sci. Technol. 50(8): 4422.

Jr, Stahl Rg. 1991. The genetic toxicology of organic compounds in natural waters and wastewaters. Ecotoxicol. Environ. Saf. 22(1): 94-125.

Kang, Gwi Hyeon, Jang Jae Gyeong, Lee Ji Yeong, Mun Hyeon Su, Jang In Seob, Kim Jong Min and Kim Byeong Hong. 2004. A low BOD sensor using a microbial fuel cell. J. Korean Soc. Environ. Engi. 26.

Karube, I., T. Matsunaga, S. Mitsuda and S. Suzuki. 2009. Microbial electrode BOD sensors. Biotechnol. Bioeng. 102(3): 1535-1547.

Karube, Isao, Tadashi Matsunaga, Satoshi Mitsuda and Shuichi Suzuki. 1977. Microbial electrode BOD sensors. Biotechnol. Bioeng.19(10): 1535-1547. doi:10.1002/bit.260191010.

Kaur, Amandeep, Saad Ibrahim, Christopher J. Pickett, Iain S. Michie, Richard M. Dinsdale, Alan J. Guwy and Giuliano C. Premier. 2014. Anode modification to improve the performance of a microbial fuel cell volatile fatty acid biosensor. Sens. Actuators, B 201(4): 266-273.

Kaur, Amandeep, Jung Rae Kim, Iain Michie, Richard M. Dinsdale, Alan J. Guwy and Giuliano C. Premier. 2013. Microbial fuel cell type biosensor for specific volatile fatty acids using acclimated bacterial communities. Biosensors 47(18): 50-55.

Kharkwal, S., Y.C. Tan, M. Lu and H.Y. Ng. 2017. Development and long-term stability of a novel microbial fuel cell BOD sensor with MnO_4, catalyst. Int. J. Mol. Sci. 18(2): 276.

Kim, Jung Rae, Amandeep Kaur, Richard Dinsdale, Alan Guwy and Giuliano Premier. 2013. Detection of volatile fatty acids using MFC type biosensor with an acclimated bacterial community. J. Korean Soc. Bitechnol. Bioeng. **269-269??**: 1.

Kim, M., S.M. Youn, S.H. Shin, J.G. Jang, S.H. Han, M.S. Hyun, G.M. Gadd and H.J. Kim. 2003. Practical field application of a novel BOD monitoring system. J. Environ.Monito. Jem 5(4): 640.

Kim, Mia, Hyung Soo Park, Gil Ju Jin, Won Hui Cho, Kwon Lee Dong, Moon Sik Hyun, Ho Choi Chang and Hyung Joo Kim 2006. A Novel Combined Biomonitoring System for BOD Measurement and Toxicity Detection Using Microbial Fuel Cells. IEEE Conference on 2006, pp. 1247-1248.

Kim, Taegyu and Jong In Han. 2013. Fast detection and quantification of *Escherichia coli* using the base principle of the microbial fuel cell. J. Environ. Manage. 130(1): 267-275.

King, S.T., M. Sylvander, M. Kheperu, L. Racz and W.F. Harper Jr. 2014. Detecting recalcitrant organic chemicals in water with microbial fuel cells and artificial neural networks. Sci. Total Environ. 497-498: 527-533.

Kretzschmar, Jörg, Luis F.M. Rosa, Jens Zosel, Michael Mertig, Jan Liebetrau and Falk Harnisch. 2016. A microbial biosensor platform for inline quantification of acetate in anaerobic digestion: potential and challenges. Chem. Eng. Technol. 39(4): 637-642.

Kudo, A., Y. Fujikawa, S. Miyahara, J. Zheng, H. Takigami, M. Sugahara and T. Muramatsu. 1998. Lessons from Minamata mercury pollution, Japan—after a continuous 22 years of observation. Water Sci. Technol. 38(7): 187-193.

Lefebvre, O., A. Uzabiaga, Y.J. Shen, Z. Tan, Y.P. Cheng, W. Liu and H.Y. Ng. 2011. Conception and optimization of a membrane electrode assembly microbial fuel cell (MEA-MFC) for treatment of domestic wastewater. Water Sci. Technol. 64(7): 1527-1532.

Li, Feifang, Zhanwang Zheng, Bin Yang, Xingwang Zhang, Zhongjian Li and Lecheng Lei. 2016. A laminar-flow based microfluidic microbial three-electrode cell for biosensing. Electrochim. Acta 199: 45-50.

Liang, Cheng, Quek Soonbee and R. Cordruwisch. 2014. Hexacyanoferrate-adapted biofilm enables the development of a microbial fuel cell biosensor to detect trace levels of assimilable organic carbon (AOC) in oxygenated seawater. Biotechnol. Bioeng. 111(12): 2412.

Liu, B., Y. Lei and B. Li. 2014. A batch-mode cube microbial fuel cell based 'shock' biosensor for wastewater quality monitoring. Biosensors 62: 308.

Logan, B.E., B. Hamelers, R. Rozendal, U. Schröder, J. Keller, S. Freguia, P. Aelterman, W. Verstraete and K. Rabaey. 2006. Microbial fuel cells: methodology and technology. Environ. Sci. Technol. 40(17): 5181.

Lorenzo, M. Di, T.P. Curtis, I.M. Head and K. Scott. 2009. A single-chamber microbial fuel cell as a biosensor for wastewaters. Water Res. 43(13): 3145.

Lovley, Derek, Kelly P. Nevin, Byoung Chan Kim, Richard H. Glaven, Jessica P. Johnson, Trevor L. Woodard, Barbara A. Methè, Di Donato Jr, Sean F. Covalla and Ashley E. Franks. 2009. Anode biofilm transcriptomics reveals outer surface components essential for high density current production in geobacter sulfurreducens fuel cells. Plos One 4(5): e5628.

Maie, El-Gammal, Reda, Abou-Shanab, Irini, Angelidaki, Omar, Per Viktor and Sveding. 2017. High efficient ethanol and VFA production from gas fermentation: effect of acetate, gas and inoculum microbial composition. Biomass Bioenergy 105: 32-40.

Marsili, E., D.B. Baron, I.D. Shikhare, D. Coursolle, J.A. Gralnick and D.R. Bond. 2008. Shewanella secretes flavins that mediate extracellular electron transfer. Proc Natl Acad Sci USA. 105(10): 3968.

Mia, Kim, Hyun Moonsik, G.M. Gadd and Kim Hyungjoo. 2007. A novel biomonitoring system using microbial fuel cells. Journal of Environmental Monitoring 9(12): 1323.

Michael Berg, Hong Con Tran, Thi Chuyen Nguyen, Hung Viet Pham, Roland Schertenleib and Walter Giger. 2001. Arsenic contamination of groundwater and drinking water in Vietnam: a human health threat. Environ. Sci. Technol. 35(13): 2621-2626.

Michelle, Rasmussen and Shelley D. Minteer. 2015. Long-term arsenic monitoring with an enterobacter cloacae microbial fuel cell. Bioelectrochemistry 106 (Pt A): 207-212.

Modin, O. and B.M. Wilén. 2012. A novel bioelectrochemical BOD sensor operating with voltage input. Water Research 46(18): 6113-6120.

Moon, Hyunsoo, In Seop Chang, Kui Hyun Kang, Jae Kyung Jang and Byung Hong Kim. 2004. Improving the dynamic response of a mediator-less microbial fuel cell as a biochemical oxygen demand (BOD) sensor. Biotechnol. Lett. 26(22): 1717-1721.

Ortiz-Martínez, V.M., M.J. Salar-García, A.P. de los Ríos, F.J. Hernández-Fernández, J.A. Egea and L.J. Lozano. 2015. Developments in microbial fuel cell modeling. Chem. Eng. J. 271: 50-60. doi:https://doi.org/10.1016/j.cej.2015.02.076.

Paerl and W. Hans. 1997. Coastal eutrophication and harmful algal blooms: importance of atmospheric deposition and groundwater as “ New” nitrogen and other nutrient sources. Limnol. Oceanogr. 42(5): 1154-1165.

Pasternak, Grzegorz, John Greenman and Ioannis Ieropoulos. 2017. Self-powered, autonomous biological oxygen demand biosensor for online water quality monitoring. Sens. Actuators, B 244: 815-822.

Patil, S., F. Harnisch and U. Schröder. 2010. Toxicity response of electroactive microbial biofilms—a decisive feature for potential biosensor and power source applications. Chemphyschem 11(13): 2834-2837.

Peixoto, L., B. Min, G. Martins, A.G. Brito, P. Kroff, P. Parpot, I. Angelidaki and R. Nogueira. 2011. In situ microbial fuel cell-based biosensor for organic carbon. Bioelectrochemistry 81(2): 99-103.

Pham, T.H., N. Boon, P. Aelterman, P. Clauwaert, Schamphelaire L. De, L. Vanhaecke, Maeyer K. De, M. Höfte, W. Verstraete and K. Rabaey. 2008. Metabolites produced by Pseudomonas sp. enable a gram-positive bacterium to achieve extracellular electron transfer. Appl. Microbiol. Biotechnol. 77(5): 1119-1129.

Quek, Soon Bee, Liang Cheng and Ralf Cord-Ruwisch. 2015. In-line deoxygenation for organic carbon detections in seawater using a marine microbial fuel cell-biosensor. Bioresour. Technol. 182: 34-40.

Quek, Soon Bee, Cheng Liang and Ralf Cord-Ruwisch. 2015. Microbial fuel cell biosensor for rapid assessment of assimilable organic carbon under marine conditions. Water Research 77: 64-71.

Raposo, F., R. Borja, J.A. Cacho, J. Mumme, K. Orupõld, S. Esteves, J. Noguerol-Arias, S. Picard, A. Nielfa and P. Scherer. 2013. First international comparative study of volatile fatty acids in aqueous samples by chromatographic techniques: evaluating sources of error. TrAC, Trends Anal. Chem. 51(11): 127-143.

Reguera, G., K.D. Mccarthy, T. Mehta, J.S. Nicoll, M.T. Tuominen and D.R. Lovley. 2005. Extracellular electron transfer via microbial nanowires. Nature 435(7045): 1098.

Rozendal, René A., Hubertus V.M. Hamelers, Korneel Rabaey, Jurg Keller and Cees J.N. Buisman. 2008. Towards practical implementation of bioelectrochemical wastewater treatment. Trends Biotechnol. 26(8): 450-459.

Schievano Andrea, Alessandra Colombo, Alessandra Cossettini, Andrea Goglio, Vincenzo D'Ardes, Stefano Trasatti and Pierangela Cristiani. 2017. Single-chamber microbial fuel cells as on-line shock-sensors for volatile fatty acids in anaerobic digesters. Waste Manage 71.

Schievano, Andrea, Francesca Pizza, Claudio Pino, Davide Perrino, Alessandra Colmbo and Pierangela Cristiani. 2016. Experiences of floating microbial fuel cells, supplying on-line sensors for water quality.

Schneider, G., M. Czeller, V. Rostás and T. Kovács. 2015. Microbial fuel cell-based diagnostic platform to reveal antibacterial effect of beta-lactam antibiotics. Enzyme Microb. Technol. 73-74: 59-64.

Schwarzenbach, R.P., T. Egli, T.B. Hofstetter, U. Von Gunten and B. Wehrli. 2010. Global water pollution and human health. Social Science Electronic Publishing 35(1).

Schwarzman, M.R. and M.P. Wilson. 2009. New science for chemicals policy. Science 326(5956): 1065-1066.

Shen, Y.J., O. Lefebvre, Z. Tan and H.Y. Ng. 2012. Microbial fuel-cell-based toxicity sensor for fast monitoring of acidic toxicity. Water Science & Technology A Journal of the International Association on Water Pollution Research 65(7): 1223.

Shen, Y., M. Wang, I.S. Chang and H.Y. Ng. 2013. Effect of shear rate on the response of microbial fuel cell toxicity sensor to Cu(II). Bioresour. Technol. 136(5): 707-710.

Stein, N.E., H.V.M. Hamelers, C.N.J. Buisman, A. Bergel, D. Féron and H.C. Flemming. 2010. Stabilizing the baseline current of a microbial fuel cell-based biosensor through overpotential control under non-toxic conditions. Bioelectrochemistry 78(1): 87-91.

Stein, Nienke E., Hubertus V.M. Hamelers and Cees N.J. Buisman. 2012a. Influence of membrane type, current and potential on the response to chemical toxicants of a microbial fuel cell based biosensor. Sens. Actuators, B 163(1): 1-7.

Stein, Nienke E., Hubertus V.M. Hamelers, Gerrit Van Straten and Karel J. Keesman. 2012b. Effect of toxic components on microbial fuel cell-polarization curves and estimation of the type of toxic inhibition. Biosensors 2(3): 255.

Stein, Nienke E., Hubertus V.M. Hamelers and Cees N.J. Buisman. 2012c. The effect of different control mechanisms on the sensitivity and recovery time of a microbial fuel cell based biosensor. Sens. Actuators, B 171-172(9): 816-821.

Su, L., W. Jia, C. Hou and Y. Lei. 2011. Microbial biosensors: a review. Biosensors 26(5): 1788.

Tilman, D., J. Fargione, B. Wolff, C. D'Antonio, A. Dobson, R. Howarth, D. Schindler, W.H. Schlesinger, D. Simberloff and D. Swackhamer. 2001. Forecasting agriculturally driven global environmental change. Science 292(5515): 281.

Tran, Phuong Hoang Nguyen, Tha Thanh Thi Luong, Thuy Thu Thi Nguyen, Huy Quang Nguyen, Hop Van Duong, Kim Byunghong and The Pham Hai. 2015. Possibility of using a lithotrophic iron-oxidizing microbial fuel cell as a biosensor for detecting iron and manganese in water samples. Environ. Sci.: Processes Impacts 17(10): 1806.

Tront, J.M., J.D. Fortner, M. Plötze, J.B. Hughes and A.M. Puzrin. 2008. Microbial fuel cell biosensor for in situ assessment of microbial activity. Biosensors 24(4): 586-590.

Urgun-Demirtas, M., B. Stark and K. Pagilla. 2006. Use of genetically engineered microorganisms (GEMs) for the bioremediation of contaminants. Crit. Rev. Biotechnol. 26(3): 145-164.

Wang, X., N. Gao and Q. Zhou. 2013. Concentration responses of toxicity sensor with Shewanella Oneidensis MR-1 growing in bioelectrochemical systems. Biosens. Bioelectron. 43(1): 264.

Wang, Y., S.E. Kern and D.K. Newman. 2010. Endogenous phenazine antibiotics promote anaerobic survival of Pseudomonas Aeruginosa via extracellular electron transfer. J. Bacteriol. 192(1): 365-369.

Webster, D.P., M.A. Teravest, D.F. Doud, A. Chakravorty, E.C. Holmes, C.M. Radens, S. Sureka, J.A. Gralnick and L.T. Angenent. 2014. An arsenic-specific biosensor with genetically engineered Shewanella Oneidensis in a bioelectrochemical system. Biosensors 62(20): 320-324.

West, E.A., A. Jain and J.A. Gralnick. 2017. Engineering a native inducible expression system in Shewanella Oneidensis to control extracellular electron transfer. ACS Synth. Biol. 6(9): 1627-1634.

Wu, F., Z. Liu, S.G. Zhou, Y.Q. Wang and S.H. Huang. 2009. Development of a low-cost single chamber microbial fuel cell type BOD sensor. Environ. Sci. 31(7): 1596.

Wu, Wen Guo, K.L. Lesnik, Shou Tao Xu, Lu Guang Wang and Hong Liu. 2014. Impact of tobramycin on the performance of microbial fuel cell. Microb. Cell Fact. 13(1): 91.

Wu, Y.Q., S.X. Du and Y. Yan. 2011. Ultraviolet spectrum analysis methods for detecting the concentration of organic pollutants in water. Spectroscopy & Spectral Analysis 31(1): 233.

Xu, G.H., Y.K. Wang, G.P. Sheng, Y. Mu and H.Q. Yu. 2014. An MFC-based online monitoring and alert system for activated sludge process. Sci Rep 4: 6779.

Xu, Zhiheng, Bingchuan Liu, Qiuchen Dong, Yu Lei, Yan Li, Jian Ren, Jeffrey Mccutcheon and Baikun Li. 2015. Flat microliter membrane-based microbial fuel cell as 'on-line sticker sensor' for self-supported in situ monitoring of wastewater shocks. Bioresour. Technol. 197: 244-251.

Xu, Zhiheng, Yucheng Liu, Isaiah Williams, Li Yan, Fengyu Qian, Zhang Hui, Dingyi Cai, Wang Lei and Baikun Li. 2016. Disposable self-support paper-based multi-anode microbial fuel cell (PMMFC) integrated with power management system (PMS) as the real time 'shock' biosensor for wastewater. Biosensors 85: 232.

Yamashita, T., N. Ookawa, M. Ishida, H. Kanamori, H. Sasaki, Y. Katayose and H. Yokoyama. 2016. A novel open-type biosensor for the in-situ monitoring of biochemical oxygen demand in an aerobic environment. Sci. Rep. 6: 38552.

Yang, Weiyang, Xuejian Wei, Arwa Fraiwan, Christopher G. Coogan, Hankeun Lee and Seokheun Choi. 2016. Fast and sensitive water quality assessment: a ML-scale microbial fuel cell-based biosensor integrated with an air-bubble trap and electrochemical sensing functionality. Sens. Actuators, B 226: 191-195.

Yao, Xin. 2002. Evolving artificial neural networks. Proceedings of the IEEE 87(9): 1423-1447.

Yuan, W., N. Yang and X. Li. 2016. Advances in understanding how heavy metal pollution triggers gastric cancer. Biomed Res Int 2016(1): 7825432.

Zeng, L., X. Li, Y. Shi, Y. Qi, D. Huang, M. Tade, S. Wang and S. Liu. 2017. FePO$_4$ based single chamber air-cathode microbial fuel cell for online monitoring levofloxacin. Biosensors 91: 367-373.

Zhang, Yifeng and Irini Angelidaki. 2011. Submersible microbial fuel cell sensor for monitoring microbial activity and BOD in groundwater: focusing on impact of anodic biofilm on sensor applicability. Biotechnol. Bioeng. 108(10): 2339-2347.

Zhang, Yifeng and Irini Angelidaki. 2012. A simple and rapid method for monitoring dissolved oxygen in water with a submersible microbial fuel cell (SBMFC). Biosensors 38(1): 189-194.

Zhao, S., P. Liu, Y. Niu, Z. Chen, A. Khan, P. Zhang and X. Li. 2018. A novel early warning system based on a sediment microbial fuel cell for in situ and real time hexavalent chromium detection in industrial wastewater. Sensors 18(2): 642.

Zheng, Tao, Yu Shang Xu, Xiao Yu Yong, Bing Li, Di Yin, Qian Wen Cheng, Hao Ran Yuan and Yang Chun Yong. 2015. Endogenously enhanced biosurfactant production promotes electricity generation from microbial fuel cells. Bioresour. Technol. 197: 416.

Zhou, Tuoyu, Huawen Han, Liu Pu, Xiong Jian, Fake Tian and Xiangkai Li. 2017. Microbial fuels cell-based biosensor for toxicity detection: a review. Sensors 17(10): 2230.

Sustainability Assessment of Microbial Fuel Cells

Surajbhan Sevda[1], Swati Singh[1], Vijay Kumar Garlapati[2], Swati Sharma[2], Lalit Pandey[1], T.R. Sreekrishnan[3] and Anoop Singh[4]*

[1] Department of Bioscience and Bioengineering, Indian Institute of Technology Guwahati, India
[2] Department of Biotechnology & Bioinformatics, Jaypee University of Information Technology (JUIT), Waknaghat, HP - 173234, India
[3] Department of Biochemical Engineering and Biotechnology, Indian Institute of Technology Delhi, New Delhi, India
[4] Department of Scientific and Industrial Research (DSIR), Ministry of Science and Technology, Government of India, Technology Bhawan, New Delhi - 110016, India

Introduction

An integrated technique where a biological catalyst is used to convert chemical energy into electrical energy (or vice versa) is termed as bio-electro-chemical system (BES). A biocatalyst stands for any microbe, its secondary metabolite, or its enzymes. In 1911, M.C. Potter first discovered electricity generation by bacteria. Later in 1934, another discovery of production 35V potential by utilizing microbial half-cell was made. By the end of 19th century, scientist B.H. Kim reported mediator-less electron transport across the electrode. Till date, various ongoing research on the exploitation of this ability of microbes to generate tremendous power is being performed worldwide.

Conventional fuel cell works on exploiting chemical catalyst in terms of metal electrodes, which are highly expensive, susceptible of corrosion, surface poisoning, high catalyst loading, and selective electrolyte properties, making it a complicated, non-economic, and ecological threat when disposed. In comparison to conventional fuel cells, BES works under mild reaction condition using inexpensive substrate, making it a very promising technology in the near future. Recent research focuses on the use of bio-electro-chemical technique for the production of energy (electrical or chemical) from waste

*Corresponding author: apsinghenv@gmail.com

materials as substrate with the help of biological catalyst, making this process affordable, feasible, and environment friendly.

In this chapter various types of BES, their application, highlighting majorly on wastewater treatment using MFC, and economic aspects of implementation of this technology at industrial scale have been discussed. Depending on the mode of operation and the type of end product formed, bio-electrochemical technologies can be broadly classified as:

1. Microbial fuel cell (MFC)
2. Microbial electrolysis cell (MEC)
3. Microbial electro synthesis (MES)
4. Microbial solar cell (MSC)
5. Microbial desalination cell (MDC)
6. Enzymatic fuel cell (EFC)

Microbial Fuel Cell

Microbial fuel cell is the most studied BES, which produces electricity by extracting chemical energy from complex organic substrates (Fig. 16.1). Microbial fuel cell utilizes electroactive microbes as biocatalysts in an anaerobic anodic chamber to generate electricity. Presence of these microorganisms at anode drags the electron generated from the oxidation of complex organics towards the electrode causing generation of electric current. These electroactive microbes form a biofilm across the anode electrode. Under anaerobic condition, the biofilm utilizes complex organic substrate, producing electrons that are shuttled to the cathode. Microbial fuel cell differs from conventional fuel cell as it utilizes biotic microbes as catalyst, under neutral pH and temperature range of 15°C to 45°C (Min et al. 2008, Larrosa-Guerrero et al. 2010). The added advantage is its neutral impact on the environment.

Various research has been done at lab scale using synthetic organic substrate or model microbial consortia as catalyst for production of electricity in MFC. However, it is still unsuccessful for real-time implementation. This is due to insufficient knowledge about the mechanism and kinetics of electroactive bacteria. With present imaging and electrochemical technique, this phenomenon is understood only up to a certain extent (Beyenal and Babauta 2015, Artyushkova et al. 2015). To understand the affinity of microbes towards anode, their mechanism of biofilm formation and the coordination between various species forming biofilm, our present technology is insufficient. In addition, real wastewater as a substrate comprises complexity of inherent microbial consortia, which comprises both electroactive and non-electroactive microbes. Studying such a complex mixture under varying day-to-day environmental parameters is complicated and a major reason of failure of MFC at industrial platform.

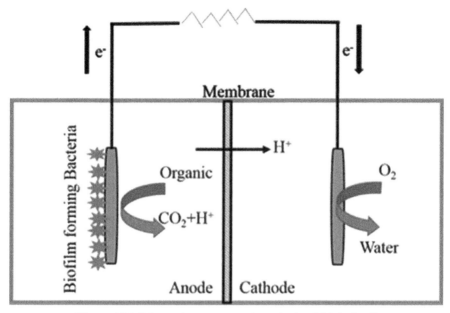

Figure 16.1. Schematic representation of microbial fuel cell.

Microbial Electrolysis Cell

Microbial electrolysis cell is a technique for hydrogen production using oxidation of organic matter by microbes. The first discovery of MEC occurred simultaneously and independently in 2005 at Penn State University and Wageningen University. Since then, the MEC research has been carry forward by different researchers (Liu et al. 2005, Rozendal and Buisman 2005). In general, an MEC involves the anaerobic oxidation of organic matter at anode by an electroactive bacteria-producing CO_2, electrons, and protons. The electrons are shuttled towards cathode and the protons are allowed to pass through a proton selective membrane, which separates anodic chamber from cathodic chamber. The transported electrons then react with shuttled electrons to form H_2 gas at cathode (Fig. 16.2). However, this process requires an external voltage supply (~0.2-0.8 V). Hence, a typical microbial electrolysis cell produces H_2 at cathode under biologically controlled condition of pH 7, 30°C, and 1 atm pressure with an external energy input (Liu et al. 2005). However, the energy input in an MEC is lesser than that of a typical electrolysis unit (1.23-1.8 V).

Microbial Electrosynthesis Cell

A microbial electrosynthesis cell (MES) involves the investment of power to drive thermodynamically unfavorable reactions or to enhance the rate of reaction by using microbes. These days MES is being used effectively for the production of O_2, water, caustic solution, glutamic acid, as well as hydrogen peroxide. In this way, MES is used to convert toxic waste material into value

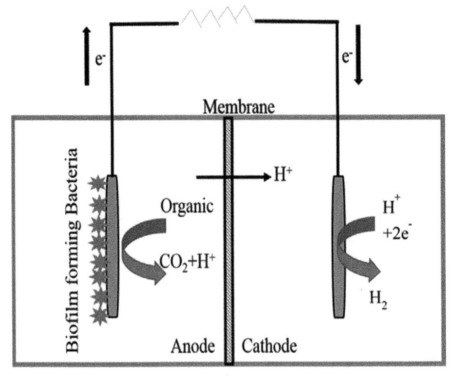

Figure 16.2. A microbial electro synthesis unit.

added product. In a typical MES, microbes or their enzymes are immobilized as catalyst on anode for electrosynthesis. Biocatalysts in the form of enzymes, are preferred over chemical catalyst for the electrosynthesis process because of their high specificity; however, the use of entire microbe as catalyst has some added advantages. These advantages include self-regeneration and flexibility in substrate and process parameters. With these advantages, microbes also have a drawback of initial time of adaptation in the fuel cell and consumption of some part of substrate for initial growth. Studies report that microbes when used as a whole, showed decrease in the electrode over potential as compared to enzymes (Lowy et al. 2006, Rabaey et al. 2008). In addition, microbes were used to reduce CO_2 using electric current in an MES (Nevin et al. 2010). Hence, MES involves the reduction or oxidation of various organic compounds using application of potential across the electrode in the presence of microbes as catalyst.

Microbial Solar Cell

Microbial solar cell is a technique used to convert solar energy to synthesis chemical in the presence of microbial catalyst. Catalysts used for MSC are photoautotrophic (which also includes higher plants) that are capable of harvesting solar energy. These catalysts synergistically work with electroactive microbes to drive the electrons towards the anode for the generation of electricity. The basic steps in a synergistic MSC include:

- Photosynthesis
- Transfer of organic substrate to the anodic chamber
- Oxidation of organic substrate by electroactive bacteria
- Reduction of oxygen at cathode

MSC are generally classified based on their mode of capture of solar energy and their mode of electron transfer across the electrode. Till date, three types of MSC are studied, which include (i) a higher plant for capturing solar energy and its transfer via rhizodeposition (they are also called as plant microbial fuel cell, PMFC), (ii) a phototrophic biofilm which diffuses the energy to adhered electrodes, and (iii) photobioreactor-based MFC. There is very little information in the understanding of the mechanism of transfer of solar energy to chemical energy via these systems (Logan and Regan 2006, Lovley 2008, Richmond 2008).

In a PMFC, the plant transfers its solar energy in the form of rhizospheric depositions, which are chemically sugars, polymeric carbohydrates, organic acids, dead cell material, and certain enzymes (Schamphelaire et al. 2008, 2010). These plant exudates comprises about 20-40% of plant's photosynthetic productivity, which are consumed by electroactive bacteria, adhered on anode for the production of electricity (Lovley 2008). A typical PMFC involves the plant's root to grow in anodic chamber where a continuous feed is generated *in situ* for anodic bacteria to produce electricity *in situ*. In a PMFC study, 50 mW/m² of power was generated in a month by *Spartina anglica* (Timmers et al. 2010). The microbes involved in PMFC are usually *Desulfobulbus, Geobacteraceae, Natronocella, Beijerinckiaceae, Rhizobiales, Rhodobacter* and sometimes *Geobacter sulfurreducens* (Kaku et al. 2008, Schamphelaire et al. 2010). MSC-based phototrophic biofilm forming microbes include microbes of genus *Chlorophyta, Cyanophyta* and *Canoperate* (Lovley 2008). A mixture of these microbes with some electroactive bacteria are usually found on the anodic chamber of photobiofilm-based MSCs. In case of photobioreactors-based MSCs, algae zooplankton is used to harvest solar energy. Such systems are usually used in case of harvesting energy from coastal marine ecosystem (Girguis et al. 2010, Reimers et al. 2007). However, all the MSCs formed till date are operating under pilot scale. Due to unavailability of data and techniques, such as inability to calculate coulombic efficiency to standardise accurate calculations, the success of MSC are only at lab scale.

Microbial Desalination Cell (MDC)

It is an extrapolation of the principle involved in native MFC with an added application of treating wastewater along with the production of electricity. Similar to MFC, it comprises an anodic chamber maintained at anoxic condition separated from cathodic chamber using a cation and anion exchange membrane, incorporated in the middle (Luo et al. 2012, Sevda et al.

2015). An MDC utilizes the inherent electroactive microbes present in sludge or wastewater for the oxidation of organic content to produce electron in line, which are transferred to the anode, where they are directed to cathode via an external electric wire. In the cathodic chamber, these electrons combine with proton exchanged via membrane from anode, completing the circuit and forming clean water. The potential difference between the two electrodes produces electricity, which drives the desalination process of inlet water (Sevda et al. 2017, Sevda and Abu-Reesh 2017a). A typical MDC requires 180-231% less power to drive desalination of 5-30 g/L salt solution, in comparison to conventional desalination techniques (Sevda and Abu-Reesh 2017b, 2018, Wang and Ren 2013). During this process, based on oxidation of sludge and waste matter, electrons are produced that are driven towards an external circuit towards the cathode, where they bind with hydrogen ion and oxygen molecule, forming pure water. This creates a potential difference across the two electrode dragging cations towards the cathode via cation exchange membrane and anions towards anode through anionic membrane, leaving the middle chamber devoid of salts (upto 99%) for letting out desalted water (Forrestal et al. 2012). However, MDC is efficient only when the salt concentration is higher than electrolytic concentration to surpass the ohmic resistance. A lower salt concentration leads to lower desalination rate (Yuan et al. 2012). Hence, an MDC can produce water and generate electricity by treating wastewater without any extra energy input. It is a highly affordable and profitable technique for bioremediation of wastewater in the coming future.

Enzyme Mediated Fuel Cell

This type of BES utilizes biological catalyst in the form of enzymes for the oxidation of substrate and simultaneous generation of electricity. Use of enzymes as catalyst miniaturizes fuel cell, due to which various EFC of micrometer measurements have been developed (Gellett et al. 2010, Katuri et al. 2011). Specificity of enzyme narrows down the reaction to comparatively mild conditions i.e., neutral pH, and standard temperature. In addition, unlike MFC, no mediator is required in EFC, rather the enzymes are immobilized on inert electrode eliminating the need for separators and other housing parts of conventional fuel cell (Heller 2004, Gellett et al. 2010, Katuri et al. 2011). Due to its high specificity and selectivity, an EFC displays 1.24 V potential under 100% columbic efficiency when glucose is oxidised to CO_2 and H_2O at 298 K. Various research is done to immobilize multiple enzymes on the electrode to increase the oxidation rate of glucose, extracting up to 24 electrons (Simon et al. 2002). In a typical EFC, the reduction of substrate involves the presence of a co-factor, a non-proteinaceous and electroactive component, which directs the electron from the enzyme to the electrode (or vice versa). In a typical glucose oxidation, the common cofactors involved are NAD (Nicotinamide adenine dinucleotide), FAD (Flavin adenine dinucleotide), and PQQ (pyrroloquinoline quinone).

The transfer of electrons from the enzyme to the electrode is classified as (i) direct electron transfer (DET), and (ii) mediated electron transfer (MET). When the cofactor is tightly bound to the enzyme, it causes direct electron transfer between the enzyme and electrode (Calabrese Barton et al. 2004, Cracknell et al. 2008, Ludwig et al. 2010). However, studies report that appreciable DET occurs only when the enzyme and electrode are in the vicinity of 2 nm (Marcus and Sutin 1985). Along with this, the orientation of the active site also plays a critical role; often the close proximity of electrode blocks the substrate to bind to the enzyme.

Due to these challenges, small electroactive molecules are used to shuttle the electrons to the electrode, termed as mediated electron transfer (MET). In a MET, the redox potential of mediator should be more positive when used at anode for biocatalysis, and more negative during cathodic biocatalysis is required for the shuttling of electron via mediator (Gallaway and Calabrese Barton 2008). However, the maximum potential output from EFC has been reported as 0.2-0.3 V indicating the need of more improvisation by stacking of cells and stabilization of power to manifest high power density (Sakai et al. 2009). Recent studies also focused on printing of electrode component to maximize volume, that in turn enhances overall power output. In a comparative study of printing aldol, dehydrogenase showed better results in comparison to glucose oxidase, GOx (Jenkins et al. 2012). Hence, we can conclude that high reaction rate and specificity along with low over-potential and miniaturization makes EFC a portable fuel cell for power generation as well as a biosensor for biochemical reactions.

Hypothesis of Wastewater Treatment Using Microbial Fuel Cell

Rapid urbanization, population growth, and anthropogenic activities impede a negative impact on environmental pollution, especially wastewater streams along with the energy crisis. The domestic and industrial wastewater streams mainly comprise organic substances, toxic chemicals, xenobiotic compounds, heavy metals, nitrogen, and phosphorus (Choudhury et al. 2017). The wastewaters from the domestic, agricultural, and industrial plants are the predominant causes for impairment of drinking waters through the surface water eutrophication, hypoxia and algal-booms impairment. To tackle the uncontrollable issue, wastewater treatment is the common practice throughout the world to protect the potable sources of drinking water along with the environmental protection. The existing wastewater treatment processes (WWTPs) are mostly energy- and chemical-drive processes which are coupled with the high investments without any observable profits (Gude 2016, Sevda et al. 2018). The energy component of wastewater mainly lies in organic, nutritional (nitrogen and phosphorus), and thermal forms which was equivalent to ~1.79 kWh/m^3, ~0.7 kWh/m^3, and ~7 kWh/m^3, respectively (McCarty et al. 2011). Usually, the WWTPs utilized for carbon and nitrogen removal have an energy demand of 0.5-2 kWh/m^3, which accounts for the 1-4% of the nation's average daily electricity consumption (McCarty et al. 2011). Moreover, during the operation of WWTP, various greenhouse gases (GHGs), such as CO_2 and N_2O will release to the atmosphere along with the different volatile compounds. Apart from the investment, energy- intensives, and GHGs emission, the end problem with the WWTPs mainly lies in the excess produced sludge, which further needs a separate disposal operation. The electricity production through WWTPs is also accompanied with the emission of GHGs i.e., production of one kWh electricity is accompanied with the emission of 0.9 kg CO_2 (Wang et al. 2010).

The existing WWTPs for the domestic waters works on well-established activated-sludge process, also not at all chemical- and energy-intensive and needs high capital and maintenance costs. Aeration is the main prerequisite in case of activated-sludge-based WWTPs, which accounts for around 75% of total plant energy costs. The 60% of the operational costs are a tie-up with the treatment and sludge disposal (Sustarsic 2009). On the other hand, anaerobic digestion (AD) is the option for dealing with the high strength wastewaters from industries. Compared to the activated sludge process, AD has a potential for energy savings through biogas production (Visvanathan and Abeynayaka 2012). In spite of biogas production, the AD suffers from the separation and purification aspects of produced biogas (CH_4) (Stillwell et al. 2011). Based on the above facts, the existing WWT systems are not at all sustainable, which further needs developing a newer energy capturing technologies with sustainable footprints (Khandan et al. 2014). Any technology which captures the most energy potential of wastewater with net energy producer than the consumer is the viable note-worthy technological breakthrough in case of wastewater treatment (Logan 2008). The above issues paved the way for looking for the superior wastewater treatment systems design, which encompasses the recovering resources with less energy utilization and maximum energy recovery rather than the removal of pollutants.

Microbial fuel cell (MFC) technology is one of the sustainable approaches where direct conversion of waste to electricity and chemical products is possible (Logan and Rabaey 2012). In MFC, the chemical energy will be converted to electricity through the oxidation of biodegradable substrates with the help of exo-electrogenic bacteria (Montpart et al. 2014). Hence, production of electrical energy from waste streams along with the wastewater treatment through MFC technology is a viable, sustainable approach to tackle the burden of excessive wastewater stream (Qin et al. 2017, He et al. 2017). The operation of MFC mainly lies with the action of exo-electrogenic bacteria at anodic chamber, which oxidizes organic matter, present in wastewater streams, which ends up with the attaining of carbon (energy) and electrons. After utilization of little part of the carbon for the growth, the exoelectrogenic bacteria transfer the electrons to the external electron shuttle via the conductive biofilm matrix. The transferred electrons are utilized at the cathodic chamber to reduce the oxygen, which closes the electric circuit for electricity production (Rabaey and Verstraete 2005). Apart from sustainability, the MFC technology is further advantageous for wastewater treatment over other conventional technologies concerning enhanced conversion efficiency with low solid-waste generation (Park et al. 2017).

The other positive attributes of MFC technology include handling of low-strength wastewaters (which will not suit for activated sludge and AD processes) and operation at ambient temperatures (Corbella et al. 2016). The technology will not need any further gas treatment and energy inputs for aeration and have widespread applicability even in locations with insufficient electrical infrastructures (Qin et al. 2017). The mixed cultures present in wastewater streams serve as the source of inoculums for exo-electrogenic bacteria utilized in MFC (Kim et al. 2007). According to a recent study, a power density of 4200 mW/m^3 can be produced by MFC along with the complete removal of chemical oxygen demand (COD) and other pollutants (Zinadini et al. 2017). Overall, utilization of MFC technology for wastewater treatment has energy, environmental, economic, and operational benefits. The energy benefits of MFCs for wastewater treatment include direct power generation, no need of aeration, and low sludge yield. The footprints of MFCs on the environment include water reclamation, low carbon traces, and no need for any special sludge disposal mechanisms. Low operational costs coupled with the revenues associated with the energy and value-added product, and no further downstream operations contribute to the economic benefits of MFC technology for wastewater treatment. The operational benefits of this so-called MFC technology lie in the utilization of readily available microorganisms as the source of inoculums with tolerance for different environmental stress and amenable to real-time monitoring of the process (Li et al. 2014).

Different Applications of Bioelectrochemical System

Bio-electro-chemical system is a technique that employs the ability of microorganisms to catalyse the substrate using redox enzymes for the production of energy in either chemical or electrical form. It is majorly used for bioremediation of waste substrate along with the production of various value added chemical products, as discussed below.

Treatment of Domestic Wastewater and Production of Value Added Products

Unlike conventional wastewater treatment techniques, the use of BES system decreases the overall sludge produced after the treatment, decreasing the overall cost of treatment, as no new sludge treatment unit is required. Along with waste minimization, BES produces electricity by using electroactive microbes that can utilize organic content present in the wastewater for the production of electrons based on their redox enzymatic activity. In the same system, various value added products are also synthesized with few slight modifications in the overall system, keeping the process economic and feasible. Few such chemical compounds synthesized by BES technique are hydrogen, acetate, methane, hydrogen peroxide, caustic soda, etc. (Liu et al. 2005, Rabaey et al. 2010, Modin and Fukushi 2012).

BES biosensor

In recent studies, BES has been used for the testing of toxicity level of water as well as its net biochemical oxygen demand (BOD). It provides a real time, on site, and rapid analysis of analyte with high specificity and selectivity. BOD is the parameter to quantify the net organic content in the substrate. The electric current produced by an MFC (mediator-less) is directly proportional to the net organic content. Hence, an extrapolation of net current generated can provide an insight of the BOD value of the sample. Various reports have verified this relation of BOD with the current generated (Chang et al. 2004, 2005). In the same way, if all the experimental parameters are kept constant, the rate of consumption of substrate in the presence of toxin varies directly in terms of overall current generated. Hence, a small decrement in the polarization curve indicates the presence of toxins, leading to inhibition of microbial consumption of organic matter. Various studies have concluded the applicability of MFC-based biosensor for the toxicity analysis (Stein et al. 2012a, b).

Recovery of nutrient and heavy metals using BES

Recovery of nutrient across the membrane is solely dependent on the phenomenon of charge neutrality. Across the ion exchange membrane, there is an active transport of ions, which is accumulated near the oppositely charged electrodes from where they can be further recovered. It is applicable for the removal of nutrient from agricultural and domestic wastewater. Various studies have been performed for the removal of ammonia and phosphate ions from the waste stream using MFC employed with ion exchange membrane (Liu et al. 2018).

In a study, Kim et al. (2008) studied the recovery of ammonium ion (NH_4^+) from the animal wastewater using BES, where NH_4^+ ion got accumulated near the cathode upon passing through cation exchange membrane under the application of electric current. He concluded that redox reaction by the microbes of MFC are the driving force for the separation of ions and simultaneous electricity production. A combination of MFC with precipitation unit was used for the separation and recovery of phosphorus and ammonia from municipal wastewater with an efficiency of 94.6% for phosphorus and 20.6% for ammonium (Zang et al. 2012). Similarly, various studies based on ammonia recovery using MFC have been carried out using different type of substrate (Tao et al. 2014).

Microbes can oxidise or reduce heavy metals, thereby changing their solubility in the waste stream causing them to separate from the waste stream. In a study, Tao et al. (2014) studied the metal recovery from fly ash leachate using a combination of BES with electrolysis unit. They reported the metal removal efficiency as 98.5% for Cu, 95.4% for Zn, and 98.1% for Pb. Similarly, in another study performed by Modin et al. (2012), municipal solid incinerated ash leachate was utilized for bio-electrochemical recovery of Cu by 84.3%, Pb by 47.5%, Cd by 62%, and Zn by 44.2%. Hence, BES system can be a great resource for the recovery of essential heavy metals and nutrient along with power generation and bioremediation of waste stream.

Economic Assessment of Bioelectrochemical System

The conventional wastewater treatment system's overall cost depends on two factors, first is the cost of setupof the system and second is the operational cost. The first part is the one-time investment for building the new plant, and second one is operational cost or day-to-day cost for the smooth operation of the process. If the same capacity wastewater plants run in anaerobic and aerobic mode, the operational cost is more for aerobic mode due to the cost of sparing high amount of air. As the BES system considers the wastewater a resource and gets extra energy while treating it, so we need to keep total operational and maintenance costs minimal so that the entire process is able to produce a net amount of energy that can be stored and used for further applications. The entire BES consists of membranes, electrodes, and other related materials. The life cycle assessment (LCA) is used to evaluate the environment burden, effects associated with BES process by finding out the quantifying energy, materials used, and wastes released to the environment. In the LCA process, in the initial step, goals and scope definition are described, followed by life cycle inventory. In the third stage life cycle, impact assessment is carried out for the entire process and based on these three stages, life cycle interpretations are performed and results are compared to the existing systems. In the final assessment of the entire process, transportation of raw materials is also needed. In the BES, the reactive medium is defined as composition of electrode (e.g., granular graphite), electroactive bacteria that act as electrical conductor for potential difference used to optimize bacterial reductive potential. The BES system showed a promising environmental performance, mostly because of the lighter environmental burden of graphite-based electrodes and other carbon-based electrode. For estimating the better performance of a BES system, three technologies named LCA, integrating dynamic simulation (DS), and techno-economic assessment (TEA) were combined and developed a system for wastewater treatment by removal of chemical oxygen demand (COD) (Shemfe et al. 2018).

The above model is made by following assumptions, such as (i) Biofilm was assumed as a conductive matrix, (ii) Electron transfer occurs via direct conduction mechanism through the biofilm, and (iii) Temperature and pH were considered constant throughout the system. The estimated cost of generation of HCOOH was €0.005–0.015 g^{-1} for its generation of 0.094–0.26 kg yr^{-1} and a COD removal rate of 0.038–0.106 kg yr^{-1} (Shemfe et al. 2018).

The last decade is considered a golden period for progress in MFC, MEC, and other related technologies. In this era, the basic focus was given to prove the proof of the concept for MFC, bioelectricity generation, hydrogen production, and chemical production, while treating wastewater in the system. Now the basic hurdle is how smoothly these technologies can be scaled up and need to find out a scheme that whether at the initial stage these technologies can be synergized with the conventional technologies and the main focus will be to improve the overall process. This type of integration can avoid loss of material and energy from wastewater stream and result in the combined system increasing the overall efficiency, environmental and economic performance. The new thermodynamic modeling framework for MFC, MEC, and other BES was able to predict the multicomponent physico-chemical behavior, best configuration, and technical feasibility for resource recovery from wastewater streams (Sadhukhan et al. 2016). The BES developed as the most efficient system for low strength wastewater and lignocellulosic biomass into hydrogen/chemical or bioelectricity generation. The main component in these systems are material for electrodes, catalysts, and separator with innovative design, to make these systems more attractive and economical compared to the conventional system (Pant et al. 2012).

The main working force in BES is the interaction between electrodes and electroactive microbes and this technology is self-energy efficient in terms of net energy needed and production. In general, the power production defined as normalized of electrode area (kWh/m^2) and anode electrode volume (kWh/m^3), the nitrogen recovery (kWh kg/N), salt removal during desalination (kWh/kg), and chemical production (kWh/kg). The key power consumption in the entire process are pumping system (feeding pumps/recirculation) and external power source in the case of MEC. The analysis shows the comparison with various studies that the energy balance is positive due to the higher value of produced hydrogen and chemicals during the process (Zou and He 2018). The microbial desalination cell (MDC) shows better performance with applied external potential and lower total energy consumption (Sevda et al. 2015, 2017, Sevda and Abu-Reesh 2017 a, b, 2018).

Before industrial commercialization of BES technology, all economic assessment needs to be done and also compared to the same scale conventional processes. For making economic assessment, the investment decisions, such as internal rate of returns and new present value are the two main points to consider for its commercialization. Figure 16.3 shows the basic flow diagram of various stages of MFC for treating wastewater streams and bioelectricity generation. The overall economic assessment was conducted with two different conditions (cathode with and without Pt, respectively). The MFC process was also compared to activated sludge process, and all the initial assessment shows that the MFC is a more attractive option. When a full setup of MFC system is installed, further there will be an annual growth in the rate of electricity production, and so MFC's total revenue will be increased with time.

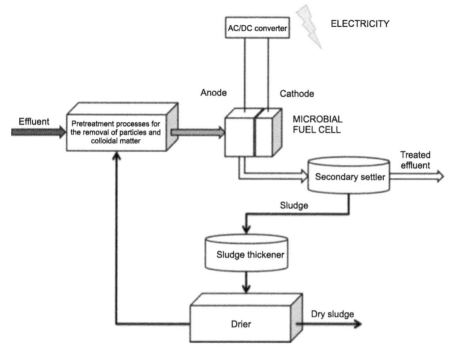

Figure 16.3. The scheme of the MFC (BES) experimental status.
(Source: Trapero et al. 2017)

In overall summary, the economic assessments show the implementation of MFC technology is a promising alternative compared to the activated sludge, and it has more potential economic benefits (Trapero et al. 2017). The BES generates a net energy gain as the technology can utilize the chemical energy present in the wastewaters. In the BES technology, several alternative electron donors were investigated, which shows promise for future application. In the last decade, great advancements have been made into upscaling of this technology. Environmental and economic analysis showed that the main advantage of this technology is in its positive environment impact.

Conclusion

The BES technology generates energy from organic waste by converting chemical energy matters directly to bioelectricity, biohydrogen, and other valuable chemicals. The BES are supported by electroactive microbes and these bacteria have a special ability to transfer electrons out of their cell. There is high range of actual waste available for use in BES; the organic waste present in agriculture waste, wastewater, human waste, distillery wastewater, paper and pulp wastewater, waste from food production, and many more can be utilized by BES system. For scale up of this technology, initial work is reported and its commercial uptake is expected to be driven by reliability and cost. At the current stage of BES development, if scale up is done with only bioelectricity, the end product is not viable due to the overall cost, such as the cost of wires, current collector, and cost of running pumps. Moreover, chemicals such as caustic soda, H_2O_2, ethanol, acetic acid are more economical and valuable in case the chemical generation is stronger, and these products can be used in the other locally available industries along with wastewater treatment. The development and use of BES technology for treating wastewater could be useful to help reducing the overall cost of wastewater treatment, biohydrogen generation, and valuable chemical production.

References

Artyushkova, K., J.A. Cornejo, L.K. Ista, S. Babanova, C. Santoro, P. Atanassov and A.J. Schuler. 2015. Relationship between surface chemistry, biofilm structure, and electron transfer in Shewanella anodes. Biointerphases 10(1): 019013.

Beyenal, H. and J.T. Babauta. 2015. Biofilms in bioelectrochemical systems: from laboratory practice to data interpretation. John Wiley & Sons. New Jersey, USA.

Calabrese Barton, S., J. Gallaway and P. Atanassov. 2004. Enzymatic biofuel cells for implantable and microscale devices. Chem. Rev. 104(10): 4867-4886.

Chang, I.S., H. Moon, J.K. Jang and B.H. Kim. 2005. Improvement of a microbial fuel cell performance as a BOD sensor using respiratory inhibitors. Biosens. Bioelectron. 20(9): 1856-1859.

Chang, I.S., J.K. Jang, G.C. Gil, M. Kim, H.J. Kim, B.W. Cho and B.H. Kim. 2004. Continuous determination of biochemical oxygen demand using microbial fuel cell type biosensor. Biosens. Bioelectron. 19(6): 607-613.

Choudhury, P., U.S.P. Uday, N. Mahata, O.N. Tiwari, R.N. Ray, T.K. Bandyopadhyay and B. Bhunia. 2017. Performance improvement of microbial fuel cells for waste water treatment along with value addition: a review on past achievements and recent perspectives. Renew. Sust. Energ. Rev. 79: 372-389.

Corbella, C., M. Garfí and J. Puigagut. 2016. Long-term assessment of best cathode position to maximise microbial fuel cell performance in horizontal subsurface flow constructed wetlands. Sci Total Environ. 563-564: 448-455.

Cracknell, J.A., K.A. Vincent and F.A. Armstrong. 2008. Enzymes as working or inspirational electrocatalysts for fuel cells and electrolysis. Chem. Rev. 108(7): 2439-2461.

Forrestal, C., P. Xu, P.E. Jenkins and Z. Ren. 2012. Microbial desalination cell with capacitive adsorption for ion migration control. Bioresour. Technol. 120: 332-336.

Gallaway, J.W. and S.A. Calabrese Barton. 2008. Kinetics of redox polymer-mediated enzyme electrodes. Journal of the American Chemical Society 130(26): 8527-8536.

Gellett, W., M. Kesmez, J. Schumacher, N. Akers and S.D. Minteer. 2010. Biofuel cells for portable power. Electroanalysis 22(7-8): 727-731.

Girguis, P.R., M.E. Nielsen and I. Figueroa. 2010. Harnessing energy from marine productivity using bioelectrochemical systems. Curr. Opin. Biotechnol. 21(3): 252-258.

Gude, V.G. 2016. Wastewater treatment in microbial fuel cells: an overview. J. Clean. Prod. 122: 287-307.

He, L., P. Du, Y. Chen, H. Lu, X. Cheng, B. Chang and Z. Wang. 2017. Advances in microbial fuel cells for wastewater treatment. Renew. Sust. Energ. Rev. 71: 388-403.

Heller, A. 2004. Miniature biofuel cells. Phys. Chem. Chem. Phys. 6(2): 209-216.

Jenkins, P., S. Tuurala, A. Vaari, M. Valkiainen, M. Smolander and D. Leech. 2012. A mediated glucose/oxygen enzymatic fuel cell based on printed carbon inks containing aldose dehydrogenase and laccase as anode and cathode. Enzyme Microb. Technol. 50(3): 181-187.

Kaku, N., N. Yonezawa, Y. Kodama and K. Watanabe. 2008. Plant/microbe cooperation for electricity generation in a rice paddy field. Appl. Microbiol. Biotechnol. 79(1): 43-49.

Katuri, K., M.L. Ferrer, M.C. Gutiérrez, R. Jiménez, F. del Monte and D. Leech. 2011. Three-dimensional microchanelled electrodes in flow-through configuration for bioanode formation and current generation. Energ. Environ. Sci. 4(10): 4201-4210.

Khandan, N., T. Selvaratnam and A.K. Pegallapati. 2014. Options for energy recovery from urban wastewaters. *In*: 9th Conference on Sustainable Development of Energy, Water, and Environmental Systems. SDEWES2014.0596-1.

Kim, J.R., S. Cheng, S.E. Oh and B.E. Logan. 2007. Power generation using different cation anion and ultrafiltration membranes in microbial fuel cells. Environ. Sci. Technol. 41: 1004-1009.

Kim, J.R., Y. Zuo, J.M. Regan and B.E. Logan. 2008. Analysis of ammonia loss mechanisms in microbial fuel cells treating animal wastewater. Biotechnol. Bioeng. 99(5): 1120-1127.

Larrosa-Guerrero, A., K. Scott, I. Head, F. Mateo, A. Ginesta and C. Godinez. 2010. Effect of temperature on the performance of microbial fuel cells. Fuel 89(12): 3985-3994.

Li, W.W., H.Q. Yu and Z. He. 2014. Towards sustainable wastewater treatment by using microbial fuel cells-centered technologies. Energy Environ. Sci. 7(3): 911-924.

Liu, H., S. Grot and B.E. Logan. 2005. Electrochemically assisted microbial production of hydrogen from acetate. Environ. Sci. Technol. 39(11): 4317-4320.

Liu, T., X. Chen, X. Wang, S. Zheng and L. Yang. 2018. Highly effective wastewater phosphorus removal by phosphorus accumulating organism combined with magnetic sorbent MFC@ La $(OH)_3$. Chem. Eng. J 335: 443-449.

Logan, B.E. 2008. Microbial Fuel Cells. John Wiley & Sons, Hoboken, NJ.

Logan, B.E. and K. Rabaey. 2012. Conversion of wastes into bioelectricity and chemicals by using microbial electrochemical technologies. Science 337(6095): 686-690.

Logan, B.E. and J.M. Regan. 2006. Microbial fuel cells—challenges and applications. Environ. Sci. Technol. 40(17): 5172-5180.

Lovley, D.R. 2008. The microbe electric: conversion of organic matter to electricity. Curr. Opin. Biotechnol. 19(6): 564-571.

Lowy, D.A., L.M. Tender, J.G. Zeikus, D.H. Park and D.R. Lovley. 2006. Harvesting energy from the marine sediment–water interface. II: Kinetic activity of anode materials. Biosens. Bioelectron. 21(11): 2058-2063.

Ludwig, R., W. Harreither, F. Tasca and L. Gorton. 2010. Cellobiose dehydrogenase: a versatile catalyst for electrochemical applications. Chem. Phys. Chem. 11(13): 2674-2697.

Luo, H., P. Xu, P.E. Jenkins and Z. Ren. 2012. Ionic composition and transport mechanisms in microbial desalination cells. J. Membrane Sci. 409: 16-23.

Marcus, R.A. and N. Sutin. 1985. Electron transfers in chemistry and biology. BBA – Rev. Bioenergetics 811(3): 265-322.

McCarty, P.L., J. Bae ad J. Kim. 2011. Domestic wastewater treatment as a net energy producer—can this be achieved? Environ. Sci. Technol. 45(17): 7100-7106.

Min, B., Ó.B. Román and I. Angelidaki. 2008. Importance of temperature and anodic medium composition on microbial fuel cell (MFC) performance. Biotechnol. Lett. 30(7): 1213-1218.

Modin, O. and K. Fukushi. 2012. Development and testing of bioelectrochemical reactors converting wastewater organics into hydrogen peroxide. Water Sci. Technol. 66(4): 831-836.

Modin, O., X. Wang, X. Wu, S. Rauch and K.K. Fedje. 2012. Bioelectrochemical recovery of Cu, Pb, Cd, and Zn from dilute solutions. J. Hazard. Mater. 235: 291-297.

Montpart, N., E. Ribot-Llobet, V.K. Garlapati, L. Rago, J.A. Baeza and A. Guisasola. 2014. Methanol opportunities for electricity and hydrogen production in bioelectrochemical systems. Int. J. Hydrogen Energ. 39(2): 770-777.

Nevin, K.P., T.L. Woodard, A.E. Franks, Z.M. Summers and D.R. Lovley. 2010. Microbial electrosynthesis: feeding microbes electricity to convert carbon dioxide and water to multicarbon extracellular organic compounds. MBio. 1(2): e00103-10.

Pant, D., A. Singh, G.V. Bogaert, S.I. Olsen, P.S. Nigam, L. Diels and K. Vanbroekhoven. 2012. Bioelectrochemical systems (BES) for sustainable energy production and product recovery from organic wastes and industrial wastewaters. RSC Adv. 2: 1248-1263.

Park, Y., H. Cho, J. Yu, B. Min, H.S. Kim, B.G. Kim and T. Lee. 2017. Response of microbial community structure to pre-acclimation strategies in microbial fuel cells for domestic wastewater treatment. Bioresour. Technol. 233: 176-183.

Qin, M., E.A. Hynes, I.M. Abu-Reesh and Z. He. 2017. Ammonium removal from synthetic wastewater promoted by current generation and water flux in an osmotic microbial fuel cell. J. Clean Prod. 149: 856-862.

Rabaey, K. and W. Verstraete. 2005. Microbial fuel cells: novel biotechnology for energy generation. Trends Biotechnol. 23: 291.

Rabaey, K., S. Bützer, S. Brown, J.R. Keller and R.A. Rozendal. 2010. High current generation coupled to caustic production using a lamellar bioelectrochemical system. Environ. Sci. Technol. 44(11): 4315-4321.

Rabaey, K., S.T. Read, P. Clauwaert, S. Freguia, P.L. Bond, L.L. Blackall and J. Keller. 2008. Cathodic oxygen reduction catalyzed by bacteria in microbial fuel cells. ISME J. 2(5): 519.

Reimers, C.E., H.A. Stecher, J.C. Westall, Y. Alleau, K.A. Howell, L. Soule, H.K. White and P.R. Girguis. 2007. Substrate degradation kinetics, microbial diversity, and current efficiency of microbial fuel cells supplied with marine plankton. Appl. Environ. Microbiol. 73(21): 7029-7040.

Richmond, A. 2008. Handbook of microalgal culture: biotechnology and applied phycology. John Wiley & Sons, USA.

Rozendal, R. and C. Buisman. 2005. Process for producing hydrogen. Patent WO2005005981.

Sadhukhan, J., J.R. Lloyd , K. Scott, G.C. Premier, E.H. Yu, T. Curtis and I.M. Head. 2016. A critical review of integration analysis of microbial electrosynthesis (MES) systems with waste biorefineries for the production of biofuel and chemical from reuse of CO_2. Renew. Sust. Energ. Rev. 56: 116-132.

Sakai, H., T. Nakagawa, Y. Tokita, T. Hatazawa, T. Ikeda, S. Tsujimura and K. Kano. 2009. A high-power glucose/oxygen biofuel cell operating under quiescent conditions. Energy Environ. Sci. 2(1): 133-138.

Schamphelaire, L.D., A. Cabezas, M. Marzorati, M.W. Friedrich, N. Boon and W. Verstraete. 2010. Microbial community analysis of anodes from sediment microbial fuel cells powered by rhizodeposits of living rice plants. Appl. Environ. Microbiol. 76(6): 2002-2008.

Schamphelaire, L.D., L.V.D. Bossche, H.S. Dang, M. Höfte, N. Boon, K. Rabaey and W. Verstraete. 2008. Microbial fuel cells generating electricity from rhizodeposits of rice plants. Environ. Sci. Technol. 42(8): 3053-3058.

Sevda, S. and I. Abu-Reesh. 2017a. Improved petroleum refinery wastewater treatment and seawater desalination performance by combining osmotic microbial fuel cell and up-flow microbial desalination cell. Environ. Technol. https://doi.org/10.10 80/09593330.2017.1410580.

Sevda, S. and I. Abu-Reesh. 2018. Improved salt removal and power generation in a cascade of two hydraulically connected flow-microbial desalination cells. J. Environ. Sci. Health. A. Tox. Hazard. Subst. Environ. Eng. 53(4): 326-337.

Sevda, S. and I. Abu-Reesh. 2017b. Energy production in microbial desalination cells and its effects on desalination. J. Energy. Environ. Sust. 3: 53-58.

Sevda, S., I.M. Abu-Reesh and Z. He. 2017. Bioelectricity generation from treatment of petroleum refinery wastewater with simultaneous seawater desalination in microbial desalination cells. Energ. Convers. Manage. 141: 101-107.

Sevda, S., H. Yuan, Z. He and I.M. Abu-Reesh. 2015. Microbial desalination cells as a versatile technology: functions, optimization and prospective. Desalination 371: 9-17.

Sevda, S., T.R. Sreekrishnan, N. Pous, S. Puig and D. Pant. 2018. Bioelectroremediation of perchlorate and nitrate contaminated water: a review. Bioresour. Technol. 225: 331-339.

Shemfe, M., S. Gadkari, E. Yu, S. Rasul, K. Scott, I.M. Head, S. Gu and J. Sadhukhan. 2018. Life cycle, techno-economic and dynamic simulation assessment of bioelectrochemical systems: a case of formic acid synthesis. Bioresour. Technol. 255: 39-49.

Simon, E., C.M. Halliwell, C.S. Toh, A.E. Cass and P.N. Bartlett. 2002. Immobilisation of enzymes on poly (aniline)–poly (anion) composite films: preparation of bioanodes for biofuel cell applications. Bioelectrochemistry 55(1-2): 13-15.

Stein, N.E., H.V. Hamelers and C.N. Buisman. 2012a. The effect of different control mechanisms on the sensitivity and recovery time of a microbial fuel cell based biosensor. Sens. Actuators B Chem. 171: 816-821.

Stein, N.E., H.V. Hamelers, G. van Straten and K.J. Keesman. 2012b. Effect of toxic components on microbial fuel cell-polarization curves and estimation of the type of toxic inhibition. Biosensors 2(3): 255-268.

Stillwell, A.S., C.W. King, M.E. Webber, I.J. Duncan and A. Hardberger. 2011. The energy-water nexus in Texas. Ecol. Soc. 16(1): 2.

Sustarsic, M. 2009. Wastewater treatment: understanding the activated sludge process. CEP 105, 26-29.

Tao, H.C., T. Lei, G. Shi, X.-N. Sun, X.-Y. Wei, L.-J. Zhang and W.-M. Wu. 2014. Removal of heavy metals from fly ash leachate using combined bioelectrochemical systems and electrolysis. J. Hazard. Mater. 264: 1-7.

Timmers, R.A., D.P. Strik, H.V. Hamelers and C.J. Buisman. 2010. Long-term performance of a plant microbial fuel cell with Spartina anglica. Appl. Microbiol. Biotechnol. 86(3): 973-981.

Trapero, J.R., L. Horcajada, J.J. Linares and J. Lobato. 2017. Is microbial fuel cell technology ready? An economic answer towards industrial commercialization. Appl. Energ. 185: 698-707.

Vilajeliu-Pons, A., C. Koch, M.D. Balaguer, J. Colprim, F. Harnisch and S. Puig. 2018. Microbial electricity driven anoxic ammonium removal. Water Res. 130: 168-175.

Visvanathan, C. and A. Abeynayaka. 2012. Developments and future potentials of anaerobic membrane bioreactors (AnMBRs). Membr. Water Treat. 3: 1-23.

Wang, X., Y. Feng, J. Liu, H. Lee, C. Li, N. Li and N. Ren. 2010. Sequestration of CO_2 discharged from anode by algal cathode in microbial carbon capture cells (MCCs). Biosens. Bioelectron. 25(12): 2639-2643.

Wang, H. and Z.J. Ren. 2013. A comprehensive review of microbial electrochemical systems as a platform technology. Biotechnol. Adv. 31(8): 1796-1807.

Yuan, L., X. Yang, P. Liang, L. Wang, Z.-H. Huang, J. Wei and X. Huang. 2012. Capacitive deionization coupled with microbial fuel cells to desalinate low-concentration salt water. Bioresour. Technol. 110: 735-738.

Zang, G.-L., G.-P. Sheng, W.-W. Li, Z.-H. Tong, R.J. Zeng, C. Shi and H.-Q. Yu. 2012. Nutrient removal and energy production in a urine treatment process using magnesium ammonium phosphate precipitation and a microbial fuel cell technique. Phys. Chem. Chem. Phys. 14(6): 1978-1984.

Zinadini, S., A. Zinatizadeh, M. Rahimi, V. Vatanpour and Z. Rahimi. 2017. High power generation and COD removal in a microbial fuel cell operated by a novel sulfonated PES/PES blend proton exchange membrane. Energy 125: 427-438.

Zou, S. and Z. He. 2018. Efficiently "pumping out" value-added resources from wastewater by bioelectrochemical systems: a review from energy perspectives. Water Res. 131: 62-73.

Index

Editors Biography

Dr. Lakhveer Singh is presently working as a Senior Research Associate at Biological and Ecological Engineering department, Oregon State University, USA. Prior to his joining Oregon State University, he worked as an Assistant Professor from Dec. 2013-July 2017 at University Malaysia Pahang, Malaysia. His main areas of research are Bioenergy, Water treatment, Bioelectrochemical systems, Nanomaterial synthesis for energy and water applications and Bioreactors development. He has published 44 papers in prominent journals in Elsevier's, Springer, Wiley, and has authored 4 book chapters. His works have been cited 682 times with an h index of 15 and an i10 index of 19 to till date. In addition, he has edited one book: (i) **Waste Biomass Management – A Holistic Approach** (2017) and presently editing two books. He is presently Editor-board member of the following journals: International Journal of Energy Engineering, Scientific and Academic Publisher, USA, International Journal of Engineering Technology and Science, and Insight- Energy Science. He is a member of the following scientific societies: International Water Association (IWA) and **Asian Council of Science Editors** (ACSE). He also holds four patent filing application for his research. He has been a reviewer for various journalspublished by Elsevier, Wiley, RSC and Springer.

Dr. Durga Madhab Mahapatra is presently working as a Courtesy Faculty at the Department of Biological and Ecological Engineering (BEE), School of Engineering, Oregon State University, USA. Prior to his joining Oregon State University, he worked as a Research Associate at the Indian Institute of Science (IISc), Bangalore. His main areas of research are Algal Biofuels and Bioproducts, Biomass and Bioenergy, Water and Wastewater Treatment, Industrial Process Engineering and Bioreactors Design and Analysis. He has extensive experience in research and industry oriented projects. He has more than 50 publications in peer-reviewed journals, scientific reports, book chapters and invited articles. His works have been cited 704 times with an H index of 13. He is presently editing two books. He has won many awards in national and international forums, exhibitions, conferences and is a recipient of the prestigious Amulya and Vimala Reddy Award for

Sustainable Technologies and Gandhian Young Technological Innovation Award. He is a member of American Society of Agricultural and Biological Engineers (ASABE), Indian Science Congress (ISC), Asia Pacific Society on Algal Biotechnology and has been a reviewer for various journals published by Springer, Elsevier, RSC and Wiley.

Printed and bound by CPI Group (UK) Ltd, Croydon, CR0 4YY

24/10/2024

01778307-0011